Student Poster Sessions 2017

Core Programming Area at the 2017 AIChE Annual Meeting

Minneapolis, Minnesota, USA
29 October – 3 November 2017

ISBN: 978-1-5108-5813-8

Printed from e-media with permission by:

Curran Associates, Inc.
57 Morehouse Lane
Red Hook, NY 12571

Some format issues inherent in the e-media version may also appear in this print version.

Copyright© (2017) by AIChE
All rights reserved.

Printed by Curran Associates, Inc. (2018)

For permission requests, please contact AIChE
at the address below.

AIChE
120 Wall Street, FL 23
New York, NY 10005-4020

Phone: (800) 242-4363
Fax: (203) 775-5177

www.aiche.org

Additional copies of this publication are available from:

Curran Associates, Inc.
57 Morehouse Lane
Red Hook, NY 12571 USA
Phone: 845-758-0400
Fax: 845-758-2633
Email: curran@proceedings.com
Web: www.proceedings.com

TABLE OF CONTENTS

Some abstracts in this book refer to images that were not made available by the publisher

510514 Infrared Spectra and the Coverage of Molecules on Surfaces of Solid Catalysts

Abstract:

Fourier transform infrared (FTIR) spectroscopy, a powerful tool for probing reactive species in reaction conditions, has limited applications to studies of catalytic reactions and materials. This is due to the dipole-coupling effects, a phenomenon where surface molecules cross-communicate with surrounding molecules at high surface coverage, affecting the spectra and leading to inaccurate interpretations of spectroscopic measurements. This project studied the vibrational frequency and intensity of surface carbon monoxide molecules at different gas coverage on the surface of Pt clusters. Using both experimental and computational methods, we successfully established evidence of dipole coupling at high surface coverage on Pt clusters.

Copyright © American Institute of Chemical Engineers. All rights reserved.

511073 Catalytic Activity, Stability, and Selectivity of Single Atom Pt on Nanoporous CuO_x in the CO Oxidation and PROX Reactions

Abstract:

Currently, the removal of CO from automobile exhaust is accomplished by the oxidation of CO in catalytic converters. Noble metals catalysts are used on different supports but the problem is that they are rare and expensive(1). The goal is to have highly efficient catalysts for the reaction at low costs. The same reaction known as PROX (Preferential CO Oxidation) in a hydrogen- rich stream is important for PEMFC (Proton-exchange membrane fuel cells) which uses a Pt anode. The challenge is to attain a highly selective CO oxidation reaction to purify the inlet feed from CO as best as possible and prevent the poisoning of the Pt anode(2). This project shows that successfully synthesized nano-porous (np) Cu by the sacrificial support method(3), and Pt single atoms embedded in the surface of np-Cu by galvanic replacement is an excellent catalyst with high surface area and superior catalytic performance for both the dry CO oxidation reaction and the PROX reaction. In these oxidative reaction environments, the Cu will be oxidized to CuO_x. Pt/np-CuO_x with a Pt loading of 0.1 at.% has shown a better catalytic conversion over the np-CuO_x control catalyst at all conditions examined. It also showed good stability and maintained its activity during long-term steady- state experiments of 45 h. The Pt/np-CuO_x catalyst also showed high selectivity when used for the PROX reaction as the selectivity stayed at 100% at a temperature as high as 140 °C. This means that the single atom Pt /np-CuO_x catalyst used in this study has a good combination of high activity, stability and selectivity for the CO oxidation and PROX reaction under widely different tested conditions.

References

[1] Li, Shanlong et al. "Copper Doped Ceria Porous Nano-structures towards a Highly Efficient Bi-functional Catalyst for Carbon Monoxide and Nitric Oxide Elimination". *Chem. Sci.* 6.4 (2015): 2495-2500. Web. 13 Oct. 2016.

[2] Liu, Jilei et al. "Tackling CO Poisoning with Single-Atom Alloy Catalysts". *J. Am. Chem. Soc.* 138.20 (2016): 6396-6399. Web. 13 Oct. 2016.

[3] Shan, Junjun et al. "Selective non-oxidative Dehydrogenation of Ethanol to Acetaldehyde and Hydrogen on Highly Dilute NiCu alloys", *Appl. Catal. B: Environ.* 205 (2017): 541-550.

Copyright © American Institute of Chemical Engineers. All rights reserved.

511129 Zeolites Based Catalytic Processes for Furfuryl Alcohol Dehydration: Activity Test, Characterization and Deactivation of Catalysts

Abstract:

Alcohol dehydration and condensation reaction have been extensively investigated. Currently more attention has been paid to alternative renewable sources due to the increase demand of energy and depletion of fossil fuels. As a lignocelluloses derived alcohol, furfuryl alcohol (FA, $C_5H_6O_2$) is an attractive platform chemical that can be used in the production of chemicals and fuels. FA dehydration and condensation reaction into oligomers and polymer have also been investigated because of a wide range of applications, such as nanocomposite carbon, carbon nanotubes, and fuel blending components. In this work, furfuryl alcohol dehydration by heterogeneous acid catalysts was studied under 100°C at atmospheric pressure for 3 hours and 6 hours. The comparison was established among three zeolite catalysts (H-MFI, H-MOR, and H-BEA), having the similar Si/Al ratios but having different pore architectures, therefore provided different catalytic activities. By reacting the above zeolites with FA, the FA underwent alcohol dehydration while coke was formed on zeolites. The study of coke formation has brought lots of interest since it results in decreased catalytic activities and increased byproducts. FT-IR spectroscopy were applied to characterize the zeolites before and after reactions. Coke formed on zeolites were characterized by using Raman and TGA. The FA dimers and trimers were identified by GC-MS and GC-FID was used to determine the conversion and selectivity of oligomers. It was found that 6 hours reaction yielded a higher conversion than 3 hours reaction; however, the selectivity differs due to the difference in structure and pore size of zeolites.

Copyright © American Institute of Chemical Engineers. All rights reserved.

511157 Tuning High-Performance YSZ-Based Intermediate Temperature Solid Oxide Fuel Cells: Engineering SOFC Cathode Structures

Abstract:

Solid Oxide Fuel Cells (SOFCs) are solid-state electrochemical devices that operate at high temperature (above 700C) and are capable of efficiently converting the chemical energy stored in the bonds of fuel molecules (i.e., H_2, CO, hydrocarbons) to electrical energy. SOFCs are composed of a cathode compartment that reduces gas-phase O_2 to oxygen ions. The oxygen ions diffuse through a ceramic oxygen ion conducting electrolyte to the anode, where the fuel is oxidized. One of the limiting factors with SOFCs is the operation at very high temperatures (~800C) to achieve reasonable electrochemical rates. Many efforts have been reported in lowering the operating temperature of these systems. In this work, we demonstrate that significant performance improvements at intermediate temperatures can be achieved by improving the cathode-electrode kinetics. We show that incorporation of nanostructured layered mixed metal oxides in porous scaffolds of ion-conducting oxides (i.e. YSZ) at the cathode can lead to significant power density improvements at temperatures of 550-600C.

Copyright © American Institute of Chemical Engineers. All rights reserved.

509967 Operando Raman Spectroscopic Analysis of Silica Supported V-Nb Catalysts for Tuning Redox Properties

Abstract:

We report that the redox properties of traditional silica supported vanadia catalysts can be tuned through the addition of niobia. Heterogeneous catalysis has widespread use in industry for numerous processes. Vanadium over silica is a traditional precursor metal oxide catalyst usually employed in the reduction of alkanes to form terminal olefins. In order to improve the conversion rate to olefins, it is desirable to balance the reductive and oxidative properties of the catalyst so that the processes of catalytic action (alkane reduction) and active site regeneration (vanadium oxidation) occur closer to the same rate.

Inclusion of a secondary component in the catalyst alters dispersion and monolayer coverage of vanadium, altering the reaction kinetics of oxidation and reduction for the catalyst. However, not much is known about how a second metal in ternary systems like this influences the physical and chemical properties of the vanadium oxide active site. For a series of vanadia catalysts with different amounts of a secondary component, niobia, a state of the art methodology was utilized to perform Operando Raman spectroscopy to monitor the rate of change of the catalyst state at the surface. This was followed by mass spectroscopy using pulses of gaseous reagent in separate experiments to analyze the bulk of the catalyst. Thus, the redox properties of the surface and the bulk of the catalyst were analyzed simultaneously, providing new insight into the complete redox behavior. From this, the most effective catalyst composition of those tested was determined.

Copyright © American Institute of Chemical Engineers. All rights reserved.

510962 Synthesis and Characterization of Ordered Mesoporous Resin (OMR) for Pharmaceutical Synthesis

Abstract:

Ordered Mesoporous Resin (OMR) is an organic polymer with applications in catalysis, energy storage, sensors, adsorption and separation process. Current variants of OMR are produced using solvents such as ethanol as well as template (surfactants such as F-127). While these methods generate a porous structure, they produce waste and require an additional heating step to remove the template. Additionally reducing the use of solvents decreases the toxicity of the manufacturing process.

Through our research, a highly porous OMR has been produced using no solvents or templates but instead by using a hyperbranched polyethyleneimine. Pictures of the OMR were taken using a scanning electron microscope (SEM) and show a highly porous structure. Then, Brunauerâ€"Emmettâ€"Teller theory (BET) was used to calculate surface area measurements to confirm the porosity of the OMR structures. Due to the porous nature and relative stability of the OMR it is highly suitable to be used as a structure to be modified for use in catalysis. Further development of a acid functionalized and palladium doped OMR are being analyzed for use in catalysis.

The polymer can be functionalized with acids which will optimize the catalysis step of creating pharmaceuticals. The initial aim of creating a porous yet stable structure of OMR is necessary to allow for an increased rate of reaction and to provide substantial surface area available for attaching functional groups. Having functional groups attached to the polymer will speed up the process, minimize the amount of waste produced, and will reduce the number of steps taken to create pharmaceuticals.

The other variant of OMR under development is doped with small quantities of Palladium. This is done with the goal of producing the high surface area OMR but with palladium uniformly distributed throughout the structure. This palladium doped catalyst has potential applications for the pharmaceutical industry to catalyze cross-coupling reactions. For this reaction, the palladium doped OMR is a heterogeneous catalyst thus eliminating the need for a difficult separation step to remove palladium from the mixture.

Copyright © American Institute of Chemical Engineers. All rights reserved.

510207 A Kinetic Investigation of Solvent Effects on Fischer Esterification for Biobased Chemical Production

Abstract:

The production of biomass-sourced esters by joint biological-chemical synthesis is a current subject of interest, as these high-value chemicals have applications as fuel additives, polymers, pharmaceuticals, and more. A key aspect of the process is the acid-catalyzed esterification of platform intermediates, but solvent effects on the reaction kinetics remain unclear. Previous work on the esterification of butanol and butyric acid suggested that a change in solvent from THF to toluene decreases the reaction order, potentially caused by site-blocking and asymmetric changes in activities of the reactants, products, and kinetically-relevant transition states. This study improved on the experimental procedures used in previous work to better capture the initial rate regime and extend the range of targeted concentrations. Experiments were run in isothermal, well-mixed batch reactors using Amberlyst-46 as an acid catalyst. Concentrations were quantified using GC-FID. Rates were determined by linear fitting of catalyst-normalized butyl butyrate concentrations over the reaction timescale. Fractional orders of about 0.7 were found for both reactants in the 0.1 to 60 mM concentration range. Rate order plots displayed no visible curvature, suggesting minimal influence of site blocking. Nonlinear regression modeling showed that adsorbed butanol was the most plausible abundant surface intermediate, although differences between model fits were small in comparison to the uncertainty present, suggesting that the kinetics data alone cannot be used to make a definitive assignment of which species is abundant on the surface. Future work intends to further extend the concentration range and reduce catalyst loading to limit conversion for high-leverage experiments. Characterization of potential solvent effects on selective may also be an important avenue of study.

Copyright © American Institute of Chemical Engineers. All rights reserved.

510753 Catalytic Hydrodeoxygenation of Renewable Oils to Diesel-Range Hydrocarbons Using Ir-ReOx/SiO2 Catalyst

Abstract:

Hydrodeoxygenation (HDO) of renewable plant and waste cooking oils to long-chain alkanes is a sustainable and promising process for the production of carbon neutral diesel products. We found that an Ir-ReOx/SiO2 catalyst is effective for the HDO of such oils for conversion to long-chain alkanes under very mild conditions . Near quantitative carbon yields (>80 wt%) in alkanes is obtained from various renewable oils (palm oil, soybean oil, rapeseed oil, sunflower oil and waste cooking oil) at 180 °C and 2 MPa H2. A reaction pathway for the HDO of renewable oils is proposed. The first step is the hydrogenation of C=C double bonds to form saturated triglycerides. Then the saturated triglycerides undergo hydrogenolysis to form a mixture of esters, acids and propane. The hydrogenation of the carboxylic group via aldehyde produces a fatty alcohol, which undergoes a sequential direct hydrogenolysis to long-chain alkanes with high carbon efficiency and selectivity over the Ir-ReOx/SiO2 catalyst. Partial decarbonylation of the aldehyde intermediates to alkanes with a single carbon loss is also observed, but minimized.

Copyright © American Institute of Chemical Engineers. All rights reserved.

509743 Effects of Interzeolite Transformation on the Aluminum Distribution in Zeolite Frameworks

Abstract:

The distribution of aluminum atoms within zeolite frameworks, both in terms of local arrangement and tetrahedral-site (T-site) location, determines their structural, ion-exchange, and catalytic properties. Here, we study different methods for interzeolite transformation of faujasite (FAU) zeolites into chabazite (CHA) zeolites, and their effects on the framework aluminum distribution in CHA. Previous literature reports have demonstrated that the common structural building units, such as the double six-member ring (D6R) building units in FAU and CHA, are an important structural feature that enables the conversion of one zeolite topology into another. Here, we employ the highly symmetric, single T-site zeolites (FAU and CHA) to study framework Al arrangement independently of T-site location effects that are also present in lower-symmetry zeolite frameworks. Two hydroxide mediated synthesis procedures were studied, one utilizing the organic-structure directing agent (OSDA) N,N,N-trimethyl-1-adamantylammonium cation, and the other utilizing only inorganic alkali cations as structure-directing agents. Both procedures have previously been reported to facilitate the interzeolite conversion of FAU into CHA, however neither has been studied to understand how local Al arrangements are influenced by the interconversion synthesis route. Powder X-ray diffraction patterns (XRD) and argon adsorption isotherms are used to probe changes in zeolite topology during the conversion of FAU to CHA. Elemental analysis with atomic absorption spectroscopy (AAS) was used to characterize the total Al content of each zeolite sample, and temperature programed desorption (TPD) of ammonia was used to quantify the total number of BrÃ¸nsted acid sites. Aqueous-phase divalent cobalt (Co^{2+}) ion-exchange isotherms were measured for each topology, and saturation Co^{2+} uptakes on each sample were used to determine the total number of paired Al sites ($Al\text{-}O\text{-}(Si\text{-}O)_x\text{-}Al$; x = 1,2 paired and x â‰¥ 3 isolated) in FAU and CHA. Diffuse reflectance UV-Vis spectroscopy (DRUV-VIS) was used to ensure that no cobalt oxides were present and only bare Co^{2+} was exchanged onto the zeolite. We discuss how changing the conditions used for interzeolite conversion can lead to differences in the framework Al distribution in the CHA product zeolites compared to each other and the parent FAU zeolites.

Copyright © American Institute of Chemical Engineers. All rights reserved.

511064 Continuous Flow Microreactor

Abstract:

Independent researchers studying chemical processes face a challenge due to its lack of equipment for small scale experiments. In industry, the resources are available to run chemical reactions on a large scale in huge tanks with essentially unlimited resources. To efficiently and consistently induce chemical reactions on a small scale, a continuous flow microreactor is a simple solution that can provide this desired effect. The main concept that proves a microreactor to be an excellent method for running small-scale reactions is the turbulence generated in small diameter pathways and bends that achieve a good mix of reactants. In addition, if a volatile reaction is desired this reduces the risk involved as the amount of reactants in use at any time in continuous flow microreactor is significantly low compared to batch production.

The design of the micro reactor is relatively simple with an inlet leading into a narrow winding pathway up until the outlet. The two main concerns around the design are visibility of the reaction and rapid prototyping to enable quick turnaround for varying experiments. To address visibility, the microreactor is made by adding polydimethylsiloxane (PDMS) to molds of the top and bottom half of the design. To achieve rapid prototyping, the main mode of production implements 3D printing of the molds.

The production of a microreactor begins with the design of the negative of the top and bottom half of it using CAD software. Then, the designs are exported into a file that can be imported into a 3D printer. Once the molds are produced the next step is mixing a 1:10 ratio of curing agent and PDMS monomers to enable the PDMS to set. After this, the molds and PDMS are baked until it is set.

The original design was intended for a specific experiment, but the scope of this project scales far beyond the initial experiment. The rapid prototyping nature of 3D printing can allow for unique reactors for different experiments or quick redesigns if a flaw is found post production. Also, many modifications can be made to the original design to extend its usefulness. These could include parallel pathways to run either cold or hot fluid through the system to create a temperature change to enhance the reaction or adding multiple inlets to combine multiple fluids rather than coating the pathway in reactant as is done in the initial experiment.

Copyright © American Institute of Chemical Engineers. All rights reserved.

509990 Investigating the Kinetics of Polymerization Required for Bulk-Mediated Alignment of Liquid Crystal Elastomers

Abstract:

Liquid crystal elastomers (LCEs) are lightly cross-linked polymer networks that exhibit liquid crystal order combined with rubber elasticity. LCEs are thermally responsive and if well aligned, they can elongate by over 300% when transitioning from the isotropic (unaligned) to nematic (aligned) state. Previous approaches to induce alignment restrict the LCEs to thin films with limited geometries. For expanded applications, more geometries are desired. This study presents a strategy for developing bulk-mediated alignment of LCEs utilizing the photodimerization of poly (vinyl cinnamate) during simultaneous polymerization via a base-catalyzed thiol-acrylate Michael addition (TAMA) reaction. It is expected that the alignment process will be sensitive to an assortment of reaction conditions and rates, and therefore, the foundational TAMA reaction must be thoroughly investigated. Experiments with varying amounts and types of catalysts and reagents were monitored using time resolved FT-IR spectroscopy. It was found that the rate of the reaction increases and then saturates when catalyst concentration is increased. Preliminary data suggests that the cinnamate pendants dimerize in the system when irradiated with UV light, but more experiments are needed in order to demonstrate the viability of bulk-mediated alignment. If successful, this alignment technique may be instrumental in developing responsive elastomers having complex three dimensional geometries with applications ranging from rewritable braille to soft muscle actuators.

Copyright © American Institute of Chemical Engineers. All rights reserved.

510689 Effect of Particle Size on Activity of Nickel Based Tri-Reforming Catalysts

Abstract:

As the impacts of climate change continue to worsen, global demand for a reduction in greenhouse gas emissions will continue to grow. Tri-reforming of methane is a practical technique to neutralize CO_2 emissions directly from the stack with a desirable H_2/CO product ratio. The process synergistically utilizes dry reforming (DR), steam reforming (SR), and partial oxidation of methane (POM). Tri-reforming has been demonstrated to be feasible on an industrial scale at the Korean Gas Corporation (KOGAS) to produce dimethyl ether. One of the key advantages of tri-reforming is that it is thermodynamically favorable in comparison to DR. Typically, Ni-based catalysts are preferred for tri-reforming due to a relatively low cost and high activity. The activity of Ni catalysts can be improved by developing materials that demonstrate high resistance to oxidation and coke formation. This could be achieved by controlling the Ni particle size and enhancing the metal-support interaction. In this work, two unique catalysts, Ni-CeO_2 nanoparticle and NiO-CeO_2@SiO_2 coreshell, were synthesized to investigate the particle size effect on tri-reforming activity. Modification of particle size may provide control over the surface area of active site exposure. The results showed that core-shell structures had less deactivation and comparable activity. Particle size was modified by altering calcination temperature and metal ratios. Ni particle size has been shown to range between 13 and 52 nm on the synthesized catalysts. This particle size could be maintained after the reaction. The two-step calcination process used in the preparation of Ni-CeO_2 catalysts, in comparison to the one-step calcination process, does not enhance the tri-reforming activity due to a weak metal-support interaction. In the future, an investigation of similar materials that could include metals such as Zr, Ti, and Mg will be conducted. Additionally, the impact of NO_x and SO_2 on catalytic activity during a tri-reforming reaction will be investigated.

Copyright © American Institute of Chemical Engineers. All rights reserved.

510091 Perovskites for Photocatalysis: Determination of Band Structure

Abstract:

As our earth continue to be faced with the major issue that is global warming it is imperative to develop technology to reduce and reverse this trend. Although much research has been done into photocatalysts due to their potential for providing us with cheap, green, and sustainable energy through the photoreduction of carbon dioxide, the technology is still greatly plagued by inefficiencies. Many studies have focused on metal oxides, however their large band gaps (e.g TiO_2â€™s band gap of 3.2eV) and band edge positions cause them to be inefficient for this application. Perovskites, hybrids of two or more metal oxides, with the general formula ABO_3, could be the answer. The goal of this project is to determine the potential of perovskites to serve as an alternative to current metal oxide photocatalysts due to their adjustable band edges and band gaps, made possible by varying the metal composition.

In this study, $LaCr_{1-x}Fe_xO_3$ (where x = {0, 0.25, 0.5, 0.75, 1}) perovskites were synthesized via the Pechini process. XRD (x-ray powder diffraction) was used to verify catalyst structure. Optical band gaps were determined through DRS (diffused reflectance spectroscopy) UV-vis spectroscopy. Band edges were obtained using an electrochemical cell, and Mott-Schottky analysis. Values were then compared to both the CO_2 reduction and H_2O splitting potentials to determine if these materials could be suitable as photocatalysts. It was found that through the replacement of Cr ions in $LaCrO_3$ with Fe ions that the band gap does not show a linear tread, but rather exhibits the smaller band gap of $LaFeO_3$. We find that both $LaFe_{0.25}Cr_{0.75}O_3$ and $LaFe_{0.75}Cr_{0.25}O_3$ show potential for CO_2 photoreduction and H_2O splitting in the visible light range with band gaps of ~ 2.0eV and appropriate band edges. It is demonstrated that the perovskite materials present a tunable band edge and adjustable band gap allowing for the potential for use as photocatalysts. Further research is needed to verify the effectiveness of these materials in the actual CO_2 photo-reduction.

Copyright © American Institute of Chemical Engineers. All rights reserved.

510177 Valorization of Bioelectrochemically Derived Hydrogen Peroxide through Catalytic Sulfoxidation

Abstract:

In this work, we demonstrate a combined bioelectrochemical and inorganic catalytic system for resource recovery from wastewater. We designed a microbial electrolysis cell (MEC) for continuous hydrogen peroxide (H_2O_2) production and used this bioelectrochemically-derived H_2O_2 as a green oxidant for sulfoxidation, a representative oxidation reaction used industrially for chemical synthesis and oxidative desulfurization of transportation fuels. We continuously operated a MEC equipped with a gas diffusion electrode cathode for six months, achieving a peak current density above 1.4 mA/cm^2 with 60% acetate removal efficiency at an anodic coulombic efficiency of 67%. During operation, we evaluated several cathode buffers for solubility and pH compatibility with downstream catalytic systems and identified a 200 mM phosphate buffer that allowed us to achieve high effluent H_2O_2 concentrations (3.0 g/L) at a low energy input (1.6 Wh/g H_2O_2) and pH (10) in 24-hour batch tests. The MEC-derived H_2O_2 solution was used directly as an oxidant for a representative catalytic sulfoxidation of 4-hydroxythioanisole over a solid niobium(V)-silica catalyst. We achieved complete selective conversion of 50 mM 4-hydroxythioanisole to 4-(methylsulfinyl)-phenol on a 0.5 mol% catalyst loading in 100 minutes in aqueous media. Reactivity was within 20% for batch experiments performed with commercial H_2O_2 and batch and CSTR reactors operated with MEC catholyte. Our results demonstrate a new and versatile approach for valorization of wastewater through continuous production of H_2O_2 and its subsequent use as a green oxidant.

Copyright © American Institute of Chemical Engineers. All rights reserved.

510515 Investigation of Platinum Alloys for Light Alkane Dehydrogenation

Abstract:

Economical and efficient conversion of feedstock to product is central in designing a sophisticated and sustainable process strategy. Specifically, the recent discovery of shale gas sourced from horizontal drilling and hydraulic fracturing of naturally occurring shale formations has dramatically increased the abundance of light alkanes and has consequently reduced the price of natural gas to an all-time low. Shale gas reserves are predicted to have an impact on the production of chemicals by providing a cheap and abundant feedstock supply. A majority of shale gas is comprised of methane with the balance being a mixture of light alkanes like ethane or propane. These light alkane species are also a major byproduct of many industrial hydrocarbon processing pathways like petroleum cracking. Often, these light alkanes are flared which releases greenhouse gases, but by utilizing them as feedstock to produce olefins via alkane dehydrogenation, they stand to increase the profitability and sustainability of this process. Olefins are versatile chemical intermediates that are used to produce a diverse collection of value-added chemicals in a variety of industries including pharmaceuticals, bulk and fine chemicals, and fuels. Industrially, a supported platinum alloy catalyst is typically employed, but the overall process suffers from the formation of carbonaceous species on the catalyst surface. This work synthesizes, characterizes, and analyzes promising novel bi-metallic alloy catalyst materials for alkane dehydrogenation predicted by Density Functional Theory (DFT) calculations. Hydrotalcite-like materials are used to support the bi-metallic alloys because of their thermal and chemical stability, high surface area, and moderate basicity which reduces coking tendencies. Thermocatalytic activity and selectivity is quantified by the conversion of ethane to ethene via alkane dehydrogenation in a continuous flow system. While the formation of alkene has been studied, the mechanism for coke formation is unknown and this project culminates in a fundamental understanding of how carbonaceous species interact with catalyst surfaces which will provide useful insight for industrial applications.

Copyright © American Institute of Chemical Engineers. All rights reserved.

510634 Optimization of Spikey Gold Nanoparticles and 2D Gold Nanosheets Syntheses

Abstract:

Due to their unique optoelectronic properties, gold nanoparticles (Au NPs) can be used in a variety of applications including electronics, photodynamic therapy, sensors [especially surface enhanced Raman spectroscopy (SERS)], biological probes and imaging, diagnostics, and heterogeneous catalysis. Also, Au NPs can be incorporated into the $LiFePO_4$ cathode of a lithium ion battery to increase electrical conductivity as they do not alter the crystal structure of the cathode nor readily oxidize[1]. This work involves optimizing the anisotropic shape of spikey Au NPs through varying the chemical parameters, such as the reducing agent and colloidal stabilizer or surfactant. Au NPs in 2D shapes, similar to graphene, are intriguing materials for next generation electronics and optoelectronics. Their unique morphology readily allows for incorporation in nanoelectronic devices and layered heterostructures. Using polyvinylpyrrolidone (PVP) as a colloidal stabilizer, ethylene glycol as a reducing agent, and dimethyl sulfoxide (DMSO) as a shape directing agent, micron-sized 2D nanosheets can be selectively synthesized. We optimized a previously-reported synthesis that had many impurity particles to develop an improved version with a higher yield of 2D Au nanosheets.

Copyright © American Institute of Chemical Engineers. All rights reserved.

Enhanced Photocatalysis of TiO2 with a Pt Co-Catalyst

Trevor R. Fisher[a], Advisor: Rajesh Shende[b]

[a] Department of Chemical, Biological and Materials Engineering, University of Oklahoma, Norman, OK 73069 U.S.A.

[b] Department of Chemical and Biological Engineering, South Dakota School of Mines and Technology, Rapid City, SD 57701 U.S.A.

Introduction

Titanium dioxide (TiO_2) is an affordable photocatalyst for generating hydrogen gas from water, organic waste, and UV-visible light [1]. An electron-hole pair is generated in TiO_2 when it absorbs light. When a water molecule interacts with the electron hole, it splits into hydrogen and oxygen gas [2]. However, as a pure substance, TiO_2 can not generate hydrogen gas at an economically viable rate [3]. To improve hydrogen generation, TiO_2 must be modified in order to:

- Increase visible light absorption
- Stabilize electron-hole pairs

Increasing visible light absorption provides more energy to generate electron-hole pairs. Enhanced light absorption can be achieved by doping with impurities or exposing a more reactive surface of TiO_2 [4].

It is also important to stabilize the electron-hole pairs once they are generated. Longer lifetimes of electron-hole pairs increase the positive charge on the surface of TiO_2, leading to more favorable conditions for splitting water [5].

Experiments have shown that the photocatalytic rate can be enhanced by coupling TiO_2 with a platinum (Pt) co-catalyst to promote electron hole generation [6]. This study will use *ab initio* density functional theory (DFT) modeling to support that Pt enhances the photocatalytic properties of TiO_2.

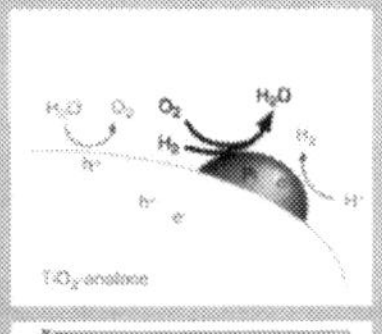

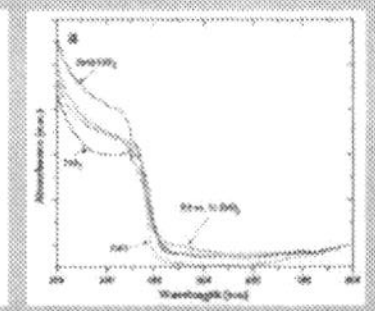

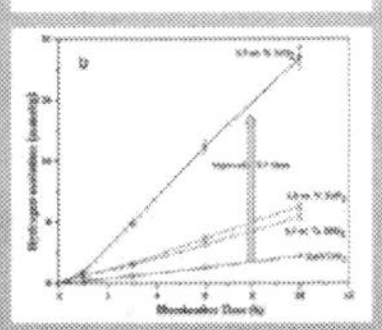

Figure 1 (Top Left) Water decomposition into hydrogen and oxygen gas on a Pt-TiO_2 surface [2]. (Top Right) UV-visible light absorption spectrum for TiO_2 coupled with ZrO_2 and ZnO [6]. (Bottom Left) Hydrogen generation from different ratios of ZrO_2 to TiO_2 suspended in ethanol [6].

Objectives

The objectives of this project are to:

- Construct low-energy conformations for [101] and [001] slabs of anatase TiO_2;
- Create an interface between a Pt crystal and the TiO_2 slabs;
- Calculate how the electron density for Pt and TiO_2 change when they interface;
- Support experimental evidence that Pt increases electron hole stability in TiO_2 with theoretical modeling.

Methods, Results & Discussions

1. Structures:

DFT Parameters: Software – Materials Studio; Exchange Correlation Functional – GGA-PBE; Pseudopotentials – "Ultrasoft"; Kinetic Energy Cutoff – 450 eV; k-points – 11x11x11 (bulk) / 11x11x1 (slabs); Energy Convergence Tolerance – 1.0e-5 eV/atom

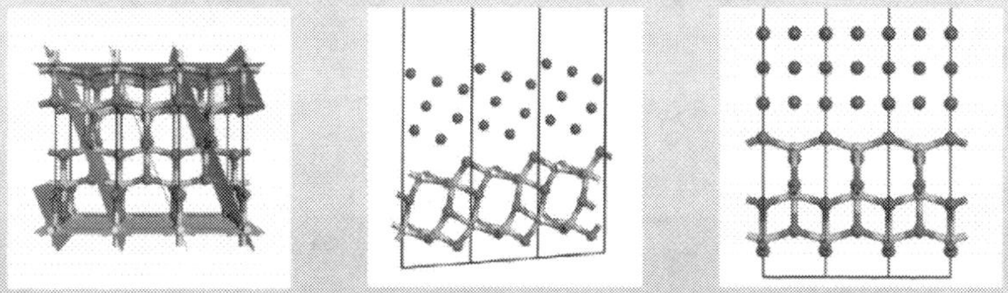

Figure 2. Relaxed structures for bulk anatase TiO_2 (Left), (101) slab (Middle), and (001) slab (Right) with Pt adsorbed to the surface. Ti, O, and Pt atoms are colored grey, red, and blue, respectively. The planes on the bulk structure indicate the exposed surfaces of the (101) (Blue) and (001) (Orange) slabs.

	(101) Slab	(001) Slab
Clean Slab Surface Energy	0.407 J/m²	0.944 J/m²
Pt Energy of Adsorption	0.12 J/m²	-0.24 J/m²

2. Electron Density Difference

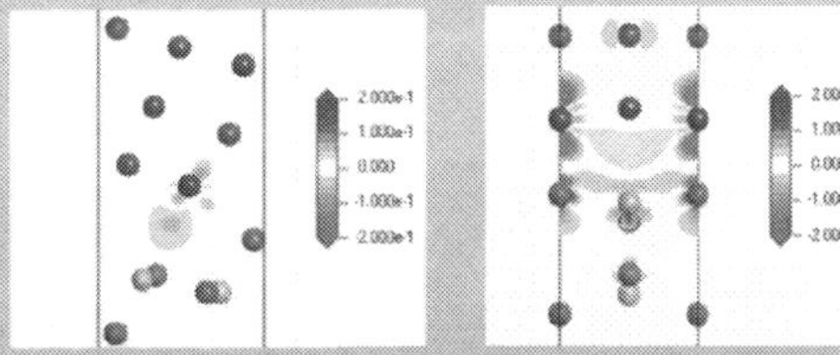

Figure 3. The electron density of the combined Pt-TiO_2 slab system relative to the electron density found in the isolated Pt and TiO_2 sub-systems. Blue, positive values represent a region in the combined system with higher electron density than in either Pt or TiO_2 sub-system, whereas red values indicate a lower electron density.

Larger changes in electron density are seen in the (001) slab than in the (101) slab.

In the (001) slab, electron density significantly decreases around the interface Pt atoms, and the interface O atoms have orbitals with increasing and decreasing electron density. The surface of the TiO_2 looses electron density while the Pt surface gains electron density.

Only slight changes in electron density are seen in the (101) slab, which indicates weak bonding with the Pt. The Pt atom closest to the TiO_2 surface loses electron density, and the region between the Pt and the closest Ti atom increases in electron density.

3. Partial Density of States

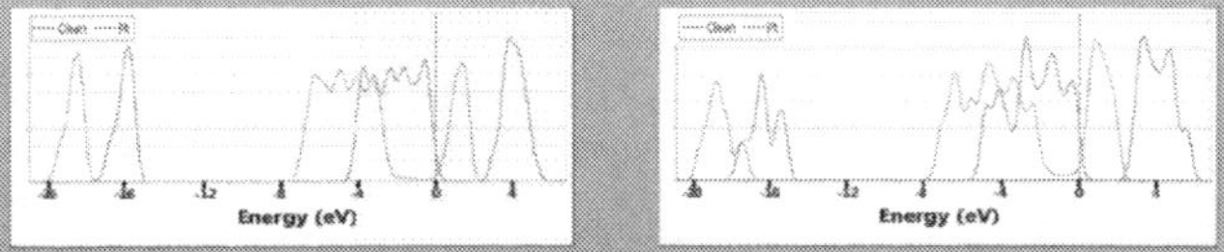

Figure 4. The partial density of states of only the Ti and O atoms in the (101) slab (Left) and the (001) slab (Right). The isolated TiO_2 slab PDOS is in blue, and the Pt-TiO_2 slab PDOS is in red.

The TiO_2 PDOS for the (101) and (001) slabs are similarly affected by the interaction with Pt. The presence of Pt shifts the Fermi level near the valence band closer to the conductance band. This p-to-n type semiconductor transition indicates that electrons are more likely to carry charge around the Pt-TiO_2 surface, and electron holes predominantly carry charge around a clean TiO_2 surface.

The electron states are also delocalized by Pt and occupy band gap states, which can lead to increased visible light absorption [4].

Conclusions

The Pt-TiO_2 interface is able to enhance photocatalysis on TiO_2 by:

- Stabilizing electron-hole pairs by spacing out regions of higher and lower electron density;
- Promoting a p-to-n semiconductor transition to separate electron-hole pairs;
- Increasing band gap states to increase visible light absorption.

Future Plans

- Explore alternative conformations of Pt over the (101) slab to find a more energetically stable system with stronger bonding interactions and a lower Pt adsorption energy
- Incorporate a GGA-U functional to more accurately study the electronic states near the band gap
- Create supercells to look at the interactions between Pt-TiO_2 regions and clean TiO_2 regions

References

1. I. Rosetti, "Hydrogen production by photoreforming of renewable substrates", *ISRN Chem Eng*, **2012**
2. R. Abe, K. Sayama, H. Arakawa, "Significant effect of iodide addition on water splitting into H_2 and O_2 over Pt-loaded TiO_2 photocatalyst: suppression of backward reaction", *AIST*, **2003**
3. C-H. Liao, C-W. Wang, J.C.S. Wu, "Hydrogen Production from Semiconductor-based Photocatalysis via water splitting", *Catalysts*, **2012**:2:490-516
4. G. Liu, C. Sun, S.C. Smith, L. Wang, G. Q. Lu, H-M. Cheng, "Sulfur doped anatase TiO_2 single crystals with a high percentage of {001} facets", *JCIS*, **2010**:349:477-483
5. S. Jensen, D. S. Kilin. "Electronic properties of nickel-doped TiO_2 anatase", *J. Phys.* **2015**:27
6. A. M. Hussein, R. V. Shende, "Enhanced hydrogen generation using ZrO_2-modified coupled ZnO/TiO_2 nanocomposites in the absence of noble metal co-catalyst", *International Journal of Hydrogen Energy*, **2014**:39:5557-5568

Acknowledgements

This project was funded by the National Science Foundation Division of Engineering Education and Centers grant no. 1461190. We thank South Dakota School of Mines and Technology for access to supercomputing facilities (gamow) to run DFT calculations. Special thanks to Dr. Kevin Hadley and other participating faculty in the TEAMS REU at South Dakota School of Mines and Technology for involving undergraduates in computational modeling.

511082 Electroless Plating of Copper on Silicon Wafer Via a Seedless, Environmentally Friendly Process

Abstract:

Abstract:

Copper is a soft, malleable, and ductile metal with very high thermal and electrical conductivity. The copper thin film was made by an ELETROLESS PLATING method, to directly grow a copper film on the silicon substrate without using any seed layer. The mechanism and kinetics of the growth of the copper film were studied. Ascorbic acid was employed as the reducing agent, which is a cheaper and non-hazardous material comparing to hydrazine or formaldehyde as the commonly used reducing chemical. In addition, the effort also been made to produce a nanoparticles solution of copper through a similar process and can be used to fabricate a copper thin film directly by spin coating. The copper thin film we made can be potentially used as a conductive circuit, RFID tags, or front-contact metallization for silicon heterojunction solar cells.

Copyright © American Institute of Chemical Engineers. All rights reserved.

509684 The Effect of Mordant on Carmine Adsorption Thermodynamics and Kinetics

Abstract:

The effect of mordant on the adsorption kinetics and thermodynamics of carmine dye were studied by analyzing the dyeing process at 25, 55, 75 and 85°C at a constant initial dye concentration by measuring the absorbance using UV- visible adsorption spectroscopy at λ_{max} 500 nm. The adsorption kinetics were found to follow a pseudo first-order kinetic model for all the mordants tested, as a high correlation coefficient (R^2) was obtained for the first-order model and the activation energies obtained for all the mordants ranged between 10.4-13.3 kJ/mol. It was also determined that the presence of mordant had a statistically significant effect on the rate constant. Langmuir and Freundlich adsorption isotherms were applied to describe the experimental equilibrium isotherms. The results indicate that Freundlich model provides the best correlation. Thermodynamic parameters: free energy (ΔG^o), enthalpy (ΔH^o) and entropy (ΔS^o) were also determined for carmine dyeing. The thermodynamic parameters showed that for all mordants, adsorption is spontaneous and thermodynamically favorable. It was deduced that the dyeing process is likely to be controlled by a physisorption process for both adsorption and kinetic studies.

Copyright © American Institute of Chemical Engineers. All rights reserved.

Understanding intramolecular cooperativity in acid-base silica-supported organocatalysts

Jingwei "Judy" Xie, Nathan C. Ellebracht, and Christopher W. Jones*
School of Chemical and Biomolecular Engineering
Georgia Institute of Technology, Atlanta, GA

MOTIVATION

Homogeneous organometallic catalysts:
- Expensive
- Homogeneous
- Unrecyclable

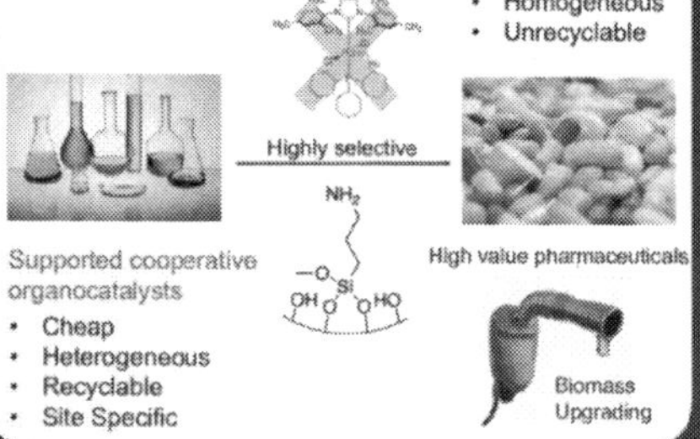

Supported cooperative organocatalysts
- Cheap
- Heterogeneous
- Recyclable
- Site Specific

SILICA SUPPORTED CATALYSTS

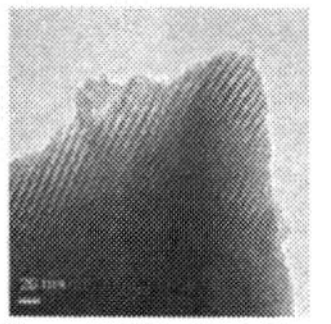

Figure 1. TEM of SBA-15

Mesoporous silica (SiO_2): SBA-15

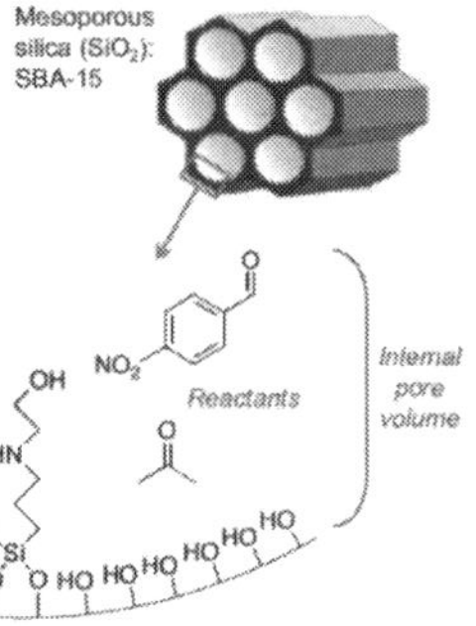

Scheme 1. Functionalized SBA-15 surface

COOPERATIVE CATALYSIS

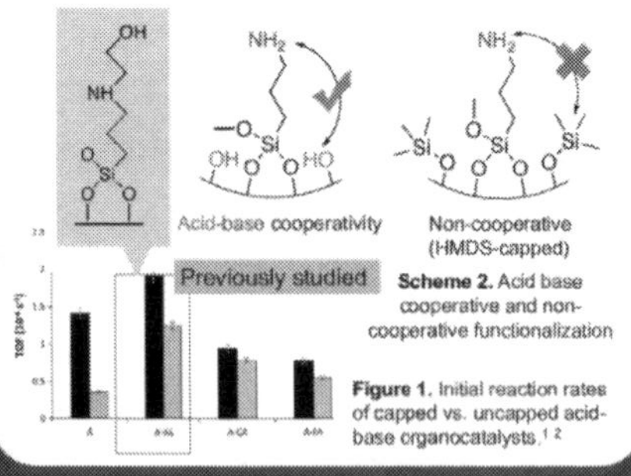

Scheme 2. Acid base cooperative and non-cooperative functionalization

Figure 1. Initial reaction rates of capped vs. uncapped acid-base organocatalysts.[1,2]

CATALYST SYNTHESIS & APPROACH

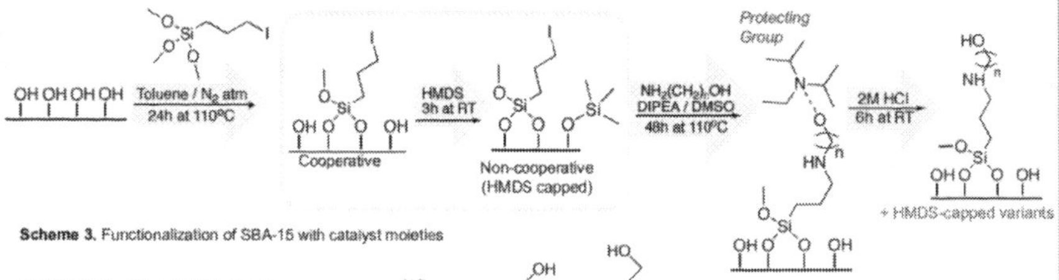

Scheme 3. Functionalization of SBA-15 with catalyst moieties

Longer linker length hypothesized to facilitate intramolecular cooperativity.

Transition state stability hypothesized to impact specific preferable linker length in promoting intramolecular interaction.

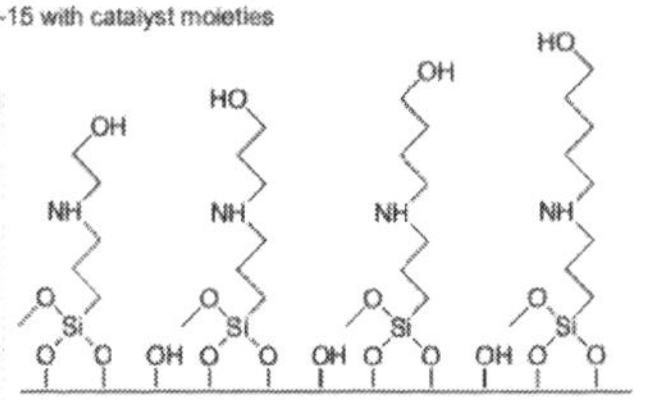

Scheme 4. Chemical structure and naming convention of catalysts

Catalyst Site Loading (mmol/g)				
Material	SBA-AA2C	SBA-AA3C	SBA-AA4C	SBA-AA5C
Uncapped	0.25	0.25	0.24	0.24
Capped	0.26	0.24	0.28	0.21

Table 1. Catalyst site loadings calculated from Elemental Analysis conducted by Atlantic Microlab (Norcross, GA) and estimated from Thermogravimetric Analysis

RESULTS & CONCLUSIONS

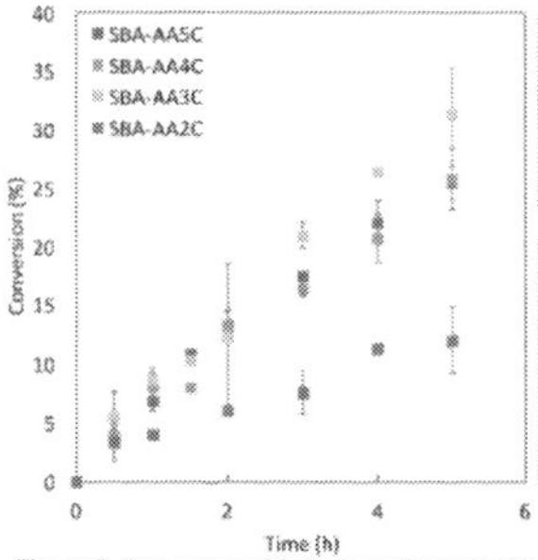

Figure 2. Conversion of 4-nitrobenzaldehyde with uncapped catalysts

The 3~5 carbon linker lengths offered comparable rates.

The 2 carbon linker length resulted in low rate, receiving no benefit from attached alcohol group.

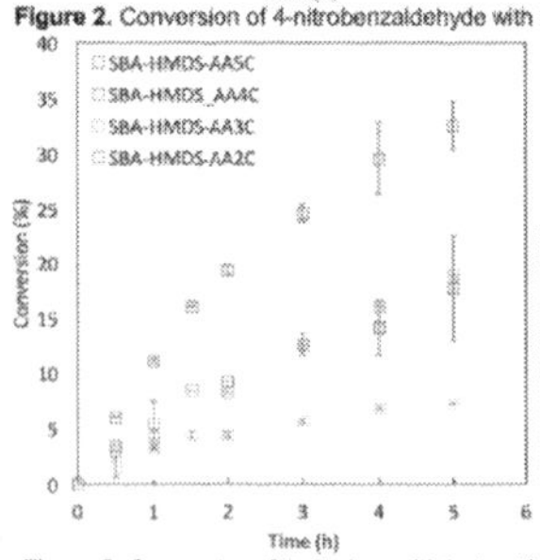

Figure 3. Conversion of 4-nitrobenzaldehyde with capped catalysts

The capped 5 carbon material resulted in higher rates, including observed improvements from capping.

KINETICS EXPERIMENTS

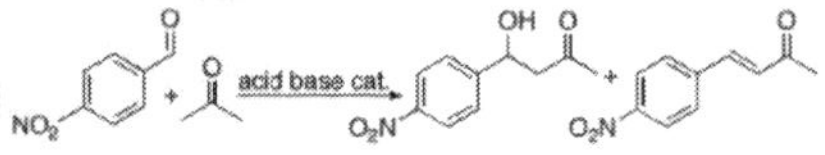

Scheme 5. Aldol condensation of 4-nitrobenzaldehyde and acetone

- 30mM [4-nitrobenzaldehyde] in [acetone]
- Internal standard [structure]
- 45°C under reflux
- 15mol% active sites
- Products analyzed by Gas Chromatography

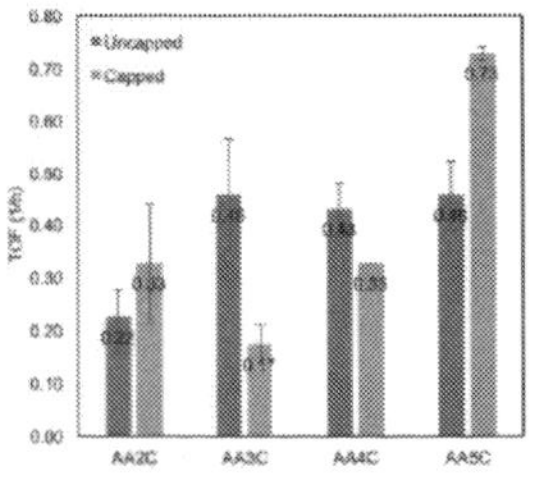

Figure 4. Turnover frequencies (TOFs) of the studied catalysts

Error bars: Standard deviation from individual experiments

The capped 2 and 5 carbon materials provided higher rates than the uncapped counterparts

FUTURE WORK

- Duplicate Elemental Analysis on current materials to obtain more statistical certainty in amine loading data and run further kinetics experiments using updated loading information
- Perform additional characterizations methods such as N_2 physiosorption, Infrared Spectroscopy and Thermogravimetric Analysis to confirm structure and prepare for publishing

APPLICATIONS

- Extend to a study of more industrial relevant reactions

Scheme 6. Cross aldol condensation of furfural and acetone

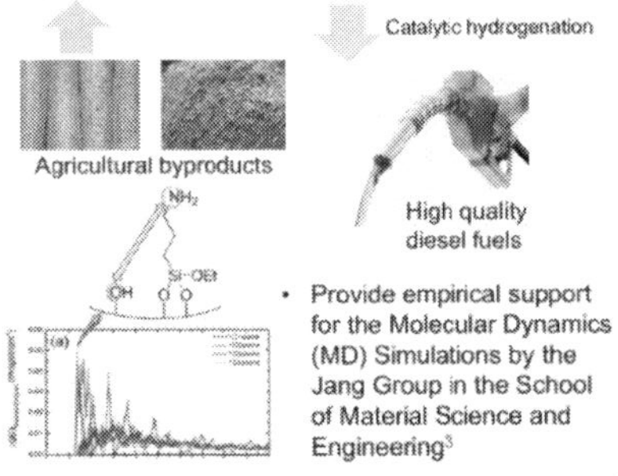

- Provide empirical support for the Molecular Dynamics (MD) Simulations by the Jang Group in the School of Material Science and Engineering[3]

REFERENCES

[1] J. Lauwaert, E. G. Moschetta, P. Van Der Voort, J. W. Thybaut, C. W. Jones, G. B. Marin. Spatial arrangement and acid strength effects on acid-base cooperatively catalyzed aldol condensation on aminosilica materials. J. Catal. 325 (2015) 19-25.
[2] N. Brunelli, C.W. Jones. Tuning acid–base cooperativity to create next generation silica-supported organocatalysts. J. Catal. 308 (2013) 60–72
[3] K. C. Kim, E. G. Moschetta, C. W. Jones, S. S. Jang. Molecular Dynamics Simulations of Aldol Condensation Catalyzed by Alkylamine-Functionalized Crystalline Silica Surfaces. J. Am. Chem. Soc. 2016, 138, 7664–7672.

ACKNOWLEDGEMENTS

This work is supported by the Undergraduate Research Opportunities Program at Georgia Tech and Department of Energy Office of Basic Energy Sciences through Catalysis Contract Number DEFG02-03ER15459. The author would like to thank the Jones Group and PhD Candidate Nathan Ellebracht for the guidance.

UROP U.S. DEPARTMENT OF ENERGY

510850 Electrocatalytic Oxygen Reduction Reaction on Oxide and Nitride Systems

Abstract:

The goal of this research project was to explore various cathodic catalysts that may exhibit an enhancement or equivalent properties compared to that of a platinum catalyst, which is the industry standard as a cathodic catalyst in alkali solutions. This research endeavor could lead to an inexpensive alternative for electrochemical reactions requiring an electrocatalyst. These materials were synthesized via hydrothermal process and then characterized with SEM and XRD methods. In order to perform the electrochemistry experiments oxygen reduction reactions were conducted via cyclic and rotary disk electrode voltammetry. Materials that were examined in this project were various crystalline structures of mesoporous tungsten oxide (WO3). WO3 exhibited an onset potential of approximately -275mV. For the oxide and nitride systems, cobalt was used in the presence of reduced graphene oxide. Cobalt oxide systems exhibited fair effectiveness as a electrocatalyst (onset potential of -300mV), but cobalt nitride demonstrated superior capabilities as a electrocatalyst having an onset potential of -250mV. The data gathered was then compared to both literature and experimental values of a platinum cathodic catalyst to gauge on whether these systems could be used as comparable electrocatalysts.

Copyright © American Institute of Chemical Engineers. All rights reserved.

510880 Synthesis and Optimization of Oil Soluble FKM Seal Compatible Copper Passivators

Abstract:

Copper and its alloys are widely utilized in automotive and industrial applications because of its frictional properties and high ductility. Unfortunately, a number of additives used in these industries can be corrosive to copper. To mitigate this, copper passivators are often coupled with these additives. While there are effective copper passivators on the market, many are limited in application due to limited solubility in oil, being detrimental to commonly utilized fluoroelastomer (FKM) seals, and being reactive with numerous other components. Additionally, many of these passivators are quite costly. Due to this, we sought to develop a cost-effective and atom economical copper passivator that was both oil soluble and FKM seal compatible. Sixteen triazole derivatives were synthesized and tested for copper passivation, demulsibiltiy, and solubility. Derivatives were synthesized by alkylating the nucleophilic nitrogen on benzotriazole with a branched alkyl group. All oxygen based groups were used to minimize the presence of reactive amines, theoretically reducing FKM seal compatibility issues. Optimization of this synthesis was carried out to maximize atom economy and minimize cost. Asymmetrically branched benzotriazole derivatives provided the most favorable conversions and the best solubility. Derivatives which utilized shorter aldehydes and longer, unbranched alcohols produced the smallest quantity of by-product. Specific limitations regarding the alkyl chain length of both the alcohol and aldehyde were also identified. The cheapest derivative synthesized was estimated to cost less than 34% of that of Irgamet 39, an effective and commonly utilized copper passivator. The purity of these derivates, after optimized reaction parameters were implemented, ranged from 90-94%. These reaction parameters also halved the reaction time necessary for comparable product conversions. Optimized derivatives were determined to be completely stable, remaining within Â± 3% of the derivativeâ€™s product purity at the time of synthesis. Currently, we are running static seal testing against bisphenol AF cured Type II fluoroelastomers to assess each derivativesâ€™ seal compatibility. In the future we aim to determine mechanism behind FKM seal degradation.

Copyright © American Institute of Chemical Engineers. All rights reserved.

510990 Electrochemical Reduction of CO2 in a Biphasic Continuous-Flow Electrolysis Reactor

Abstract:

Electrochemical conversion of carbon dioxide (CO_2) to carbon monoxide (CO) can simultaneously mitigate anthropogenic greenhouse gas emissions and produce useful industrial chemicals. When combined with a renewable energy source, a reactor can recycle CO_2 into CO, which is the carbon source of synthesis gas, in a sustainable fashion. Through Fischer-Tropsch processing, syngas is processed into a range of liquid hydrocarbons that can be used as fuels or precursor materials. Electrochemical fuel generation processes are advantageous because they usually occur at mild conditions, near ambient pressure and temperature, in comparison to the thermochemical processes. In addition, the time scale for adjusting the applied voltage, which controls production rate, is small and, allows the process to be synced with periods of low cost electricity. These benefits along with decreasing energy prices, have contributed to a growing interest in synthesizing catalysts for energy-efficient CO_2 conversion to CO.

Pursuant to this goal, significant development has been made on identifying electrocatalysts that are of high activity, stability, and selectivity towards CO over the competing hydrogen (H_2) evolution side reaction. My research group has screened an array of precious metal catalysts and identified carbon supported gold nanoparticles as a superior material, achieving over 90% coulombic efficiency. However, we conducted these experiments in a small-scale electrochemical H-cell designed for catalytic studies. A more practical application is to deposit the catalyst onto a gas diffusion layer and then characterize the electrode performance in a reactor that has higher product yield. Drawing from fuel cell literature and electrolyzer designs, we developed a continuous-flow electrochemical reactor with gas phase reactant delivery, where the anode and cathode are bridged by the flowing electrolyte and the gaseous reactant permeates directly to the active catalyst site. Using this analytical platform, we aim to quantify performance of the system as a function of electrode preparation, cell configuration, and operating conditions. Optimization of the flow cell can inform future catalyst and electrode development, and serve as a bridge between academic pursuit and industrial scalability.

Copyright © American Institute of Chemical Engineers. All rights reserved.

509228 Mechanistic Studies of Thermal Decomposition of Nickel-Gallium Layered Double Hydroxides (Ni-Ga LDHs)

Abstract:

Layered double hydroxides (LDHs) represent one of the most technologically promising catalysts due to their low cost and relative ease of preparation. The large number of possible combinations available due to the numerous synthetic methods, as well as a broad variety of metals that could be used, make the LDHs suitable for a variety of catalytic applications. The purpose of this project is to explore the structure and the properties of nickel-gallium layered double hydroxides (Ni-Ga LDHs), and to investigate their potential application in catalytic dry reforming of methane. Therefore, the ultimate goal would be to explore the feasibility of using Nickel-Gallium LDHs as stable catalysts to facilitate the transformation of carbon dioxide and methane into syngas, which is a useful mixture of carbon monoxide and hydrogen. Syngas production currently relies on coal gasification, a harmful process for the environment. The choice to use nickel and gallium as the major components was based on the fact that nickel and gallium are highly selective and efficient to activate methane and carbon dioxide respectively.

A series of nickel-gallium layered double hydroxides (LDHs) with three different nickel-gallium molar ratios (3:1, 1:1, 1:3) were synthetized by hydrothermal method and sol-gel method. The hydrothermal method requires prolonged heating at a relatively high temperature (above 100 °C) in autoclaves, while the sol-gel method requires overnight stirring at a moderate temperature (around 50-60 °C).

A variety of characterization methods, including thermogravimetric-differential thermal analysis (TG/DTA), mass spectrometer (MS), and Fourier transform infrared spectroscopy (FTIR), were applied to investigate physical structures of the prepared nickel-gallium LDHs, as well as the decomposition mechanisms. Mechanistic studies revealed that the decomposition processes underwent three separate steps, namely dehydration (around 200 °C), followed by dehydroxylation (around 300 °C), and then decarbonation (around 400 °C). Furthermore, different heating rates (2.5, 3.5, 5, 6.5, and 7.5 °C/min) were applied to calculate the activation energies for different steps regarding carbon dioxide and water, and elucidate kinetics models. The average value of activation energy ranged from 60 to 90 kJ/mol for water, and from 130 to 200 kJ/mol for carbon dioxide. As-prepared samples were also calcined in air at three different temperatures (200, 260, and 355 °C), and then characterized by thermogravimetric-differential thermal analysis (TG/DTA). Results suggested the presence of water in both the interlayer region and in the layer surface, as well as in the layer itself.

Copyright © American Institute of Chemical Engineers. All rights reserved.

509397 Electrocatalytic Generation of Ozone on Nickel- and Antimony-Doped Tin Oxide Electrodes for Water Purification

Abstract:

Ozone is recognized as a safe and powerful disinfectant in water that leaves no harmful residuals because it rapidly decays into oxygen. The current method of ozone generation is cold corona discharge, which is unfavorable because it creates harmful nitrogen oxides in air and yields low concentrations of ozone when dissolved in water, lowering operational efficiency. A possible solution to this problem is generating ozone electrochemically, which can be done by choosing catalytic electrode materials that are much more selective in oxidizing water into ozone than oxygen. Nickel- and antimony-doped tin dioxide (NATO) has proven to be an electrode material active for ozone generation; however, its current efficiency (i.e. selectivity of ozone instead of oxygen) has not exceeded 50% in literature [1] and the material has poor stability.

Typically, NATO electrodes are synthesized as a sol-gel from a coating of metal precursors decomposed thermally in air. In this study, we electrodeposit thin films of NATO to control catalyst loading and compare the effects of synthetic routes on the ozone generation reaction. From baths containing either chloride or nitrate ions, we deposit either metallic or oxide films. With chloride ions, Sn(II) is reduced directly to Sn(0) on the substrate, and then is thermally oxidized to SnO2. With nitrate ions, reduction of nitrate ions increases local pH and causes SnO2 to precipitate onto the substrate. The films are characterized by X-ray diffraction and Scanning Electron Microscopy/Energy Dispersive X-ray analysis, and ozone selectivity is measured with spectroelectrochemistry. We observe that thermal oxidation yields higher current efficiencies than electrochemical oxidation. Annealing time does not affect ozone current efficiency, but high annealing temperatures increase both SnO2 crystallinity and ozone current efficiency. Increasing the concentration of Sb and Ni in the electrodeposition bath increases both the amount of dopant present and the ozone current efficiency. This electrodeposited NATO synthesis method has yet to be optimized, but achieves ozone current efficiencies of 35-45%, comparable to the sol-gel method. Efforts to relate the structural and material properties of NATO to ozone selectivity and activity are ongoing.

Sources:

[1]. Christensen, P. A. â€œRoom Temperature, Electrochemical Generation of Ozone with 50% Current Efficiency in 0.5M Sulfuric Acid at Cell Voltages < 3V.*â€ Ozone: Science and Engineering*, no. 31, 2009, pp. 287â€"293., doi:10.1080/01919510903039309.

Copyright © American Institute of Chemical Engineers. All rights reserved.

509508 Imogolite Nanotubes As a Catalyst for Biomass Conversion

Abstract:

Rising energy costs and concerns regarding climate change necessitate the development of efficient and sustainable processes for the production of fuels and chemicals. Catalysts are an attractive solution to minimize the cost and environmental impact of these processes. One important process is biomass conversion, which can be employed to produce 5-hydroxymethyfurfual (HMF). HMF is a valuable intermediate for the sustainable production of biofuels, plastics, and chemicals. It is produced by the cascade reaction of glucose to fructose to HMF. Several intriguing catalysts for the isomerization of glucose to fructose have been identified, including immobilized enzymes, triethylamine, and zeolites. While these catalysts have sparked immense interest in this area, further research is necessary to identify novel catalysts for the isomerization reaction that are cost effective and environmentally friendly. The purpose of this research is the study of imogolite nanotubes as a catalyst for the isomerization of glucose to fructose. Imogolite nanotubes offer a highly tunable platform and the capability to design an effective heterogeneous catalyst. The successful synthesis of imogolite nanotubes has been confirmed. Preliminary catalytic tests on the imogolite nanotubes show promising activity for the isomerization reaction. Further work includes the determination of the active site of imogolite that is responsible for the catalytic activity. Extensive characterization of the materials to determine dimensions, composition, and acid-base properties gives insight on the active site. An opportunity to improve the selectivity of the imogolite catalyst for fructose lies in modification of the nanotube structure and composition through changes to the synthesis procedure. Effective modifications to the structure and properties of imogolite have been identified and performed, including methyl functionalization, precise control over dimensions, and heteroatom substitution. The effects of these modifications on surface properties and catalytic activity aid in identification of the active site and enable further tuning of catalytic activity. The results of preliminary tests establish that imogolite has the potential to be applied as an effective isomerization catalyst. Through further investigation of the nature of the catalytic activity, imogolite could be a novel catalyst that aids in the efficient, eco-friendly, and sustainable production of fuels and chemicals from biomass.

Copyright © American Institute of Chemical Engineers. All rights reserved.

510676 Reductive Amination of Furfural and Hydrogenation of Furonitrile

Abstract:

Amines are widely used in the manufacture of pharmaceuticals, agricultural chemicals, polymers, and surfactants. However, the majority of amines today are produced via petrochemical means. Thus, there is an environmental and economic driving force to produce amines from renewable resources, such as biomass. Ammonia is the ideal choice for a nitrogen source, as it is inexpensive and readily available. However, the use of ammonia produces reactive primary imines. The reactions were performed in a 150 mL stirred batch reactor under continuous stirring with a temperature controller. The reactor was purged with hydrogen, and then filled to the desired hydrogen pressure for reaction. The formation of these reactive amines is not able to detected by gas chromatography, as they will form other minor products. Another equally challenging factor in reductive amination is that primary amines are notoriously difficult to control, as they are better nucleophiles than ammonia, and easily form less valuable secondary amines. The effect of the metal used as the catalyst was tested, with pre-reduced Ruthenium being found as the best producer of the primary amine and Palladium was the best producer of the secondary amine. Other important variables in this experiment include temperature, amount of ammonia in reactor, amount of water in reactor, and pressure of hydrogen used. A key element of this reaction is the location of the reactants on either the catalyst surface or in solution and a change in temperature changes the profile of the solution. The amount of ammonia in the reactor is relevant because it is the nitrogen source. The presence of water in the reactor has an interesting effect. It forces the equilibrium in the balance between the secondary amine and primary amine toward the primary amine. However, this effect needs to be balanced with the fact that water and furfural are not miscible. Lastly, the amount of hydrogen in the reactor is needed to form the imines into amines. If this hydrogen is not present, hydrofuramide is formed and the material is lost. Further work suggested on this project include spectroscopy data of the adsorption of Furfurylamine on the catalyst surface and to obtain results at a wider range of temperatures.

Copyright © American Institute of Chemical Engineers. All rights reserved.

511156 Modification of Metal Oxides with Extreme Points of Zero Charge By Non-Thermal RF Plasma

Abstract:

The effect of non-thermal RF plasma modification on the point of zero charge (PZC) of magnesia and silica has been explored. It has been previously established that plasma modification effectively lowers the PZC of alumina, an effect attributed to adsorbed gas species. The purpose of this study is to build on that work, answering the question, "Can non-thermal RF plasma moderate more extreme PZC?"

In this work, samples were modified with air and oxygen plasma, before PZC analysis. In order to further explore the effect of this treatment, Fourier-transform infrared spectroscopy (FTIR) studies were performed, to assess potential changes to adsorbed species.

Ultimately these supports will be decorated with catalytic metals via strong electrostatic adsorption (SEA), the efficacy of which (in terms of overall loading, particle size, and bond strength) is dependent on the material's PZC. Effective modification could open up previously unattractive supports for use with SEA.

Copyright © American Institute of Chemical Engineers. All rights reserved.

510454 Longer Wavelength Azo Dyes for Incorporation in Photomechanical Polymer Networks

Abstract:

Azo dyes have long been studied for their potential uses in photomechanical materials, which convert light energy directly into mechanical work for use as wireless actuators and switches. Most azo dyes undergo reversible *trans* to *cis* isomerization at ultraviolet to blue-green wavelengths. These comparatively high energy wavelengths are more prone to photodegrade materials or to be competitively absorbed by other components of a device or material, and they are incompatible with mammalian tissue. Thus, it is desirable to develop new azo-based photomechanical systems which absorb at lower energy wavelengths. Aprahamian and coworkers have reported BF_2-coordinated azo dye analogs which undergo isomerization at orange-red to near infrared wavelengths. These longer wavelength dyes minimize photodegradation and competitive absorption and are more likely to be compatible with mammalian tissue. It is necessary to install polymerizable handles on these dyes if they may be incorporated into a photomechanical material. The synthesis of several dyes is reported.

Copyright © American Institute of Chemical Engineers. All rights reserved.

510540 Hydrothermal Liquefaction of Algae Grown on Dairy Wastewaters

Abstract:

Algae-based biofuels have attracted significant research interest due to their advantages of not competing with land for food production, abilities to grow in low-quality water, higher growth rates, and strong CO2-mitigation abilities. In recent years, research has focused on hydrothermal liquefaction (HTL) of whole, wet algae biomass. HTL is a thermochemical process, considered one of the most promising techniques for valorization of wet algae biomass. HTL uses subcritical water (270-350°C and 8-18 MPa) both as a solvent and a reaction medium to convert organic biomass constituents into energy-rich bio-crude oil which can be refined into liquid fuels. HTL is most suitable for conversion of wet feedstocks including algae. In this study, we compare the oil yields and energy recoveries from the HTL of two algae polycultures used to remove excess nutrients from dairy manure effluents: a microalgae polyculture grown on brackish manure effluent in a photo bioreactor system at NMSU and a filamentous algae polyculture grown on freshwater dairy effluent as part of an algae Turfscrubber® system at ENMU.

Batch HTL conversion experiments were carried out using a 1.8 Liter autoclave reactor at reaction temperatures and retention times of 310°C and 350°C, and 30 and 60 minutes, respectively. Algae and reaction conditions were compared based on product yields and the higher heating values (HHV) of the bio-crude oils. For brackish water microalgae produced a yield of 10-24 wt. % oil, 19-38 wt. % char, and approximately 45MJ/kg for HHV. The yield for filamentous algae grown on freshwater effluent was 11-15 wt. % oil, 16-25 wt. % char, and approximately 30-44 MJ/kg. Both of these feedstocks presented high ash content, and presented a low moderate potential for energy recovery in liquid fuels. Future work will focus on bio-crude oil, and char characterization, as well as the evaluation of the dairy effluent waste water treatment and the renewable energy production combined process.

Copyright © American Institute of Chemical Engineers. All rights reserved.

510828 Synthesis of Copper Oxide-Based Nanofibers Via Electrospinning and Their Utilization in the Removal of Sulfur Compounds from Gaseous Streams

Abstract:

Electrospinning is a fast-growing synthesis technique that allows for the fabrication of one-dimensional inorganic-organic nanofibers from solutions containing a metal precursor and/or a polymer. Fibers prepared via this method may be tuned to have high surface areas, superior mechanical properties, and a tunable porosity, which makes them ideal candidates in applications involving separation, catalysis, and reactive sorption. Although the electrospinning process is simple, versatile, and adaptable, a myriad of parameters ought to be regulated and optimized to fabricate ultrafine, smooth, continuous, and defect-controlled fibers for use as sorbents in the removal of toxic sulfur compounds.

In this work, polyvinylpyrrolidone-copper (PVP-Cu) nanofibers were electrospun from a solution containing PVP (MW=1,300,000, 360,000, or 40,000 g/mol) and copper nitrate trihydrate. Four key solution properties (polymer concentration, polymer molecular weight, metal concentration, and solvent identity) were systematically varied to investigate the effect of these parameters on PVP-Cu fibers morphology. The electrospinning solutions were prepared by dissolving PVP (0.0212-0.0576 g/cm^3) in a solvent (methanol, ethanol, 1-propanol, or water). The copper precursor was dissolved in DI water while the copper nitrate to PVP weight ratio was varied from 1:4 to 1:1. The collected nanofibers were heat treated under air at 823 K for four hours at ramping rate of 2 K/min. The heat treatment was necessary to remove the PVP matrix from the fiber structure and form copper oxide (CuO). Various characterization techniques were employed in this work: thermogravimetric analysis (TGA) was used to assess the extent removal of the polymer upon heat treatment, X-Ray diffraction (XRD) was used to identify CuO phases and average crystallite sizes, scanning electron microscopy (SEM) was used to analyze nanofiber morphology, and both electron dispersive spectroscopy (EDS) as well as X-ray photoelectron spectroscopy (XPS) were used to determine the elemental composition of the fibers. Diameter size and distribution of PVP-Cu fibers were found using ImageJ software by means of converting pixels to length measurements using the SEM images scale bar.

This research effort shows that by varying the solution properties, PVP-Cu nanofibers with diameters ranging from approximately 50 to 550 nm can be formed with either ribbon-like or circular cross sections. Fused PVP-Cu nanofibers also formed in this work when the solvent was not allowed to evaporate properly. The solution viscosity, conductivity, and density as well as the solvent dielectric constant were found to be responsible for these morphological changes in PVP-Cu fibers.

After identifying the optimal solution properties ranges that yielded smooth, continuous and defect-free fibers, three CuO fibers with different mean diameters (70, 350, and 650 nm) and crystallite sizes were catalytically tested as sorbents for the removal of sulfur compounds. The fixed bed sorption reaction at room temperature demonstrated the sensitivity of fiber diameter (and thus crystallite size) on the amount of sulfur sequestered. The small reactive domains of the electrospun nanofibers and the thermodynamically favorable reaction between CuO and sulfur at low temperatures make CuO nanofibers an attractive option for the removal of sulfur compounds from gaseous streams.

Copyright © American Institute of Chemical Engineers. All rights reserved.

510967 Ru-Pt/C Catalysts Prepared By Sea and ED Methods for Direct Methanol Fuel Cells

Abstract:

The direct methanol fuel cell (DMFC) has been receiving increasing attention as a future power source suitable for small portable electronic devices such as laptops and cellphones. Currently, a Ru-Pt alloy and Pt metal are typically used as the anode and cathode catalysts, respectively, for the DMFC. Many methods exist to prepare bimetallic catalysts, and the properties and performance of the catalysts can be related to the preparation method used. In this project, we are interested in the preparation of nanometer sized Ru-Pt particles on a carbon support with different loadings of Ru and Pt. The methods of strong electrostatic adsorption (SEA) and electroless deposition (ED) were used for the preparation of Pt and Ru nanoparticles respectively. These methods utilize the manipulation of pH and temperature to affect the properties and performance of the catalyst. The particle size and distribution of the catalysts were determined by means of chemisorption and X-ray diffraction. The activities of the catalysts for the oxidation of methanol were evaluated using cyclic voltammetry. These Ru-Pt catalysts prepared by SEA and ED methods are expected to show higher activity compared to the activity of commercially available Ru-Pt catalysts prepared by alternate methods when used in the DMFC.

Copyright © American Institute of Chemical Engineers. All rights reserved.

510805 Replicating the Origin of Life on Earth: The Importance of Hydrothermal Microenvironments

Abstract:

The "primordial soup" is a widely accepted theory of the origin of life on Earth. However, this theory fails to explain how simple molecules (e.g., amino acids and sugars), diluted in prebiotic oceans, could have reached high concentrations necessary to synthesize more complex molecules, the building blocks of life (e.g., proteins and DNA). Since the discovery of the first hydrothermal vents in the deep ocean floor and their ability to provide a favorable environment for life, hydrothermal systems have emerged as an alternative to the "soup" theory. In particular, porous mineral formations near hydrothermal vents, subjected to thermal gradients from the cold deep-sea water and the hot magma underneath, could have provided microenvironments that enabled the first forms of life. In our project, we replicated the conditions for the origin of life in a laboratory scale to understand how hydrothermal microenvironments could produce high concentrations of molecules necessary to form life's building blocks. Specifically, we constructed pore-like chambers inside which convective flow can be established in a controlled way. Using this approach, we studied different possible flow trajectories that could occur inside such pores, ranging from well-organized patterns to highly disorganized or chaotic flows. The results showed that chaotic flow can quickly accumulate molecules in certain regions of the pore-like chambers, providing the necessary high concentrations for the synthesis of life's building blocks to occur. This study provides evidence for hydrothermal vents as enablers of the first forms of life in Earth's primitive oceans. Our most recent research focuses on an interesting phenomenon observed in these hydrothermal microenvironments. Two immiscible liquids inside these microenvironments form complex microstructures in a consistent and reproducible fashion due to the presence of convective flow. We are currently investigating properties of these microstructures, potential for encapsulating different liquids and species, other geometries of the pore-like chambers that could lead to their formation, and their possible applications.

Copyright © American Institute of Chemical Engineers. All rights reserved.

510988 Energy Requirements of Thermolytic Solutes in Osmotically Driven Membrane Processes

Abstract:

Reverse osmosis (RO) exists as the primary technology used to treat brackish and sea water. However, RO requires large amounts of electrical energy to overcome the natural osmotic pressure of the feed solution. Fortunately, forward osmosis (FO) is a water treatment option that recovers water from feed solutions via a draw solute introduction. Tertiary amines were studied as switchable polarity solvents (SPS) for use in a FO treatment system and can modulate polarity upon reaction with carbon dioxide. The switch reaction of dimethylcyclohexylamine (DMCHA) was studied and samples were analyzed using freezing point osmometry to determine the equilibrium constant and subsequent enthalpies and entropies of reaction. The switch reaction of cyclohexylpiperidine (CHP) was studied earlier and the data was compared with what was generated for DMCHA. At 30Â°C and 100% CO_2, DMCHA achieved a maximum salt weight percentage of 60.5% while CHP achieved a 55.8% maximum loading. vanâ€™t Hoff plots were then generated for each amine. The vanâ€™t Hoff behavior of DMCHA interestingly includes two distinct K_{eq} regions which suggests the amine undergoes two different reactions at different temperature ranges. Understanding the thermodynamics of the switch reaction is important to optimizing the SPS FO water treatment system.

Copyright © American Institute of Chemical Engineers. All rights reserved.

509988 Taylor Series for Peng-Robinson Equilibrium Properties Using Similarity Variables

Abstract:

The Peng-Robinson (PR) equation of state (EoS) is used to determine vapor pressure in addition to PVT properties. In this work, a dimensionless version of the PR equation has been simplified by combining the temperature and acentric factor into a similarity variable, *Î·*, termed the adjusted temperature. Other new dimensionless variables were defined by combining pressure and temperature into a dependent variable, *Î¨*, termed the adjusted pressure, and by defining a reduced density variable, $Ï_r$', relative to the maximum liquid density in the low temperature limit. A Taylor series expansion in the similarity variable *Î·* was used to express reduced phase densities and vapor pressure by simple polynomials. The truncation error of these series was characterized. This work resulted in a simpler method of estimating PVT properties and vapor pressure that would be predicted by the PR equation for any substance.

Copyright © American Institute of Chemical Engineers. All rights reserved.

510288 Modeling and Control of Renewable Energy Sources with Energy Storage System for Smart Grid Application

Abstract:

Even though electricity generated by fossil fuels (coal, natural gas, petroleum, and other gases) is recognized as one of the largest sources of greenhouse gas emissions, until 2016, 65 percent of the electricity generated in the United States was from these sources. Recently, a more environmental friendly energy source, renewable energy is increasingly incorporated with the electricity grid. The U.S Department of Energy mandates that by 2030, 20 percent of the total energy consumption is produced from wind energy. A major challenge of renewable energy sources is the fact that they are non-dispatchable ones, which cannot be dispatched freely to meet the electricity demands. Thus, massive energy storage system (ESS) is installed to the grid to meet the energy demand and shift the production from low-energy value periods to high-energy value periods to maximize profits.

There are two ways to improve the optimal cost of an electric grid with an ESS. The first one focuses on the placement and sizing of the ESS on the grid, and the second one focuses on the operation of the grid. These two approaches can be studied independently, but they can also be applied simultaneously to further improve efficiency. Adeodu and Chmielewski applied Economic Model Predictive Control (EMPC) and Economic Linear Optimal Control to solve the Optimal Sizing and Placement (OSP) of ESS. The focus of our PURE project in this Fall 2017 uses this OSP foundation to establish an appropriate grid operating policy. The tools that we use are EMPC and Multistage Stochastic Programming (MSP).

EMPC uses a state-space model to predict the future values of the cost of electricity generated from the current load demands and the realizations of the renewable energy generated randomly. However, since the expected values of the disturbances are zeros, EMPC ignores forecasts, so the optimal solutions can be ameliorated further. MSP allows the users to specify the scenarios of disturbances; hence it provides better controlling policy than EMPC. The disadvantage of MSP is the problem size increases geometrically with the size of future predicting horizon. To diminish the computational time of the MSP, some decomposition method such as Nested Bender Decomposition is applied.

Additionally, determining disturbance trends is critical to effective control. The disturbances, which are the load demands and the generated renewable power, are modeled using third-ordered shaping filter. The probability of each disturbanceâ€™s realization is quantified by an effective counting method. This filter simulation is run for a long time and the occurrence of each instance is counted. e.g Given 3 bins: low, mid, and high, how many times the sequence Low-Medium-High, High-High-High, High-Low-High... occur, etc.

The purpose of this project is to identify if MSP is an appropriate framework for economic dispatch on networks with energy storage. Additionally, this project will certify the hypothesis that Infinite Horizon EMP can be a reasonable approximation for the MSP.

Copyright © American Institute of Chemical Engineers. All rights reserved.

510529 Use of Type I DES for Deep Desulfurization of Diesel: Insights from Molecular Dynamics Simulations

Abstract:

With ever-increasing environmental sulfur regulations, the development of alternative desulfurisation techniques, beyonds the industrially practiced hydrodesulfurisation is essential. The current legislative requirements of sulfur content in transportation fuels is less than 10 ppm. Among the recent desulfurisation techniques, extractive desulfurisation is highly promising.The use of Deep Eutectic Solvents (DES) in the extraction of sulfur content is a new and effective alternative to the previously used solvents. Amongst several DESs, type I DES have been shown effective in almost 100% removal of refractory sulfur compounds such as: thiophenes, benzothiophenes and dibenzothiophens. In this work tetrabutylammonium chloride (TBAC), PEG, in combination with $FeCl_3$, at a ratio of 4:1:0.05 is investigated as a DES, while dodecane, and benzothiophene with dibenzothiophene were used to simulate the diesel and sulfur-based compounds, respectively. Molecular dynamics simulation is used in order to investigate the interactions between DESs, diesel, and sulfur content at different temperatures and loadings. In addition, the simulations are also performed without PEG, in order to observe the role of the polymer in the process. The results show strong interactions between DESs and sulfur based compounds in comparison with the model diesel. Moreover it is also observed that interactions are stronger with the use of PEG, however further investigation is being carried on. In summary, this study sheds light on designing efficient solvents for desulfurisation based on molecular dynamics simulation.

Copyright © American Institute of Chemical Engineers. All rights reserved.

510833 The Lifetime of Shell Anions: How Long Do Ionic Liquid Anions Remain in a Li+ Shell?

Abstract:

In recent years, ionic liquids have been considered strong candidates as electrolytes. Conventional organic electrolytes can lead to lithium dendrite growths that are known to cause fires. Ionic liquids are a favorable replacement due to their large electrochemical windows, low volatility, improved cycling, and inability to catch fire. A limiting factor of ionic liquids is the poor conductivity compared to conventional methods, which is largely influenced by lithium ion transport. Studies have shown that certain mixtures of ionic liquids have more desirable properties. In our study, we have used molecular dynamics simulations to examine various properties in several ionic liquid mixtures with the goal that learning more about these compounds at the molecular level will help us understand their unique properties at the macroscopic level.

Molecular dynamics simulations have been used in the past to measure Li+ diffusion through an electrolyte. For our study, a cubic sample of ionic liquid electrolyte was generated with various amounts of Li+ ions placed in the solution and were allowed to diffuse. The simulation can actually generate videos of such diffusion on very small timescales (i.e. picoseconds, nanoseconds). We wanted to investigate the properties of the group of anions that surround the Li+, or the anion shell, to find the amount of time that an ionic liquid anion stays within the anion shell. It is possible that there is a relationship between the fluidity of the anion shell and Li+ diffusion, as the study examines.

Molecular dynamics solutions also can give us useful information about the spatial orientation of the molecules in our system. The radial distribution function was examined to find peaks that correspond to different spatial configurations between Li+ cation and ionic liquid anion. The findings from the radial distribution function were coupled with a visual aid to illustrate these interactions for different compounds.

Copyright © American Institute of Chemical Engineers. All rights reserved.

510863 An Unbiased Classification Algorithm for Transverse Tubule Remodeling within Murine Heart Failure Models

Abstract:

Transverse tubules (TTs) are the main method of delivery of extracellular calcium into the myocyte, responsible for the initiation of calcium induced calcium release (CICR) necessary to carry out excitation contraction coupling (EC coupling) within ventricular myocytes. Disruptions in the regularly striated TT network are correlated with various degrees of calcium mishandling and are generally observed within heart failure models. Currently, TT remodeling is often judged qualitatively, subject to bias of the experimenter. To address the need to eliminate this bias, we propose a technique that utilizes: 1. Computer vision libraries. 2. Signal processing techniques such as matched filtering that provide signal-to-noise ratios, a quantitative measure of observed remodeling. Combining these methods, an unbiased classification algorithm is presented that is able to quantify both regions of TT loss and longitudinal remodeling as well as characterize whole myocytes by the magnitude of their deviation from the prototypical healthy myocyte. This allows for the expedient, and more importantly, unbiased and reliable classification of TT remodeling observed within murine ventricular myocytes.

Copyright © American Institute of Chemical Engineers. All rights reserved.

510960 Advanced Model Predictive Feedforward/Feedback Control of a Tablet Press

Abstract:

In continuous pharmaceutical manufacturing, real-time precise control of critical quality attributes (CQAs) is necessary for quality by design (QbD)-based manufacturing and real-time release (RTR) with minimum consumption of time, space, and resources. Pharmaceutical tablets can be produced through different routes with a common tablet press unit operation always placed at the end of the manufacturing process. Therefore, the tablet press is a crucial unit operation directly influencing the CQAs irrespective of manufacturing routes. Despite this, little attention has been paid to the development of an advanced efficient control system for the tablet press. Process modeling can be used as an efficient virtual experimentation tool to design, compare, and evaluate different control systems. We developed a model in Simulink (Mathworks) that includes two master control loops for tablet weight and hardness and a slave feedback loop controlling the compaction force applied to each tablet. We examined the performance of different control strategies based on proportional integral derivative (PID) control and model predictive control (MPC), as well as feedforward/feedback control. We found that a hybrid MPC-PID control strategy outperforms the PID-only control strategy. We also observed that the addition of a feedforward controller further improves the performance of the hybrid MPC-PID control strategy.

Copyright © American Institute of Chemical Engineers. All rights reserved.

511031 Modeling and Optimization of Ion Transport Membranes for Oxygen Separation from Air

Abstract:

Modeling and Optimization of Ion Transport Membranes for Oxygen Separation from Air

Allyson M. Brezler, Gaurav V. Mirlekar, and Fernando V. Lima

Submitted to Topical Area: Computing and Process Control

Department of Chemical and Biomedical Engineering, West Virginia University, Morgantown, WV

The use of high purity oxygen instead of air in fossil fuel combustion processes can increase conversion and aid in the decrease of CO_2 and NO_x emissions. Cryogenic air separation is the current technology implemented to produce oxygen, but operation requires large energy inputs that result in high costs. Ion transport membranes (ITM) for oxygen separation is an emerging technology that can be used in place of the cryogenic process to decrease costs. Ion transport membranes separate oxygen from air by selectively allowing the oxygen ions to pass through the membrane. Ceramic membranes with a perovskite crystal structure are a well-suited material for this application. In this work, the $La_{0.6}Sr_{0.4}Co_{0.2}Fe_{0.8}O_{3-d}$ (LSCF) perovskite membrane module is modeled and optimized for the implementation into a 620 MWe integrated gasification combined cycle (IGCC) power plant.

An ITM model is designed in MATLAB by performing a material balance across the membrane and applying a permeation flux equation. Preliminary results show that after optimization by employing a nonlinear programming function, the cost of a membrane required to produce 1500 mol/s of oxygen for the IGCC power plant decreased from the base case design cost. Such optimization will be further discussed to reduce the cost of the membrane. By generating a model for the cost effective membrane that is overall less expensive than current air separation units, this research will help push this technology further towards implementation into chemical and advanced energy processes.

Copyright © American Institute of Chemical Engineers. All rights reserved.

509706 Discrete Element Modeling of the Impact Breakage of a Wet Flexible Fiber Agglomerate

Abstract:

Biofuels are a source of clean and renewable energy. Understanding the impact breakage of a wet fiber agglomerate is key in the design process of a biomass reactor, leading to a reduction in material processing cost. In the biomass refinement process, long biomass fibers and their agglomerates must be broken down into a slurry for subsequent mixing and chemical reaction. The most common method for the comminution of biomass fibers is using a hammer mill. Discrete element method (DEM) simulations are performed to model the breakage of biomass fibers in a hammermill by simulating a wet fiber agglomerate impacting against a target wall. A whole spectrum of impact velocities is employed in which the agglomerate remains unbroken at a low velocity and breaks into small pieces at a high velocity. Parametric studies show that the extent of breakage of the agglomerate is increased by decreasing the liquid surface tension, liquid to solid volume ratio, contact angle, and/or liquid viscosity.

Copyright © American Institute of Chemical Engineers. All rights reserved.

510899 Design and Simulation of Pollution Control Units for Improving Sustainability

Abstract:

The United States Environmental Protection Agency (U.S. EPA) has long advocated for, and mandated regulations on chemical processes to prevent and reduce the life cycle impact the processes have on the environment. Processes in the past have released pollution to the extent that it has affected the environment, human health, and climate of the world around us. To cope with the current pollution challenges caused by the release of various chemicals into the environment, it is required to include suitable pollution control technologies for chemical processes to purify and clean waste streams in order to meet the government standards on emissions.

The design and simulation of pollution control units are critical when assessing, improving, and optimizing the process performance in terms of sustainability. As part of this project, four Pollution Control Unit (PCU) models are developed in Microsoft Visual Basic for Applications (VBA) in Excel to calculate mass and energy balances, equipment sizing, land footprint, and utility demand for a chemical process represented by a CHEMCAD simulation. To better evaluate the life cycle impact of the process regarding the land use, resources consumption and waste releases, Life Cycle Inventory (LCI) results can be obtained rapidly and accurately by the designed PCU simulators.

The chemical process for the production of acetic acid from methanol is used to test the developed PCUs. The waste streams of the acetic acid production process are connected to these units. Such streams are then treated to minimize the harmful outputs, thus increasing the process sustainability, which is reflected in the LCI calculations. Results of this work indicate that the pollutants were removed from the waste gas and mainly converted to carbon dioxide and water. For obtaining this outcome, the pollution control units are sized accordingly in order to achieve the overall LCI improvement for the process.

Copyright © American Institute of Chemical Engineers. All rights reserved.

510780 High-Throughput Molecular Simulations into the Morphology of P3HT:PCBM Blends

Abstract:

The goal of this research is to understand how temperature, solvent quality, solvent amount, and the concentrations of organic photovoltaic (OPV) components determine active layer morphology. This understanding will improve techniques for engineering OPV devices, which can be inexpensively processed from abundant materials but presently suffer from low photoconversion efficiencies. We perform molecular dynamics (MD) simulations using HOOMD-Blue accelerated with graphics processing units (GPUs) to quantify how individual molecules self-assemble into structures that influence power conversion efficiency. We simulate blends of poly(3-hexylthiophene-2,5-diyl) (**P3HT**) with [6,6]-phenyl-C61-butyric acid methyl ester (**PC61BM**) and [6,6]-phenyl-C71-butyric acid methyl ester (**PC71BM**), three of the most important molecules in OPVs. By screening hundreds of combinations of concentration, temperature, and solvent properties, we can identify the conditions that optimize their self-organization. We quantify the degree of order in the predicted morphologies with radial distribution functions, structure factors, and simulated diffraction patterns. We find morphologies in agreement with prior experimental and theoretical work, and offer suggestions for future combinatorial studies.

Copyright © American Institute of Chemical Engineers. All rights reserved.

509869 Paneled and Painless: Improving Student Interest, Confidence, and Understanding through Comics

Abstract:

Engineering and other forms of STEM education should be accessible for all who wish to learn more about topics in those fields. However, heavy jargon, lack of captivating visual aids, expensive resources, and outdated language make STEM education tedious, unappealing, even incomprehensible. Abstract concepts, such as fugacity, are only partially understood, if at all, due to lack of personal or real-world connection. To increase accessibility and make information engaging, we have produced a series of STEM-based comics, communicating difficult concepts in a visual story-telling style to introduce, solidify, and reaffirm knowledge that coincides with current STEM literature. Complicated ideas, like the refrigeration cycle, are displayed in a colorful, exciting narrative to keep readers engaged, along with visuals to break down the components into understandable pieces. Topics began in chemical engineering concepts and have since expanded to subjects such as genetics and effective education tools.

Writing the comics is completed in alignment with professors' lectures, with script edits and illustrations being performed by a student artist. Work is periodically sent back and forth to ensure the artist portrays accurate information in the visual aspect of the comic. The work is then either published to a journal, a website, or distributed to colleges and schools to be used as a curriculum supplement to whatever subject the comic accompanies. Future projects may include experiments alongside the comics or an entire textbook written like a graphic novel.

These comics have been shown to increase understanding of the particular subject matter. The humor in the comic helps readers to keep things to memory, as humor has been linked to storing information in long-term memory. Future work will include producing these comics in a variety of subjects and departments, presenting up-and-coming research in a form that can be understood by as wide an audience as possible.

Copyright © American Institute of Chemical Engineers. All rights reserved.

509873 Design, Fabrication, and Testing of a 3D Printed Centrifugal Pump

Abstract:

Additive manufacturing offers designers a rapid means of testing ideas. Other advancements in desktop 3D printing have given more power to users by allowing them to reinforce parts with fiberglass, kevlar, and carbon fiber giving parts the rigidity and strength needed for certain projects. These advancements and others like it in 3D printing show promising applications in chemical engineering and other disciplines. Some of these applications include maintenance, as well as in design and testing of new products and process equipment. In this 10 week project, a working and reliable centrifugal pump was designed using Solidworks 3D CAD software. Based on problems associated with previous pump designs such as leaking, a weak impeller drive shaft connection, a lack of assembly rigidity, DC motor damage, and a weak assembly support structure, a new list of design requirements was made. With these new design requirements in mind, a new design was drafted. A MarkForged Mark II 3D printer was used to create viable, watertight, carbon fiber and fiberglass reinforced nylon parts. Preliminary testing of pump prototypes still showed design shortcomings like air accumulation, shaft misalignment, and persistent leaking. However, all these problems were solved quickly using additional 3D printed parts, Solidworksâ€™ Finite Element Analysis tools, changing the assembly orientation to allow air to flow out, using embedded ball bearings, adding gaskets, and using packing to create seals around dynamic parts. After all design requirements were met, pump performance was measured using a testing rig to gather necessary data for a thermodynamic analysis, efficiency calculations, and pump performance curves. With a power input of 4 watts the max head height was 75 cm at a flow rate of 15 mL/s. Also the max flow rate with a power input of 4 watts was 49 mL/s. This testing rig was also used to investigate how different pump impeller shapes and designs effected overall pump performance. Lastly a final bill of materials was calculated with an estimated cost of $282 for this functional prototype. This proof of concept shows the promise, cost effectiveness, time effectiveness, and versatility of additive manufacturing in the design, maintenance, and troubleshooting of process equipment in the specific case of a common engineering application of pumping.

Copyright © American Institute of Chemical Engineers. All rights reserved.

509987 Empirical Vapor Pressure Prediction from Cubic Equations of State Using Similarity Variables

Abstract:

Vapor pressure prediction is important in a broad range of industrial applications, but general methods applying equations of state require iterative algorithms. In this work, a novel method for vapor pressure calculations was applied to the Soave-Redlich-Kwong (SRK) and Peng-Robinson (PR) equations of state. The approach used to solve this problem was to determine an empirical correlation between two similarity variables. These were an adjusted temperature variable which combined the effects of reduced temperature and acentric factor, and an adjusted pressure variable which combined reduced pressure and reduced temperature. A standard iterative vapor pressure algorithm using the EOS and a fugacity criterion was restated in terms of the new variables to generate a data set, which was fit to a variety of empirical expressions. Several promising empirical expressions were generated in this manner, and the deviation between their predictions and the numerical results from the algorithm were quantified. These expressions were capable of matching the vapor pressure predictions of SRK and PR with 0.1 percent relative deviation for values of adjusted temperature from the critical point down to 0.255. These results allow accurate estimation of vapor pressures by direct substitution into simple expressions without iterative algorithms.

Copyright © American Institute of Chemical Engineers. All rights reserved.

509991 Raman Spectroscopy for Undergraduate Laboratories

Abstract:

There are a variety of instruments that are used for quality control in industry. Each method has its own set of constraints such as long run times, extensive sample preparation, or destruction of samples. Recent advancements by MarqMetrix and Hellma, USA, have resulted in the design of the Reva Raman instrument that allows for simple, rapid, and nondestructive acquisition of Raman spectra. Raman spectroscopy can be used independently or in conjunction with other instrumental methods to analyze samples. Raman has a variety of applications such as identification, molecular feature analysis, and analytical analysis. Currently, Raman is rarely used in industry as very few people are familiar with this method. Most undergraduate chemistry major programs do not include an introduction to Raman spectroscopy throughout their four years. The objective of this project is to design labs using the Reva instrument for use at the undergraduate level.

Copyright © American Institute of Chemical Engineers. All rights reserved.

510112 Chemical Engineering Outreach Program with Food Applications and Avoiding Allergies

Abstract:

The objective of this project was to create an interactive outreach program that introduces the field of chemical engineering to elementary through high school aged students through a food chemistry group project . The design criteria for this activity included: minimal risk from allergens; ease of explanation of chemistry behind the chosen food product; that the final product of the activity could be observably different based on different inputs; that some of those differences be measurable through objective experimentation; that the activity work with a variety of age groups; and that the activity be completed in approximately 90 minutes. We were able to determine that agar-agar fruit snacks are a food product that students can make in a short amount of time without encountering common allergens. The mechanism of gelification of the fruit snacks can be described in simple or complex ways, depending on the age or interest of the student group. We were also able to determine a range of formulations that create gels with varying properties, and created basic tests that could differentiate between formulations of the product based on properties such as tensile strength, impact strength, taste, and texture. The project was run with three groups of students of varying ages and yielded positive results for both student attitudes and food product creation.

Copyright © American Institute of Chemical Engineers. All rights reserved.

510204 Student Athletes' Sleep, Trimp, and Drop Jump Patterns throughout an Athletic Season

Abstract:

Emma Harris

eah025@bucknell.edu

Bucknell University

Student Athletes' Sleep, TRIMP, and Drop Jump Patters throughout an Athletic Season

The fitness of most Olympic and high-profile professional athletes is monitored by coaches and teams using advanced tracking technologies. However, not much research exists for this type of fitness tracking in collegiate athletes, especially in women's collegiate field hockey. Investigating the use of fitness tracking methods like heart rate monitors, drop jump techniques and simple self-recording procedures allow athletes and coaches to predict and analyze performance patters. These performance patterns can identify athletes' weaknesses and strengths which allow for more efficient coaching and training. A performance pattern that is often overlooked is sleep as maintaining a healthy habitual sleep cycle can affect not only actual performance and heart rate but also overall motivation, recovery and perception of effort. Looking at the relationship between an athlete's sleep patterns and competitive performance can help coaches and athletes predict and prevent potential injuries as well as sub-par performances by individualize training methods.

The work presented here focuses on three fitness tracking metrics including reactive strength index, sleep quality, and training impact (TRIMP). The reactive strength index is recorded during a drop jump. A drop jump is when a subject steps onto a box and jumps off onto a force plate then immediately jumps upwards, without bending their knees, and then lands back on the force plate. The reactive strength index is the ratio of flight time to contact time during the jump. Sleep quality and quantity is reported using self-recording procedures where athletes report the approximate hours of sleep, number of nighttime awakenings and time to fall asleep each night. TRIMP is calculated using training duration, max heart rate and average heart rate which can be obtained from the athletes' heart rate monitors which are worn around their torsos during training. TRIMP is used by coaches and athletes as a measurement of physical exertion. This research analyzes and compares different aspects of TRIMP, reactive strength index, and sleep patterns across a given competitive season. Further investigation into the correlation between TRIMP, reactive strength index and sleep quality is important in the understanding and prediction of physical fitness and fatigue in collegiate athletes.

Copyright © American Institute of Chemical Engineers. All rights reserved.

510235 Inquiry-Based Activities to Correct Common Misconceptions in Heat Transfer and Thermodynamics

Abstract:

Students often enroll in thermodynamics courses with pre-existing misconceptions -- believing temperature, for instance, is how "hot" or "cold" something feels, conflating temperature and energy. Other frequently misunderstood concepts include rate of heat transfer versus amount of energy transferred. Interactive lab experiments have been designed to address and challenge students' misconceptions more effectively than traditional lectures. Prior work in this project demonstrated that physical experiments are the most effective at repairing student misconceptions and that simulations, while less effective, are still significantly more effective than lecture alone. Web-based computer simulations are a repeatable, cost-effective, and accessible medium for these inquiry activities compared to the financial overhead of conducting physical lab experiments. This work enabled the development of three new simulations and several modifications to older inquiry activities to make them scalable and user-friendly on mobile platforms.

Copyright © American Institute of Chemical Engineers. All rights reserved.

510794 Effect of Solids Concentration on Cloud Height in a Stirred Tank

Abstract:

In mechanically stirred vessels containing solids, the most important process results are off bottom suspension and vertical distribution of the solids. There is a large literature on off bottom suspension but little on the vertical distribution of the solids. Some studies have looked at concentration distribution with height; however a simpler, visual and useful way is to look at the interface between the solids rich area and the solid lean area called the cloud height. This paper will discuss some of the parameters that control this cloud height. Two tank sizes and one impeller type were used with the chief variables being physical properties of the fluids and solids and concentration. It was found that the liquid level, if above the cloud height, had no effect on cloud height. In addition the minimum speed to suspend solids also had no effect. The biggest effects were speed and vessel size. Physical properties had a small but significant effect. At low solids concentration below about five percent by weight there was no effect on cloud height while above that value under certain conditions increasing solids lowered the cloud height significantly.

Correlations are given for cloud height at low solids concentrations and high solids concentrations.

Advisers are Dr. Arthur Etchells and Dr. Robert Hesketh.

Copyright © American Institute of Chemical Engineers. All rights reserved.

510956 Analyzing the Effects of Desktop Learning Modules (DLMs) on Student Understanding of the Bernoulli Equation and Mass and Energy Balances

Abstract:

The use of alternative methods to aid in student learning of engineering concepts has been explored for the past several decades. One of the most common methods to supplement lecture material is hands-on learning. The desktop learning modules (DLMs), formulated by Dr. Bernard Van Wieâ€™s group from Washington State University, are hands-on apparatuses that supplement lecture material and assist student learning of a variety of engineering concepts. To determine the effectiveness of these DLMs, a Venturi module was tested on a group of forty-eight engineering students at the University of Kentucky. The students were split into two groups: the controlled group and the variable group. To begin the study, both groups were given an articulated pre-test to determine their current knowledge of the Bernoulli equation and mass and energy balances. From there, the controlled group proceeded with a typical in-class lecture. The variable group, on the other hand, filled out a worksheet while running various mini-experiments on the Venturi DLM. These two groups were then given the same post-test on mass and energy balance material. The post-test consisted of a mix of multiple choice questions and justification boxes. The justifications allow for a more psychological assessment of the students thought processes. This study will numerically compare the effectiveness of the DLMs on student understanding by using pre- and post-test data and analyze common misconceptions of mass and energy balances based off of the student answer justifications.

Keywords: hands-on learning; collaborative learning; direct learning methods; alternative learning methods; Bernoulli equation; mass and energy balances

Copyright © American Institute of Chemical Engineers. All rights reserved.

510980 A 'cards Against Humanity'-Style Card Game for Increasing Engineering Students' Awareness of Ethical Issues in the Profession

Abstract:

Ethical reasoning is an important quality for engineers to develop, as they are the ones that are tasked with designing complex systems that can impact many lives. At the university level, formal ethical reasoning is often not taught effectively. Two common approaches are the modular approach, that might be taught during a freshman class or a senior design class, or a course taught by a non-engineering practitioner, such as philosophy professor. Ethical decision making is then perceived as disconnected from the profession because of the lack of context, the brief exposure time, or both.

In this work, a card game was developed as an instructional tool to expose students to ethical decision making. A gamified approach was chosen to determine if situating ethical scenarios in a playful form would increase student engagement with the material, and allow them to take greater risks with their response choices in the context of a game. The game structure was based on the popular Cards Against Humanity. In the game, each player draws several (white) response cards, and another player (the judge) draws a single black card containing a question or other prompt. Non-judge players choose response cards from their hand, and the judge determines the "best" black card and white card combination, where "best" is left to the discretion of the judge. For this iteration of the game, many of the prompt and response cards contain well known engineering disasters or other engineering specific ethical dilemmas

To determine the effect of the game, students are asked a series of follow-up questions during a post-play debrief session, and their ethical reasoning ability can be measured both pre- and post-play using a validated instrument such as the Engineering Ethical Reasoning Instrument (EERI), developed at Purdue University.

Copyright © American Institute of Chemical Engineers. All rights reserved.

511072 Testing and Data Analysis of Meta-Material Rectifying Antennas

Abstract:

The overall objective of this research project is to rectify electromagnetic radiation at high frequencies into directional current (DC) electricity by using a bow-tie antenna coupled with a rectifying metal-insulator-insulator-metal (MIIM) diode, i.e., a rectenna. For energy-harvesting applications, these rectenna need to feature a zero-biased responsivity (ZBR) that is non-zero, nonlinear, and possess an extremely fast temporal response. On-going testing involves quantifying thermal energy conversion by rectenna devices fabricated through an inter-institutional collaboration. The tests described within this presentation includes two distinct parts: a four-point probe which tests the nonlinearity of the diode and a thermal test which examines whether the device increases in voltage output when subjected to a heat source. Of the devices that showed a response, most had a null current at zero bias, yet displayed an output voltage during thermal testing. Due to dissimilar metals used in the fabrication of the pads, the likely voltage increase is due to a thermocouple effect.

Copyright © American Institute of Chemical Engineers. All rights reserved.

510831 The James Webb Space Telescope: Contamination Control and Materials

Abstract:

The James Webb Space Telescope (JWST), expected to launch in 2018 or early 2019, will be the premier observatory for astronomers worldwide. It is optimized for infrared wavelengths and observation from up to 1 million miles from Earth. JWST includes an Integrated Science Instrument Module (ISIM) contains the four main instruments used to observe deep space: Near-Infrared Camera (NIRCam), Near-Infrared Spectrograph (NIRSpec), Mid-Infrared Instrument (MIRI), and Fine Guidance Sensor/Near InfraRed Imager and Slitless Spectrograph (FGS/NIRISS).

JWST is extremely sensitive to contamination directly resulting in degradation to performance of the telescope. Contamination control has been an essential focus of this mission since the beginning of this observatory. A particular challenge has been contamination challenges in vacuum chamber operations.

Copyright © American Institute of Chemical Engineers. All rights reserved.

511055 A Pilot Study for Investigating Human Health Impacts of Biomass-Fired Domestic Cookstoves

Abstract:

The majority of developing countries today remain reliant on biomass combustion to supply energy to their household for basic needs such as heating and cooking. The high dependency of biomass-based fuels is of concern due to the potential hazardous by-products that can be produced in the event of incomplete combustion. This will occur due to the inefficiency of domestic application systems along with insufficient ventilation. Previous studies have presented evidence of an increased health risk associated with inhalation of biomass smoke, which was found to be particularly true for women who experience daily exposure through their traditional role in the household of cooking.

Despite the efforts across the world to promote better combustion systems and cleaner fuels, immediate change has proven to be unrealistic. Therefore, an alternative intervention should be explored. This project aims to develop a pathophysiological understanding of the previously studied health consequences. We designed and built a bench-top scale combustion device purposed to simulate domestic cook stoves and a protocol was established for smoke sample collection. These collected smoke samples were then tested with a glutathione (GSH) depletion assay and endothelial permeability assay. Although a consistent result could not be expected due to the variable nature of the combustion process, a sign of glutathione depletion was observed, which is known to be associated with cell apoptosis. An endothelial permeability assay also demonstrated barrier destruction at higher dosages, while enhanced function was unexpectedly observed at a lower dosage. The smoke samples were also broken down and characterized in an effort to determine the chemical components responsible for the observed assay results via GC-MS and ICP-AES for organic and inorganic components. respectively. After separating organic components into three parts based on polarity using solid phase extraction, GC-MS analysis indicated that the samples with GSH depleting effect had higher content in the polycyclic aromatic hydrocarbon (PAH) fraction than in that of the samples that did not demonstrate GSH depletion. Studies with ICP-AES suggested copper as another potential active component for the GSH depletion observed.

Copyright © American Institute of Chemical Engineers. All rights reserved.

510599 Sustainable Approach for Green Pesticide Production in Kenya from the Croton Megalocarpus Trees

Abstract:

Sustainable Approach for Green Pesticide Production in Kenya from the Croton Megalocarpus Trees

Shelby Browning and Jeffrey Seay, Ph.D., PE

University of Kentucky College of Engineering Paducah Extended Campus

Croton megalocarpus is a nut-producing tree found all over Sub-Saharan Africa. While this tree offers many benefits to the local people such as shade, wind protection, and soil conservation, the inedible nuts it produces have long been thought of as a waste product. Currently, efforts are underway to press oil from the nuts for use as a biofuel, and the left-over seed cake has been found to be a great animal feed product. This still results in a waste product, the husks from the shelled *Croton megalocarpus* nuts. Using a process developed by the University of Kentucky Appropriate Technology and Sustainability (UKATS) research team, those wasted croton nut husks can be converted through slow pyrolysis into a product called wood vinegar. Studies suggest that diluting wood vinegar with water creates an effective non-synthetic pest repellent. The left-over biochar can also be pulverized and added to the soil as an organic fertilizer, resulting in almost a zero-waste use of croton nuts.

The focus of this research is to determine the yield of wood vinegar and biochar from *Croton megalocarpus* husks as a function of moisture content, pyrolysis temperature and the percentage of seeds remaining with the husks after the shelling operation. Response surface methodology (RSM) has been used to construct an experimental plan for the laboratory experiments. The results of this research will be used to optimize the recovery of wood vinegar and biochar from the waste husks.

Copyright © American Institute of Chemical Engineers. All rights reserved.

510604 Technological and Productive Architecture for the Cocoa Sector-Tumaco Case

Abstract:

Technological and Productive Architecture for the cocoa sector- Tumaco case

Universidad nacional de Colombia sede Bogotá. Facultad de ingeniería. Departamento de ingeniería Química y ambiental.

Author: Nadira Georgedt Aziz Corredor; ngazizc @unal.edu.co

Context.

Tumaco is a little town in the pacific coast of Colombia. In Tumaco, cocoa crops is one of the most important economic activities, representing a third of the total area planted and the source of income of 8,000 families. However, the low prices at which farmers pay grain has led the population to migrate to illicit crops as a source of higher income. This corresponds to a daunting panorama for the producing communities, notwithstanding the multiple high added value business opportunities that exist for the cocoa sector if the technologies required for the purposes are adequately incorporated.

The national government has identified a problem in the agro-industrial sector and wich is one of the most important sectors for the Colombian economy, since it generates a significant contribution to the national GDP and it seeks to promote sustained increases in productivity that significantly impact the economy.

In order to solve the problems of the sector and towards competitiveness and development, in the NATIONAL DEVELOPMNET PLAN and CONPES 3866, sectoral and territorial development policies have been proposed, which seek to increase productivity through innovation and promotion of strategic sectors. To this end, the CRCs (COMISION REGIONAL DE COMTETITIVIDAD) have been created, which structure, coordinate, formulate, evaluate and follow up on the Regional Competitiveness Plan, identifying macroprojects that generate regional impact, improving productivity and competitiveness.

Under this political panorama, it has been shown that cocoa is one of the strategic sectors for the agroindustrial development of the municipality, thanks to its high value business possibilities. In addition, cocoa is one of the potential crops for the post-conflict period, where international and national funding will be generated to encourage the change of illicit crops: coca. In the particular case of Tumaco by cacao crops. In this context, the possibilities of creating a strategic position that makes the cocoa sector of Tumaco and Nariño one of the most important centers of value is evident, increasing competitiveness both in national markets and in international markets and meeting the expectations of the community for ensure well-being in it.

Therefore, it is proposed to work specifically on technological development since it is understood as a state of collective ownership of knowledge and knowledge, applied to productive and organizational structures that generate new value propositions, which has to do with the processes and tools used by a community, with a common economic activity, to advance their capacities and respond to the demands of their natural market, it guarantees competitiveness and consequently economic development and in turn development of society.

The management of systems as complex as that of the productive sectors and even the cocoa sector in Tumaco require a theoretical and strategic framework that explains relationships, organizes and manages information. To this end, a model of technological and productive architecture for the sector must be proposed, providing the tools to integrate all the key actors and institutions, aligning them in the same high-value business vision, articulating them in search of synergies that are developed when relationships are generated of value among the actors, to configure them to find the appropriate organization that allows a sequencing and make relevant the actions carried out by the subsystems and that generates a transformation, where finally a significant change takes place in the sector, becoming a functional, sustainable system and competitive at the national level.

In the first instance, it is necessary to carry out a prospective exercise that allows the recognition of future scenarios and identify high-value business opportunities for the sector, so that sector technological development will be planned. This is based on the search for a common vision for the future, starting from the base on a human scale, allowing the identification of the purpose of the sectoral development and directing the actions towards the new reality constructed by the same local actors.

Objetives

General objective

- Propose a model of Technological and Productive Architecture to mark the economic development of the cocoa sector of Tumaco territory.

Specific objectives

- Propose a business portfolio of high value for the cocoa sector
- Identify existing technological gaps for the incorporation of added value to the production of the territory
- Generate Road Maps - Road Maps for the articulation and alignment of ID + i activities by going forward
- Propose strategies and tactics for the model implementation of the Technological and Productive Architectures for the sector
- Provide a basis for the formulation of a Strategic Plan for technological and productive development
- Promote the formulation of initiatives for productive development in the direction of business opportunities in the short, medium and long term.

Methodology

To address the different phases of the project, three methodologies will be used, as follows:

- Agrospective Methodology: Mainly oriented to the identification of innovation opportunities based on the analysis of the macro trends and their contextualization in the Colombian environment. Its principle is the construction of a vision of the future of common interest, based on the principles of development on a human scale. It allows the identification of high added value business opportunities and the definition of the value levers and the potential lines of business for the sector. Finally, it contributes elements to direct the actions towards new realities dables to build from the social alabase of producers of the territory.

This methodology is based on the analysis of the trends of Life Forms and Consumption, Drivers or business and engineering designers.

- SMOD Methodology: It allows to shape the vision for development by modifying three scenarios: the Acceptable by the market and business environment, the Desirable by the actors of a sector or territory and the agencies of development promotion and the Possible since the point of view of productive and technological capabilities. The intersection of the three scenarios gives information about the Future Dable for the cocoa sector of the territory. The objective of the prospective study based on this methodology is to define the future scenarios and establish the necessary development actions to increase the potential for the territory, through the generation and appropriation of new knowledge.

- Technological Roadmapping: The mapping of technological routes is a tool to explain, align and articulate the different R & D actions to be advanced by key players. It is based on the construction of maps or roadmaps to advance the management, monitoring and control of the different initiatives and project for development oriented to the achievement of specific technological and productive development objectives.

Copyright © American Institute of Chemical Engineers. All rights reserved.

510016 Polymorphism of Glutaric Acid Aerosols

Abstract:

Aerosols are very tiny particles (typically on the nano scale) that are present in the air. They are important to understand because of the sheer quantity of aerosols that are present in the atmosphere. Inorganic aerosols and their interactions are well-studied and understood but we still have a great deal of studying to do with regard to organic aerosols. Organic aerosols make up a large portion of the particles (20-50%) in the atmosphere by mass, a lot of these being dicarboxylic acids. Because of this, dicarboxylic acids are commonly studied in the aerosol field, with subjects of interest being their hygroscopicity and cloud condensation nuclei activity.

Glutaric acid, a water soluble dicarboxylic acid, is commonly used to model water activity in dicarboxylic acid aerosols. It exhibits dimorphism, which means that the molecule can form two different types of crystal structures. The two polymorphs of Glutaric Acid are known as the alpha and beta polymorphs. The alpha polymorph is in the metastable phase, meaning that it is less stable and after some time will revert back to the beta polymorph. The alpha polymorph is favored at higher temperatures and is more soluble than the beta phase, and has been produced under some very specific conditions. This is thought to have led to some variability with regard to the data collected in several experiments involving glutaric acid.

In this experiment, we investigated the formation of alpha glutaric acid polymorph in aerosols generated from an aqueous glutaric acid solution. The formation of both polymorphs from an aqueous solution has never been documented before. Understanding the conditions that lead to the formation of the alpha polymorph will be valuable when attempting to investigate glutaric acid as a dicarboxylic acid.

Copyright © American Institute of Chemical Engineers. All rights reserved.

510983 Sustainability Assessment and Optimization of a Pollution Reduction Process for Acetic Acid Manufacturing

Abstract:

Sustainability Assessment and Optimization of a Pollution Reduction Process for Acetic Acid Manufacturing

Selorme Agbleze[1*], Shuyun Li[1], Gerardo J. Ruiz-Mercado[2], and Fernando V. Lima[1]

Submitted to Topical Area: Sustainability

[1] Department of Chemical and Biomedical Engineering, West Virginia University, Morgantown, WV

[2] U.S. Environmental Protection Agency, Cincinnati, OH

*Undergraduate Student, presenter

The standards set by the U. S. Environmental Protection Agency (U.S. EPA) on chemical plants imply that certain non-product streams must be treated. There are guidelines for the handling and disposal of specific chemical compounds produced as by-products from such plants. Cost increases by the quantity of waste to be treated, hence optimizing a process with regards to the quantity of waste generated has a direct impact on the cost of waste treatment, with a tradeoff between variables manipulated to achieve desired specifications.

In this research, a sustainability assessment is performed employing the U.S. EPA GREENSCOPE tool for the optimization of the acetic acid production process. The goal of this optimization is to meet the same production target of acetic acid produced while reducing the amount of waste sent for waste treatment. Specific tradeoffs are analyzed in this research related to the acetic acid manufacturing process. Tradeoffs associated with the addition and removal of steps or change in process conditions necessary are considered to achieve significant waste reduction.

Preliminary results for the optimization produced a carbon dioxide and methane reduction corresponding to 84%, methanol reduction by 98%, hydrogen iodide reduction by 9% while methyl iodide is completely removed. There is also an overall electricity requirement reduction corresponding to 52%. Results for the sustainability assessment before and after the optimization will also be discussed in this presentation, including the tradeoffs between variables such as process conditions and utility requirements towards a more sustainable process operation.

Copyright © American Institute of Chemical Engineers. All rights reserved.

510025 Concentration-Dependent Kinetics of Cvoc-Ammonium Sulfate Brown Carbon Chromophores

Abstract:

The release of water-soluble volatile organic gases (VOCs) into the atmosphere results in chemical processes that occur within cloud droplets and atmospheric aerosols due to the aqueous setting. The oxidation of these volatile organic gases leads to the presence of secondary organic aerosol material (SOA). Glyoxal (GLYX) and methylglyoxal (MGLY), considered carbonyl-containing VOCs (CVOCs), have been studied as a precursor to the formation of SOAs. These water-soluble carbonyl compounds undergo Maillard-type browning reactions in the presence of ammonium salts due to its light absorbing qualities. Much is not known about the kinetics of these reactions. Herein, we discuss the potential modeling of these rate reactions across various concentrations of the CVOCs. To simulate atmospheric conditions, organics were mixed with ammonium sulfate and its light-absorbance tracked using a UV-vis spectroscopy over duration of time. The data was then modeled using Gaussian curves as its fitting parameters.

Copyright © American Institute of Chemical Engineers. All rights reserved.

510046 The Evaporation or Growth Rates of Dicarboxylic Acid Solution Droplets As a Function of Relative Humidity

Abstract:

Dicarboxylic acids are frequently observed in ambient aerosol. Beyond inorganic salts like sea salt and ammonium sulfate, oxygenated organic carbon species like dicarboxylic acids are a key to the growth of aerosols driven by increases in relative humidity (RH). The size of the aerosols or particles in the atmosphere largely determines the amount of radiative forcing they contribute to the Earth system. Radiative forcing is the difference between solar energy absorbed by the Earth versus the radiation reflected by the atmosphere and the Earth, and represents the impact of a single aerosol type or other climate forcer such as carbon dioxide where together all contributors yield the total impact on the Earth system. Evaporation of surface active dicarboxylic acids, such as malonic and glutaric acid, following water exposure are unexpected and represent a competition between size increases and decreases. Bulk scale mass measurements as a function of time are made using the Electrobalance with Variable Active Particle Humidity Control (EVAP-HC) a custom-built housing for a Cahn 2000 Electrobalance. Post evaporation composition analysis is completed with a thermogravimetric analyzer (TGA). Variables including initial solution composition and RH are controlled. The rate of mass loss is used to calculate binary diffusion coefficients using Maxwellâ€™s equations. These coefficients equal the rate at which mass, in the form of surface area, moves from the solution to gas phase and are related to the molecular size and number density. Furthermore, EVAP-HC provides an alternative method for measuring the deliquescence relative humidity (DRH), the RH where a solid particle takes up water and grows into a solution droplet. The experimentally determined linear relationship between the weight percent change over time as a function of RH for solution droplets of a specific solute is manipulated so that the weight percent change (y value) is equal to zero yielding the RH where evaporation transitions to growth (DRH) . Understanding the behavior of these dicarboxylic acid solutions via their evaporation rates, diffusion coefficients and DRHs will lead to better understanding of oxygenated organic components of aerosol and refinements in climate models to better represent the interaction with atmospheric water.

Copyright © American Institute of Chemical Engineers. All rights reserved.

509688 The Concept of Evaporation Ponds Fitted with Fresnel Lenses for Brine Disposal

Abstract:

The high salt concentration in the brine which is produced as a result of desalination processes are either returned to the sea or disposed inland. This brine has TDS as high as 36,000 mg/L. Most impacts on the marine environment arise as a consequence of this brine disposal. This increase in salt concentration, which raises the density of waters, can lead to increase of turbidity, disrupt the photosynthesis process. On the other hand, improper land disposal of brine has the potential for polluting the groundwater resources and can have a profound impact on subsurface soil properties.

If this brine is sent to evaporation pond instead, the water will get evaporated leaving behind salts which can be used for various other purposes thus eliminating the need to dispose it to sea or inland. Thus, leading to cleaner and safer environment. But the rate of evaporation is very low as compared to the output of brine of the desalination plants, and hence there is a need for a very large space to accommodate all the rejected brine.

By increasing the rate of evaporation we can decrease the space required by these evaporation ponds. The rate can be increased by using multiple fresnel lenses fitted in a grid. Fresnel lenses are thin lenses which can be as big as 1 m^2 in area. These lenses can reach a temperature of about 2000 K thus significantly increasing the evaporation rate.

This evaporation pond should be constructed of glass in semi-cylindrical shape, fitted with suction pumps on both of its semi-cylindrical bases. The floor of this pond should be constructed with high thermal conductive material such as copper for higher heat transfer. During the day, these fresnel lenses will concentrate the sun's energy and evaporate the water which is removed by the suction pumps, leaving behind the salts. In the night when it becomes non-functional, the scales and deposition of salts are removed thus maintaining the efficiency.

Keywords: brine disposal, fresnel lens, evaporation ponds

Copyright © American Institute of Chemical Engineers. All rights reserved.

509726 Henry's Constants of Persistent Organic Pollutants By a Group-Contribution Method Based on Scaled-Particle Theory

Abstract:

A group-contribution method based on scaled-particle theory was developed to predict Henry's constants for six families of persistent organic pollutants: polychlorinated benzenes, polychlorinated biphenyls, polychlorinated dibenzodioxins, polychlorinated dibenzofurans, polychlorinated naphthalenes, and polybrominated diphenyl ethers. The group-contribution model uses limited experimental data to obtain group-interaction parameters for an easy-to-use method to predict Henry's constants for systems where reliable experimental data are scarce. By using group-interaction parameters obtained from data reduction, scaled-particle theory gives the partial molar Gibbs energy of dissolution, allowing calculation of Henry's constant, H_2, for more than 700 organic pollutants. The average deviation between predicted values of log H_2 and experiment is 4%. Application of an approximate van't Hoff equation gives the temperature dependence of Henry's constants for polychlorinated biphenyls, polychlorinated naphthalenes, and polybrominated diphenyl ethers in the environmentally-relevant range 0 to 40°C.

Copyright © American Institute of Chemical Engineers. All rights reserved.

509897 Effects of Phenol and Chlorophenols on the Performance of a Continuous Flow Activated Sludge Bioreactor

Abstract:

The activated sludge process probably the most important wastewater treatment technology used for both municipal and industrial wastewaters. The process utilizes microorganisms to decompose organic matter in wastewater. The activated sludge process is frequently subjected to various inhibitory pollutants, such as toxic organics, which have inhibitory effect on the biochemical reactions and decrease the efficiency of the wastewater treatment plants. Phenols are aromatic compounds, which is released to the environment by several industrial activities, as refineries. They are particular toxic to both humans and other organisms and thus they have attracted a lot of attention and concerns. Activated sludge process, among others have been studied as a potential process for phenols removal. However, in the same time they exhibit inhibitory effects on activated sludge systems.

This study presents the experimental results on the inhibition effects phenol and chlorophenols namely 2-chlorophenol (2-CP) and 2,4,6-trichlorophenols (2,4,6-TCP) at a concentration of 80-160 ppm on the activity of urban recycle activated sludge. The inhibition the biological processes was examined in in continuous flow aerobic reactors equipped with porous membrane in order to retain the suspended solids, while allowing treated water to pass through. A synthetic influent wastewater was used. The continuous flow system was monitored by analysing the reactor mixture for total suspended solids (TSS), dissolved oxygen (DO) and pH and the treated effluent for total carbon (TC), total nitrogen (TN), phenols, NH_4^+, NO_3^- and NO_2^-. In control experiments (i.e. without phenols) the continuous flow aerobic reactors reached high levels of TC and TN removal of about 90%. The high removal of TN indicates that the reactors work under mixed aerobic-anoxic conditions. The results demonstrated that while carbon oxidation is not significantly affected by the addition of phenols in the influent wastewater nitrification and denitrification steps show considerable inhibition. Also, chlorophenols showed higher inhibition levels than phenol.

It was found that the removal of TC and TN has exhibited slight inhibition for both chlorophenols, however 2,4,6-trichlorophenols showed higher inhibition than 2-chlorophenols. The significant inhibition about 20 % in ammonia removal, nitrification rate and cell growth was observed for both 2-chlorophenols and 2,4,6- trichlorophenols.

Copyright © American Institute of Chemical Engineers. All rights reserved.

510878 Plasmonic Water Purification for Developing Countries

Abstract:

Title: Plasmonic Water Purification for Developing Countries

Authors: Justin Hayes[1], Daniel Willis[2], Dr. Kevin M. Mcpeak[2]

[1]University of Rhode Island, Kingston, RI

[2]Louisiana State University, Baton Rouge, LA

Affiliation: Cain Department of Chemical Engineering at Louisiana State University

Contacts: justin_hayes@my.uri.edu (Presenting Author), dwillis25@gmail.com, kmcpeak@lsu.edu

Topical Area: General

Abstract: In the world, there are approximately 1 billion people without access to clean water [1], and every 20 seconds 1 child will die from a pathogenic water related illness [2]. Poor citizens in developing countries can spend up to 45% of their annual income on clean water [3]. In poor rural areas, SODIS, a solar water disinfection process, takes 6-48 hours to clean 1L of water [4]. Reactive oxygen species are capable of pathogen disinfection [5]. The study focused on developing a plasmonic device which disinfects water with visible light at a low monetary cost. The plasmonic device is a flexible layered substrate which utilizes gold plasmon resonances to harness solar energy and disinfect water. Stimulated gold nanoparticle plasmons can produce hot electrons and when injected into water, can react with H^+ ions and dissolved O_2 to form hydrogen peroxide. The device was layered in the following order: 1mm thick Kapton, 100nm of aluminum via thermal evaporation, 50nm of zinc oxide spin coated and annealed, and 20nm gold nanoparticles reduced on top. To adhere the gold nanoparticles on top, the substrate was placed in a gold salt solution ($AuCl_2^-$ in solvent) and exposed to UV light. Utilizing the band gap of zinc oxide (~3.3eV), electrons were produced which reduced the $AuCl_2^-$ salt, to Au^0 and Cl^- ions. By means of Van der Waals interactions, the Au particles stay atop the zinc oxide layer. The conditions of the gold photo deposition were investigated to produce optimal sized nanoparticles and reasonable stability. Solvent used in photo deposition, UV irradiation time, and diffusion of salt solution were the variables. The substrates produced showed different stabilities when exposed to the solvents ethanol and acetonitrile, with acetonitrile yielding less stable substrates. Constant or pulsed UV irradiation times were used to deposit gold nanoparticles. Longer UV irradiation time negatively affected stability of substrates by means of zinc oxide degradation. The transport of $AuCl_2^-$ ions to the substrate surface was increased by mixing the gold salt solution during photo deposition. This increased gold deposition rate while reducing UV irradiation time. To investigate hydrogen peroxide production, the finished substrate was placed in DI water and exposed to artificial

sunlight. Hydrogen peroxide production was tested using Amplex Red. In a flow reactor, a 2x2 inch substrate produced 16.2 uM/(min*mL) of hydrogen peroxide when exposed to 400-900nm light. Other reactive oxygen species present in the system are being investigated. Materials cost of the substrate is about $0.12, which is only 0.052% of the average minimum annual income of citizens in Southeast Asia [6]. Future research will substitute pathogen infected water for the DI water, to test the system efficacy. Substrate longevity will also be investigated.

References:

- UNICEF, Progress on Drinking water and sanitation. WHO. 2012.
- Shannon et al. Nature 2008, 452, 301-310.
- "World's Poorest Spend the Most." WaterAid America. N.p., 10 Aug. 2016. Web.
- â€œHow Does It Work?â€ SODIS, sodis.ch/methode/anwendung/index_EN.
- Liu et al. Rapid water disinfection using vertically aligned MoS_2 nanofilms and visible light. Nature Nanotechnology 15 August 2016,
- "Minimum Salaries in Southeast Asia 2016." Check in Price. N.p., 31 Dec. 2016. Web..

Copyright © American Institute of Chemical Engineers. All rights reserved.

509992 Global Survey of Suspended Load Mineralogy and Heavy Metal Content in Rural Drinking Water Sources Using Point-of-Use Water Filter Kits

Abstract:

Waterborne illnesses continue to be a major global health concern. One strategy for providing clean drinking water is point-of-use water filters. Hope College has partnered with a privately-owned outdoor product company to test the efficacy of hollow-fiber membrane filters in removing bacteria, protozoa, cysts, and other water-borne microbes from rural drinking water sources. This global water-quality survey is testing water from 30 different countries. Field collection is performed by NGOs which deliver water filter kits to sites, collect and process water samples, and return the filters to Hope College for genetic analysis of microbial populations.

In addition to collecting microbial information, field kits also include polyurethane foam filters (PFFs) for capture of heavy metals. Returned PFFs will be processed and analyzed with ICP-OES for the presence of heavy metals sequestered from water. Several different protocols for the analysis of metals retained by PFFs are currently being developed and tested. These include digestion, rinsing and ashing procedures.

Returned filters also offer a unique opportunity to assess the suspended load (TSS) concentration and mineralogy of drinking water sources. Through a systematic backflush procedure, suspended load particles filtered from water can be retrieved, and the TSS load estimated by spectrophotometric measurement of the re-suspended materials. Mineralogy can be determined by a combination of PXRD and SEM analysis. Re-suspension absorbance can be correlated to model suspensions based on the composition of bedrock and substrate materials at the sampling localities or by the mineralogy of retrieved particulate matter.

Filters from many countries have been returned for analysis, including Solomon Islands, Fiji, Mexico, Liberia, India, Indonesia, Peru, Costa Rica, Haiti, Kenya, Ecuador and Senegal. Preliminary results show a range of TSS, with large differences between surface and groundwater. Minerals retrieved correlate somewhat to bedrock composition, but are not representative of the geologic substrate in all source localities.

This survey is still underway and data will be used to evaluate rural human health in the context of inorganic suspended load, the presence of heavy-metal contamination and overall microbial content of drinking water sources.

Copyright © American Institute of Chemical Engineers. All rights reserved.

510299 Wastewater Contaminant Degradation By Persulfate Activation with Nzvi Composites

Abstract:

The United States poultry industry produced 8.54 billion chickens and 238 million turkeys in 2014.[1] With a typical processing facility estimated to utilize 9 gallons of water per chicken and roughly 30 gallons per turkey, the wastewater for 2014 alone amounts to 76.9 billion and 7.14 billion gallons of wastewater for chickens and turkeys, respectively.[2] To remove contaminants from wastewater and satisfy the regulations of the Clean Water Act, poultry processors employ wastewater treatment processes. However, biological and organic contaminant levels can vary depending on the facility and the efficiency of the treatment processes used, influencing the amount of contaminants expelled into the environment. Implementing a process that degrades organic contaminants in wastewater onsite would be economically beneficial for the poultry industry. Supported nanoscale zerovalent iron (nZVI) composites activated by persulfate have shown promise in degrading organic contaminants in wastewater.[3] Because of the large surface area of the composite components and the reactivity of nZVI, nZVI composites are ideal for activation of persulfate due to increased generation of sulfate radicals.[3] These sulfate radicals are necessary to enhance and accelerate the degradation of organic contaminants.[3] In this study, we will show our continued work in developing persulfate activated nZVI composites and the implementation of these composites to degrade site-specific poultry contaminants. In particular, we will discuss the role of changing mass ratios of nZVI on a support material on persulfate activation through batch reactor experiments where chemical oxygen demand (COD) analysis and spectrophotometry are used to assess the degradation of contaminants by the synthesized composites.

1. USDA Poultry Production Data. In National Agricultural Statistics Service: 2015.
2. Process Overview. In Wastewater Treatment Manual Poultry Processing Facilities U.S. Poultry & Egg Association.
3. Yan, J.; Han, L.; Gao, W.; Xue, S.; Chen, M., Biochar supported nanoscale zerovalent iron composite used as persulfate activator for removing trichloroethylene. *Bioresour Technol* **2015,** *175*, 269-74.

Copyright © American Institute of Chemical Engineers. All rights reserved.

510368 Investigating Iodate Reduction By Microbial Communities from the Hanford Site

Abstract:

Investigating Iodate Reduction by Microbial Communities from the Hanford Site

Ayomikun Olarinoye, Tafadzwa Chigumira, Deondre Glover, Yaolin Fennell, Kimberly Jones, Patrick Ymele-Leki

Chemical Engineering Department, Howard University

This research aims to develop methods for in situ microbial reduction of radioactive Iodate (I^{129}) from radioactive waste contaminated groundwater at the Hanford Nuclear Site in Washington State. A microbial community, named â€œ*Community* 31â€, was isolated from the Hanford site. The community has yet to be characterized but was suspected by researchers at the Pacific Northwest National Laboratory (PNNL) to have Iodate reducing potential. Microbial reduction is the use of microorganism to decrease the oxidation number of a chemical species. In this case, Iodate (oxidation state of +5) was targeted to be reduced to Iodide (oxidation state -1). Community 31 was cultivated in a specially designed minimum media (Gloverâ€™s M9) that allowed for analysis of Iodate concentration with UV Spectrophotometry. A batch culture of the microbial community was exposed to an initial concentration of 100micromole/Liter of Iodate with constant shaking under anaerobic conditions. The experiment was carried out in the dark to reproduce ground water conditions and was monitored over 24 hours for growth and Iodate reduction. Samples were taken every 4 hours and measured in the UV-Spectrophotometer. Approximately 40% reduction in Iodate was recorded after 24 hours. The data suggests that Community 31 may reduce Iodate. This could have significant environmental implications as it suggests another approach for remediation of contaminated groundwater at the Hanford site.

Copyright © American Institute of Chemical Engineers. All rights reserved.

510994 Adsorption Onto Pecan Shell Biochar for Removal of N-Nitrosodimethylamine from Water

Abstract:

Sicilee Macklin

Abstract

Adsorption onto Pecan Shell Biochar for Removal of N-Nitrosodimethylamine from Water

The constant threat of water contamination promotes development of new methods and materials to effectively clean water before consumption. As a known carcinogen to animals with deteriorative effects to lungs and liver in humans, N-Nitrosodimethylamine (NDMA) has recently gained attention in the scientific community. Harmful consequences can result for living beings from exposure at low concentrations of NDMA, down to the parts per trillion level. Therefore, it is of increasing importance to remove NDMA from the water supply.

Activated carbon is commonly used to adsorb a myriad of compounds from water. In New Mexico, pecan shells are found in abundance owing to the influence of the pecan industry and are commonly thought of as a waste material. In this project, local pecan shells are pyrolyzed, some further activated to increase surface area, before testing their adsorptivity of NDMA. The current focus of this project is to utilize scintillation counting, a method used to measure the radioactivity of materials. Carbon-14-labeled NDMA is adsorbed onto the pecan shell char and the radioactivity of these samples determined via scintillation counting. The measure of radioactivity then indicates how much NDMA was adsorbed onto the pecan shell char. Various types of char pyrolyzed at different temperatures will be compared, along with a coconut shell carbon standard, to determine the best option for NDMA adsorption.

Copyright © American Institute of Chemical Engineers. All rights reserved.

511004 Nitrosodimethylamine Removal from Underground Water in NASA White Sands Testing Facility

Abstract:

N-Nitrosodimethylamine (NDMA) is a probable human carcinogen contaminating groundwater and drinking water. Some states in the U.S. limit the concentration of NDMA in water to the low part per trillion (ng/L) range. Current treatment for NDMA in contaminated groundwater is UV photolysis, an expensive and energy-intensive process for large-volume and long-term treatment applications. The objective of this work is to evaluate the feasibility of using chars and activated carbons produced from locally-available agricultural/forest residues to adsorb NDMA at treatment-relevant concentrations. Chars and activated carbons were produced from locally-sourced pecan shells adn pine bark using a range of pyrolysis temperatures (400-900°C) and activation treatments including potassium carbonate impregnation. Equilibrium adsorption isotherms of NDMA onto the prepared adsorbents were acquired using ^{14}C-labeled NDMA, 48-hour exposure times in water, and NDMA concentrations ranging from 5 to 10,000 ppt (ng/L). Liquid scintillation counting was used to quantify NDMA concentrations before and after exposure. Adsorption capacities were compared to that of a commercially-available coconut shell-derived activated carbon used for removal of organic contaminants from water. All biomass-derived treatment methods reduced the NDMA concentration in the water samples. Pecan shell adsorption results showed that K_2CO_3 activation at temperatures greater than 800°C achieved levels of adsorption closest to the commercial activated carbon.

Copyright © American Institute of Chemical Engineers. All rights reserved.

511110 Determination of a Surrogate for Contaminants in the in-Situ Degradation of 1,4-Dioxane in Groundwater By Rhodococcus Rhodochrous

Abstract:

Determination of a Surrogate for Contaminants in the In-situ Degradation of 1,4-Dioxane in Groundwater by *Rhodococcus rhodochrous*.

Gillian Williams, Grant Kresge, Eileen Lukens, Hannah Rolston, Mohammad F. Azizian, Lewis Semprini

School of Chemical, Biological & Environmental Engineering; Oregon State University, Corvallis, OR, 97330

1,4-Dioxane is a carcinogenic chemical component found in degreasers, dyes, paint strippers, and many more products pertinent to industrial processes. The chemical was used primarily in the late 1960s and 70s as an industrial chemical stabilizer for chlorinated solvents such as 1,1,1-TCA (EPA 2014). The release of chlorinated solvents has correlated to the spread of 1,4-Dioxane in groundwater. One potential remediation strategy is bioremediation. The in-situ bioremediation of this contaminant utilizes the metabolic properties of bacteria to transform carcinogenic pollutants that contaminate our drinking water sources to non-toxic substances. A model microorganism, *Rhodococcus rhodochrous*, American Type Culture Collection 21198, degrades 1,4-Dioxane through cometabolism. 21198 is fed a primary growth substrate, isobutane, which induces the production of an enzyme that fortuitously oxidizes other compounds such as 1,4-Dioxane. The purpose of this study was to determine the suitability of propylene or isobutene as a less toxic surrogate to study the in-situ degradation of 1,4-Dioxane.

Degradation tests were conducted in 27-mL batch reactors to measure the rate at which 21198 transformed various quantities of propylene or isobutene. Each reactor contained 10 mL of mineral salts media, approximately 5 mg of active 21198 grown on isobutane, and a variable volume of propylene or isobutene. 21198 was quantified as cell dry weight through total suspended solid analysis. Using a gas chromatograph equipped with a flame ionization detector, the concentration of propylene or isobutene in the batch reactor was measured over time as it was cometabolically degraded. An initial linear rate for the transformation of propylene or isobutene was determined and used to develop a Monod curve that compares the initial liquid concentration of substrate (umol/L) to the rate of transformation (umol/mg TSS*day). Monod curves are utilized in order to determine the kinetic parameters of the culture's maximum degradation rate (K_{max}) and half-saturation constant (K_s) of the contaminant.

The highest values of K_{max} expressed when degrading propylene and isobutene were 1.5 and 2 times greater than that of 1,4-Dioxane, respectively. The highest value of K_s expressed when degrading isobutene was twice that of 1,4-Dioxane. During degradation tests, 21198 was operating at or near K_{max} for the entire degradation of propylene, therefore the value of K_s for propylene was unable to be determined.

United States Environmental Protection Agency. *Technical Fact Sheet – 1,4-Dioxane.* January 2014.

Copyright © American Institute of Chemical Engineers. All rights reserved.

509702 Development of Nutrient Rich Crystals from Wastewater for the Fortification of Compost

Abstract:

Improved management of nitrogen and phosphorus in municipal solid waste (MSW) compost through nutrient-rich crystals from wastewater provides a sustainable alternative to current industrial soil amendments for gardening and agricultural practices. Current petroleum based fertilizers contribute excess nutrients such as nitrogen and phosphorous to the environment. Excess nitrogen and phosphorous cause eutrophication in waterways, biodiversity loss, contribute to global warming, and reduce air quality. Within an institution such as Lafayette College, food waste can make up a large portion of an institutionâ€™s reactive nitrogen footprint. Composting MSW, including food waste, provides a way to recycle nutrients typically lost to the environment. This research considers opportunities to add value to MSW compost while building on lessons of soil scientists. The goal is to make MSW compost a more competitive alternative to fertilizers by engineering a technique to maximize nutrient value of nitrogen and phosphorous in compost. The crystals will be comprised of recycled nutrients (nitrogen and phosphorous) from local wastewater treatment plants.

Copyright © American Institute of Chemical Engineers. All rights reserved.

509762 Effects of pH on Bentonite-Humic Acid Flocculation

Abstract:

Polymer flocculation is commonly utilized to remove colloidal particulates in solution. Flocculation (floc) performance depends on several solution properties such as pH, ionic strength, and natural organic material (NOM) content. Traditional flocculation experiments, known as jar tests, were conducted at pH levels of 3 to 11 and a range of polymer dosages to determine the flocculation efficiency by measuring turbidity. Fluorescently-tagged polymer flocs were imaged with laser confocal microscopy to determine internal floc structure. Not only does bentonite aggregate morphology change as a function of pH, but also humic acid charge density changes, which impacts the flocculation behavior. Flocculation performance of the bentonite-humic acid system differs strikingly from a bentonite only system described in previous works. Competitive flocculation of bentonite and humic acid decreases flocculation performance with increasing solution pH.

Copyright © American Institute of Chemical Engineers. All rights reserved.

510008 Effect of Particle Diameter in the Adsorption Capacity of Banana Peel Activated Carbon

Abstract:

The use of biomass as raw material to obtain activated carbon is a wide study field nowadays. This material allows the removal of contaminants from water sources such as colorants or metals. In this work, activated carbon was obtained from banana peels with the purpose of removing methylene blue of a water sample. The effect of particle diameter in the adsorption capacity was studied. In the first place, the peels are cut and rinsed with distilled water to remove any kind of dirt. Then, it is dried at 105°C for 9 hours to reduce moisture. Later, the material is ground in a mill to reduce particle size. Pyrolysis is done taking the solid to 500°C for an hour in the presence of air. The carbonized material is chemically activated with 0,5 M sulfuric acid for three hours. The carbon is sieved and divided to make adsorption tests. It was found that reducing particle size lead to an increase of adsorption capacity and efficiency, obtaining the best results with the carbon retained in the bottom. Adsorption kinetics was analyzed to the retained carbon in Mesh No. 140 with a good adjustment in pseudo-second order kinetics.

Copyright © American Institute of Chemical Engineers. All rights reserved.

510906 Engineering Struvite Precipitate Utilizing Ion Binding Peptides

Abstract:

Recycling nitrogen and phosphorous is of growing environmental importance for the agricultural industry. While these are contributing factors to issues such as eutrophication, these limited nutrients can be recovered from wastewater in the form of a nutrient rich fertilizer. Production of struvite, $NH_4MgPO_4*6H_2O$, a slow release fertilizer has the potential to provide a very feasible and economical solution. This research involves the use of a mixture of salt solutions to mimic natural wastewater levels and test the yield of precipitate at a variety of pH levels. X-ray diffraction and scanning electron microscope were utilized to analyze both composition and structure of precipitated compounds. Baseline composition and percent recoveries were established for a range of pH values, and past literature suggests low purity and recovery at neutral pH. Therefore, affinity peptides will be added to potentially enhance precipitation of struvite. A quartz crystal microbalance has been used to investigate the binding capabilities of a peptide designed to selectively adsorb to phosphate, and eventually an ammonia binding peptide will be tested.

Copyright © American Institute of Chemical Engineers. All rights reserved.

510905 Co-Occurring Nano-Particulate U(IV) and U(VI) Phases Reactivity in Abandoned Mine Wastes Under Oxidizing Conditions

Abstract:

We investigated the reactivity of co-occurring nanoparticulate U(IV) and U(VI) phases in abandoned mine wastes collected from the Jackpile Mine, Laguna Pueblo, NM, under oxidizing conditions by integrating laboratory experiments with various spectroscopy, microscopy and diffraction techniques. X-ray fluorescence analysis revealed U concentrations as high as 7000 mg/kg in the mine waste. X-ray photoelectron spectroscopy (XPS) of these mine waste samples suggested the co-occurrence of both U(IV) and U(VI). The co-occurring U-minerals were encapsulated by carbon based on electron microprobe mapping data. Data from electron microprobe and X-ray diffraction also suggest the presence of coffinite (U(IV)SiO_4) and U-phosphates (U(VI)-P-K), similar to observations made in XPS. Furthermore, the scanning transmission electron microscopy (STEM) analysis on these nano-particulate U minerals suggested that the U-P-K phase was crystalline and the predominant U phase in the mine waste. Batch experiments were carried out in oxidizing conditions to assess the reactivity of these co-occurring U-bearing mineral phases using 18 MÎ© water (DI), sodium bicarbonate ($NaHCO_3$), and sodium hypochlorite (NaOCl, bleach) over 2 weeks. DI exposure invoked U concentrations of ~ 300 Âµg/L, an order of magnitude greater than the U MCL (30 Âµg/L). Reagents consisting of sodium hypochlorite and sodium bicarbonate yielded U concentrations peaking at ~8000 Âµg/L, and 10 mM bicarbonate yielded the highest U concentrations of 24000 Âµg/L. These results suggest that the bulk of U release is attributed to desorption of U as a result of complexation with carbonate (CO_3^{2-}) to form uranyl-carbonates (U exposure to CO_3^{2-} ions) and dissolution.

Copyright © American Institute of Chemical Engineers. All rights reserved.

510129 The Characterization of Particles in Makerspaces

Abstract:

MakerSpaces are increasing popular community spaces. These spaces, designed to allow crafting, machining, and other similar â€œmakingâ€ activities, have been appearing at libraries, in schools, ranging from elementary to university level, and in community spaces, among others.

Individual devices common in MakerSpaces have been characterized by others. For example 3-D printers are a known source of small particles . We are characterizing the spaces as a whole to get information regarding likely human exposure and potential ventilation strategies to mitigate that risk.

Particle size distributions and particle concentrations are essential parameters when evaluating health impacts because they provide an idea of what kind of particles are suspended in the air and how many there are after a certain procedure was run. A Condensation Particle Counter (CPC) 3007 and Optical Particle Sizer (OPS) were used to test total particle concentration and particle size distribution (>300 nm), respectively. Both devices were set up in the MakerE at Bucknell University, which is equipped with 2 large M2 Makergear 3D printers, preheated circuit mounting bed with reflow nozzle, soldering station with fume hoods, Printed Circuit Mill, and more. There were 3 testing locations arranged in the work area (where the 3-D printers are located), at the edge of the work area, and in the lounging area outside of the work area. Results showed that while peaks in concentration at the closer sites could be higher, the farthest location showed the particles lingering in the air for longer and accumulating more when machines were run back to back in one trial.

The next step is to capture the exact size of the particles present and increase the scale of site variation. The particles will be characterized more specifically by placing the Scanning Mobility Particle Sizer (SMPS) in the work area of the MakerE. The SMPS and OPS can run for multiple days in a row which will help analyze how the particles in the room change from day to day and throughout each day. The ventilation will be varied switching off the main circulation system in the room, turning on and off the fume hoods, and opening and closing the chambers that contain the Printers. Once the MakerE is characterized we will focus on another Maker Space which is less focused on electronics and more focused on craft-type equipment.

Copyright © American Institute of Chemical Engineers. All rights reserved.

510968 Chemical Stability of High Aspect Ratio Nanoparticulate MoO3 in Biological Contexts

Abstract:

Historically, humanity has been quick to take any technological advancement and implement those improvements into our lives. Modern research has resulted in the relatively rapid creation of new man-made materials and their inclusion into industrial and household uses. However, the environmental consequences and health risks of materials are often not fully explored before economic pressures incentivize companies into implementation. One of the most well-known examples is the widespread use of asbestos in virtually all walks of daily life before its carcinogenic potential was fully detailed. Exposure to and inhalation of asbestos fibers was directly linked to the development of mesothelioma, with toxicity due to their high aspect ratio shape and recalcitrance. High aspect ratio rod-like nanoparticulate (NP) MoO3, which has found use in various chemical production and purification methods, is of interest in this historical context; when viewed under SEM, 2D NP MoO3 appears in a fibrous form similar to asbestos. Biopersistance of NP MoO3 was tested with a series of chemical stability tests in various media to determine whether the MoO3 particles have similar longevity to the very stable and persistent asbestos fibers when exposed to an inhalation pathway. Baseline comparison fluids include Nanopure water (NP) as a control, EPA Moderately Hard Water for environmental considerations, and PBS as a bulk body fluid. Roswell Park Memorial Institute (RPMI), Eagleâ€™s Minimum Essential Media (EMEM), Leibovitzâ€™s L-15 Media, Simulated Lung Fluid, and Simulated Phagolysosomic Fluid were tested to simulate the potential environments that NP MoO3 undergoing inhalation pathways might inhabit. Biopersistance was assessed through Dynamic Light Scattering (DLS) over time (run with concentrations of approximately 500 ppm), and DLS data indicated a range of stability that is qualified into three behaviors: dissolution, aggregation, and stability. These behavioral trends were confirmed with filtration over time experiments (at around 50 ppm) analyzed by Inductively Coupled Plasma Atomic Emission Spectroscopy (ICPAES). The data gathered indicates that at realistic concentrations NP MoO3 dissolved rapidly on a time scale of hours if not minutes, with the particles being stable only in nanopure water, which is not biologically relevant). NP MoO3 in simulated phagolysosomic fluid was found to undergo rapid aggregation. The outliers were attributed to pH effects rather than fluid composition; enzyme and protein concentration were controlled for by the addition of FBS into the cellular medium samples (RPMI, EMEM, L15) that were included. In conclusion, results showed that NP MoO3 is non-biologically persistent, often undergoing rapid dissolution in the various tested media. The potential for fiber based toxicity is extremely low, and any detrimental effects would be the result of dissolved ions or transformed species. Biological conditions (phagolysosomic fluid) were identified in which the NP MoO3 would be stable, opening up a possible research path.

Copyright © American Institute of Chemical Engineers. All rights reserved.

510226 E-Waste to Z-Waste

Abstract:

In 2014, 41.8 million tons of E-Waste was globally generated indicating a rise of 5 percent from 2013. The current growth rate of E-Waste production is 4-5 per cent which is expected to rise to 8 percent in 2018. Presently, only 15-20 per cent of the waste is recycled and the rest is incinerated, landfilled or exported to poor and developing countries. The figures are touching alarmingly high levels, which if not industrially treated, will be environmentally hazardous and uneconomical. From, 2001-2011, E-industry used 197-320 tons of gold, making E-Waste a 40-50 times more potential source for extraction of gold than from its ore. The conventional methods of recycling are Pyrometallurgy and Hydrometallurgy which although, economically feasible are non eco-friendly. Biometallurgy, an eco-friendly alternative, consists of leaching the E-Waste sludge using bacteria of genus *Acidthiobacillus*, *Sulfolobus*, *Acidianus* and *Leptospirillum*. *Acidthiobacillus* and Cyanogenic bacteria recover 60-90 per cent of metals depending on the methodology used. In the present study, different types of bacteria are analyzed based on their leaching efficiencies, chemical and biological properties with kinetics. Z-Waste (Zero Waste) approach has been put forward for industrial treatment of waste using Biometallurgy as base method. This review proposes a technology that converts E-Waste to Z-Waste.

Copyright © American Institute of Chemical Engineers. All rights reserved.

510254 Pathogen Removal in Moringa Oleifera Modified Sand Columns

Abstract:

As billions of people in developing regions have limited access to safe water sources, it is apparent that more affordable and available water purification methods are needed. Current water treatment methods are often expensive, manufactured from foreign materials, or unaccepted by local communities. *Moringa oleifera* trees grow rapidly in equatorial regions around the world, and their seeds contain an antimicrobial cationic peptide (MOCP) that kills bacteria by fusing cell membranes. Developing a drinking water treatment device with these seeds enables a locally-sourced and less expensive alternative to traditional water purification methods in developing world. After adsorbing MOCP onto sand through a simple mixing and rinsing procedure, the *Moringa*-functionalized sand (f-sand) carries positive charge of +20 mV, with much less organic residues in the material compared to the conventional flocculating method. The f-sand can then be packed in sand columns that can achieve higher removal of particles with similar size to pathogens, than bare sand columns. A series of f-sand column experiments using microspheres suggest the particle removal and column lifetime can be predicted using the Clean Bed Filtration theory. F-sand column achieved 6 log removal of fluorescent *E.coli* using plate counting techniques. Fluorescent *E.coli* cells were adsorbed onto f-sand after filtration, visualized by fluorescent microscopy, in comparison to no bacteria being adsorbed onto bare sand. Preliminary data indicates f-sand filter do not reach breakthrough until 100 pore volume of the column. Future tests are to determine viability of attached bacterial cells, removal of bacteriophage, and regeneration of f-sand column after breakthrough.

Copyright © American Institute of Chemical Engineers. All rights reserved.

511061 Coordinated Motility of Soil Protists in Micro-Structured Habitats: Evidence for Protist-Protist Communication and Implications for Sustainable Agriculture

Abstract:

Protists come into contact with as much as 70 percent of carbon in the carbon cycle. In the rhizosphere, the region of soil directly under the influence of growing roots, protist communities directly influence the abundance and spatial distribution of beneficial plant symbionts. However, little research has been done to understand how individual protists interact with one another and with their natural habitat, especially at physically relevant spatial scales. Of particular importance to terrestrial ecosystems and agriculture is the potential for individual protists to coordinate motility within a physically complex microenvironment. Here we measure sequential motility decisions of individual protists within a well-defined microfluidic cell culture chamber. *Euplotes vannus,* a common, motile, easy-to-culture, well-studied model ciliate was selected for the study. The experimental apparatus was a PDMS device featuring an array of source wells each connected to two target wells by a branching microfluidic channel 55 µm wide and 35µm high. Just past the branch, each channel featured a heart-shaped structure (ratchet) to promote unidirectional protist movement. In the study, sequential motility decisions of 434 *E. vannus* were tracked in 36 replicate trials. For each trial, between 10 and 40 protists were introduced into the source well and protists were continually observed as they moved freely within the device. Once an individual protist passed a threshold point in either the "Right" or the "Left" channel, a protist motility decision was recorded. Sequential protist motility decisions were analyzed using a pairwise probit regression. Protist motility decisions were significantly different from those predicted for protists displaying random motion (P=0.001), and no significant bias was found favoring either the right or the left channel. These results suggest motility decisions of protists influence subsequent motility decisions of other protists in a community. Additional analysis among sub-groups in the experiment showed coordinated motility is most prevalent among high abundance, low motility protist communities, where low motility indicates a feeding mode for protists. Correlations between motility, feeding, and reproduction have important implications on overall system function. These results will be incorporated into a computer model of protist transport in emulated soil micromodels. In the longer term, more accurate predictions of coordinated motility can improve the accuracy of microbial community models and improve performance sustainable agriculture biotechnology.

Copyright © American Institute of Chemical Engineers. All rights reserved.

510647 Assessing Herbicide and Fertilizer Drift between Conventional and Organic Farmland

Abstract:

The Spring Valley Student Farm (SVSF) and EcoGarden are two student farms at the University of Connecticut (UConn) that are operated under organic farming procedures. Adjacent are conventional farmlands that utilize synthetic herbicides and fertilizers on their crops. The objective of this study was to determine if any reasonable herbicide or fertilizer drift amassed on the organic farmland which might counteract their organic practices. Over the course of eleven weeks air and rainfall samples were collected on the SVSF and to a lesser extent EcoGarden. The approach used a combination of monitoring methods including low cost particulate matter (PM) sensors (AlphaSense OPC-N2), low-volume particulate pump monitors (3000 cm^3/min) with separate herbicide and fertilizer filters, passive rainfall collectors, and a weather station (Ambient Weather WS-1001-WIFI). These methods allowed monitoring of both dry and wet deposition onto the farmland. Eleven weeks of samples were collected and analyzed between May and August of 2017. Analysis of herbicides focused on three major brands (via their main active ingredient) being utilized: Makaze (glyphosate), Crossbow (2,4-dichlorophenoxyacetic acid), and Strut (Diglycolamine salt of 3,6-dichloro-Q-anisic acid). Analysis of total PM nitrogen from the low-volume air samplers and the rainfall samples served as a proxy for fertilizer application. PM monitors were analyzed via MatLab and compared with weather conditions to determine the relationship between total PM and drift. The air filters and water samples were analyzed by UConnâ€™s Center for Environmental Science and Engineering (CESE). Major applications of herbicides occurred on June 29th and 30th utilizing a boom sprayer while fertilizer application occurred much more frequently. Initial results indicate that herbicide drift via wet deposition was recorded at a higher rate at EcoGarden likely due to a lack of tree buffer. Drift via dry deposition was nearly negligible at the SVSF. Several hits were recorded before the first spray event indicating some sort of alternative source of herbicides is nearby to the SVSF and EcoGarden.

Copyright © American Institute of Chemical Engineers. All rights reserved.

509691 Nanotoxicity Evaluation of Doped Silicon Nanocrystals

Abstract:

Recent discoveries of high yield synthesis at el Kortshagen producing doped silicon nanocrystals (DSNCs) via non-thermal plasma reactor have sparked great research interest. DSNCs are semiconducting material with size tunable luminescent capability enabling various applications from biosensing, biomaging, florescent agents, optoelectronics, nanoelectronics, spectroscopy enchancement, and subwavelength microscopy. This versatile semiconducting material has yet to interact with the environment where living organism are found, so the environmental implications remain unknown. Hence, the proposed research entails the exploration of evaluating the nanotoxicity of DSNCs in model organism, Shewanella oneidensis MR-1 (S.oneidensis). The drop plate assay was the method used to assess the toxicity of DSNCs. TEM, XPS, and Hyperspectral imaging were used to characterize DSNCs. After experimentation, results were inconclusive to indicate the DSNCs are toxic or non-toxic.

Copyright © American Institute of Chemical Engineers. All rights reserved.

510770 Synthesis and Physiochemical Characterization of Nanocomposites for Application in the Environmental Remediation of Polychlorinated Biphenyls

Abstract:

Polychlorinated biphenyls (PCBs) are extremely persistent and hazardous organic pollutants that can be found all around the globe. Since the manufacture of PCBs was banned in the US in 1977, the scientific community has made increasing efforts to develop means for the feasible environmental remediation of these contaminants. This research focused on the synthesis and characterization of novel polymers/nanocomposites with the potential to facilitate a cost-effective and selective method of PCB removal from the environment. These novel biomaterials are composed of polystyrene crosslinked with a novel acrylated polyphenol, curcumin multiacrylate (CMA). For a subset of the samples, nanocomposite materials were synthesized with varying amounts of iron oxide magnetic nanoparticles (MNPs). It is postulated that the selective nature of these materials can be attributed to Ï€-Ï€ stacking interactions between the CMA crosslinkers and the PCB molecule due to similarity in their structures. MNPs were characterized using dynamic light scattering (DLS) to determine their size and X-ray diffraction (XRD) to confirm the crystallinity of Fe3O4. Polymers and nanocomposite materials were characterized with Fourier transform infrared spectroscopy (FTIR) and UV-visible light spectrophotometry to determine the incorporation and composition of the polyphenol. The swelling behavior in different solvents was studied through a series of swelling studies. Sorption capacity of novel biomaterials for PCBs and other model molecules were analyzed with UV-visible light spectrophotometry over various time increments.

Copyright © American Institute of Chemical Engineers. All rights reserved.

509273 Imaging Breast Cancer Metastasis over Endothelium to Bone

Abstract:

Bone is one of the common metastasis sites in patients with breast cancer. Once breast cancer cells reach bone, the result is devastating due to the vicious feedback cycle between cancer cells and bone stromal cells. But, for the cancer cells to get to the bone they must pass through the endothelium, which is normally covered by the glycocalyx; a thin layer rich in carbohydrates that protects endothelium against foreign objects, such as cancer. However, how bone metastasis is effected by the integrity of the glycocalyx is not well understood. Here, we created an in vitro system to mimic the interplay among cancer, endothelium, and bone seen in vivo. We did that by creating an endothelium (HUVEC) monolayer on top of a transwell insert (cat3464, Corning) with MDA-231 (1833 subclone) breast cancer cells seeded over the endothelium and bone derived condition medium below the transwell. The system was customized to enable direct imaging of the migration process of breast cancer cells through the endothelium. We also tested methods to impair the endothelial glycocalyx using heparinase III (Hep III) or IL-1beta. These impairments will be used to examine their effects on cancer migration in our future experiments. We concluded that (i) a tight monolayer can be achieved using the endothelial cells with a seeding density of 10,000 cell/mL and four days culture time; (ii) axial (xz) imaging of cancer-endothelium interactions is feasible using a regular inverted confocal; (iii) the optimal condition for glycocalyx impairments was .1U/mL for 2 hours using Hep III. We plan to use this in vitro system for live imaging breast cancer metastasis over endothelium to bone in both impaired and unimpaired glycocalyx systems.

Copyright © American Institute of Chemical Engineers. All rights reserved.

509870 Magnetic Nanoparticles for the Detection of Matrix Metalloproteinase-2 Activity in Tumor Models

Abstract:

Introduction: In their lifetime, around one in every seven men will get prostate cancer. This disease can be very serious, however if diagnosed can be treated. The goal of this project was to develop bioprocess-sensitive "smart" magnetic nanoparticles (MNP) which respond to overexpressed protease activity in the tumor microenvironment – and thus to tumor aggressiveness – and which can be monitored and quantified non-invasively over the entire tumor volume. This was done by using a conjugated fluorescent marker that's rate of release correlated with the activity of an enzyme linked to cancer aggressiveness.

Materials and Methods: The expression of matrix metalloproteinase-2 (MMP-2), an important tumor-associated protease, has been correlated with increasing metastatic potential in prostate cancer and is therefore an attractive target for MNP targeting. We successfully conjugated a near-infrared fluorescent marker via an MMP-2 cleavable peptide linker. This was done through a series of steps, first we attached a matrix metalloproteinase-2 (MMP-2) cleavable peptide with a fluorescent marker (FAM-GPLGVRGC) on the surface of aminated magnetic nanoparticles through NHS-amine chemistry. Thereafter, we exposed the resulting probe to various proteases of the MMP-family, mainly MMP-2, 7 and 13. The experimental data revealed that our probe demonstrated significantly higher selectivity towards MMP-2, as measured by the fluorescent intensity of the cleaved fluorescent marker.

Results and Discussion: The MNP response to the different MMP proteases indicated whether the specific nanoparticles could differentiate each protease as shown by the raw results of the experiments conducted in lab, the fluorescent response of each proteases was measured over time. The fluorescence was then measured using a microplate reader set at excitation/emission wavelengths of 495/525 nm and then set to take measurements every two minutes. This yielded promising results as seen in data, as after 160 minutes, a notable fluorescence intensity was only seen from the MMP-2 protease. This seems to affirm the correct conjugation of the peptide linker for fluorescence and the nano-particles seem to now can differentiate the MMP-2 specifically.

Conclusions: A conjugated near-infrared fluorescent marker via an MMP-2 cleavable peptide linker displayed sensitivity to MMP-2 activity; and the rate of release of the fluorescent marker correlated strongly with MMP-2 activity. In addition, the probe displayed significantly higher specificity towards MMP-2 compared to other members of the MMP family, specifically MMP-7,9, and 14.

Acknowledgements: Graduate Student Tareq Anani helped with data processing and MNP creation. Funding sources include Auburn University Biomedical Engineering Lab.

Copyright © American Institute of Chemical Engineers. All rights reserved.

510114 Optimization of ATAC-Seq Protocols for Adherent Cells

Abstract:

Keshia Capers

Abstract

Optimization of ATAC-seq protocols for Adherent Cells

Introduction: Previous studies have shown that tissue mechanics (e.g., varying matrix stiffness) can enhance human embryonic stem cell lineage commitment into mesoderm lineages, however, the contribution of chromatin reorganization in this stiffness-mediated enhancement of lineage commitment has not been explored. Thus, we aim to provide a platform to measure changes in chromatin accessibility within cells adhered to matrices of variable stiffness. To do so, we hope to utilize Assay for Transposase-Accessible Chromatin with high throughput sequencing (ATAC-seq) protocols, however, there is a lack of literature utilizing ATAC-seq for adherent cells in order to see the open chromatin landscape.

Objective: Our Goal is to optimize ATAC-seq protocols using adherent cells through pre-assay formaldehyde fixation. We hypothesize that ATAC-seq protocols using fixed cells should yield similar results to normal assay conditions.

Design: For this experiment, we plated human embryonic stem cells (HuES64) to tissue culture plastic dishes. We preformed the ATAC-sec protocol on 50,000 stem cells. Stem cells were maintained in growth medium for 3 days until they reached up to 90% confluency. The primary methods used were ATAC-seq, SPRI bead purification, PCR, qPCR, and TapeStation analysis

Results: Our results section is based off of qPCR and TapeStation. We had 6 samples; 3 were washed, 2 were fixed and 1 was washed-unfixed. For qPCR, the results tell us how many more cycles needed to be ran in PCR. Our range was 6-10 extra PCR cycles to ensure we fully amplified our DNA samples. For TapeStation results, it told us how many base pairs we had in each six samples. Our base pair range was 600 to 100 base pairs per sample.

Conclusion: In conclusion, we were able to show that ATAC-seq can function with fixed cells. After successfully performing TapeStation analysis, samples will be sent for sequencing. Sequencing will determine if there are any possible differences in ATAC-seq output between fixed and normal cells due to transposase accessibility to fixed chromatin or if there is truly no bias from the fixation process. Fixation of cells worked well thus far with the ATAC-seq protocol, but if it can be performed directly on adherent cells is yet to be determined.

Future Work: Perform fixed ATAC-seq protocol on fixed adherent cells and run qPCR and TapeStation analysis to verify success of protocol. Perform fixed ATAC-Seq protocol on cells grown on different matrix stiffnesses. Process sequencing data using computational tools and identify regions with different chromatin landscape.

Copyright © American Institute of Chemical Engineers. All rights reserved.

510526 The Viability of Lyophilized HeLa Cells

Abstract:

Introduction: There are numerous applications of human cells in the treatment of trauma and disease. The most prominent being the need of blood for transfusions and skin cells for burn victims. In the United States, fresh blood and skin cells are easily accessible; however, in many other countries, including countries where the US military is stationed and on spacecraft sent to outer space, blood is in limited supply. Consequently, blood must be preserved by cryopreservation due to the tremendous need for blood by military and civilians alike. Skin cells on the other hand must be stored in liquid media and are not preserved by freezing. In places where technology and resources are limited, the traditional methods of storing and preserving these cells makes them virtually unavailable. Therefore it is of interest to preserve cells in a way that does not include freezing or liquid suspension. A dry powdered form of cells would allow blood and skin to be transported effortlessly around the globe and stored without hassle in third world countries, military bases, and on NASA spacecrafts. The aim of the research was to determine if cells could be sprayed onto a thin layer of plastic, frozen, and lyophilized while remaining viable.

Materials and Methods: HeLa cells were cultured, trypsinized, and resuspended in CrySOfree DMSO-free freezing media (Sigma-Aldrich). The cell suspension was deposited into a 2mL spray bottle that dispensed a constant concentration of 40,000 cells/spray. The HeLa cells were sprayed into a 48-well plate at a density of 40,000 cells. In addition, a control was prepared by pipetting cells at the same density as the sprayed cells. The plate was then placed in a styrofoam container and frozen at -80oC with a cooling rate of -1oC/min. After freezing, the cells that had been sprayed into the wells were lyophilized while the control was immediately incubated. The sprayed cells were incubated after being dry frozen. A cell proliferation assay (MTS, Promega) was preformed on the cells, after a 24hr incubation period, using a UV-Vis Spectrometer set to an absorbance of 490nm. After a 5 day incubation period, photographs were taken using a confocal microscope and a manual count of the cells was preformed. The cells were again trypsinized, sprayed, frozen, and dry frozen. The day immediately following the dry freezing, the cells were photographed using SEM.

Results: The cell proliferation assay showed that the control had an absorbance of 0.802 while the freeze dried cells had an absorbance of 0.413. The absorbance, which is directly proportional to the viability, indicated that the sprayed and dry frozen cells remained viable. Although the number is 50% absorbance compared to the control it is important to note that the traditional method of freezing cells will occasionally result in 0% viability. Therefore, it is a significant finding that the cells remained viable after the great amount of shock the freeze dried cells underwent. The confocal microscope photograph of the cells showed elongated cells that remained connected together, demonstrating that the cells had not only successfully gone through the G1 phase but were also able to reproduce. The manual cell count gave roughly 2,000,000 cells. Before freezing and dry freezing there had been 320,000 cells. Therefore, through a manual cell count it was also demonstrated that the cells were viable and reproducing. The SEM photographs taken showed an intact cell membrane and nucleus indicating that no damage to the individual cells had occurred.

Conclusions: Based on the preliminary results, it can be stated that spraying cells onto plastic and freeze drying does not kill the cells. They can successfully pass through the G1 phase and go through the entire cell cycle. Reproduction was not inhibited and the cells continued to maintain all the characteristics they had had before freeze drying had occurred. Future work can continue the process with other cell lines to decide if this will be the preferred method of dry powdered packaging of cells. Increasing the viability of the freeze dried cells as well as additional research into the structure of the cell after undergoing multiple rounds of freeze drying should be continued as well.

Acknowledgments: Beu Oropeza, Luis Solis, Julio Rincon, and Dr. Renato Aguilera, UTEP Biological Sciences Dept. for supplying the HeLa cells used in this project.
Research reported in this paper was supported by the National Institute of General Medical Sciences of the National Institutes of Health under linked Award Numbers RL5GM118969, TL4GM118971, and UL1GM118970. The content is solely the responsibility of the authors and does not necessarily represent the official views of the National Institutes of Health.

Copyright © American Institute of Chemical Engineers. All rights reserved.

Fluorescent Biosensor Probes Derived from Operons for Visualizing Calcium Ion Influx in Neural Activity

Simol Shah, Chemical Engineering BSE; Adrienne Scheck, Research Professor
Barrow Neurological Insitute, Arizona State University

Abstract

Engineered protein biosensors are used in combination with live cell fluorescence microscopy to reveal spatial kinetics and locality of intracellular processes as they occur. However, in biosensory technology, engineering sensors for molecules without a natural binding protein prove to be difficult and time-consuming. The research has focused on engineering the arabinose operon protein to gauge expression behavior of Ara promoters via increased sensor response time due to receptor-ligand kinetics. Through the combined efforts of photon fluxes resulting from fluorophores of fluorescent proteins and the infinitesimal rate constant of ligand binding, the research is a pioneer study for tailor-made medicine. An intensity-based calcium ion-sensing fluorescent reporter with such indicator platforms, optical probing of neoronal circuits is achieved through intracellular stimulation of a host cell, producing Ca 2+ signals in synaptically coupled followers.

Objectives and Physical Phenomenon

Absorption-emission spectra specifically yielded conduction on two binded versions of the calcium-saturated, Ca2+-free and indo-1. Synapses associated with glial cells are reported as analytes, which bind to the bioreceptor, in this case the L-arabinose operon binded to AraC protein. Electrical transduction records amount to the absorption, refraction index, fluorescence, phosphorescence, reflectivity, and wavelength, detected through common interfaces like the LDR, photodiodes, phototransistors, and CCDs. Linear scans describing the action potential-evoked Ca2+ influx in axonal varicosities of CA1 interneurons aligned with the spontaneous activity in astrocytes in the same CA1 astrocytes loaded with green fluorescence deliver the results of automatic astrocyte detection imaging available through software.

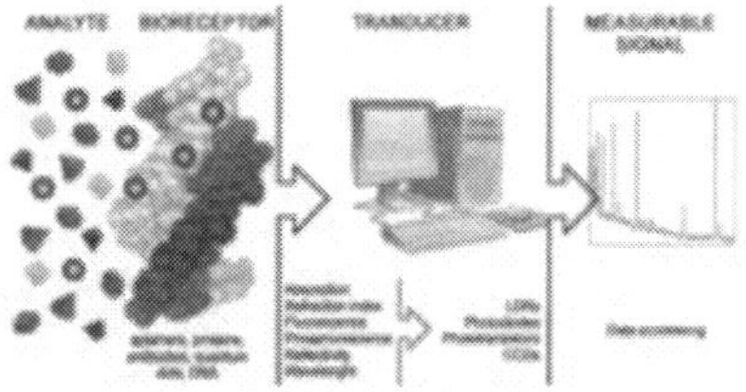

Indicator loading of specific cell populations is possible with sensors whose function is dependent on the presence of a protein. This is because cell type-specific protein expression can be achieved by the use of appropriate regulatory sequences (promoters). This concept had been proposed many years ago in conjunction with the bioluminescent calcium-sensitive reporter protein aqueorin but the light output of bioluminescent probes is rather limited and consumes a co-factor that needs to be applied to the preparation . Genetically encoded fluorescent probes became a powerful approach after the molecular cloning of Aequorea victoria green fluorescent protein (GFP) and similar proteins from corals, and their bright spectrally and optically improved variants . A now broadly used approach to generate fluorescent protein based sensors involves the molecular fusion of a conformation-sensitive GFP-based reporter protein to a second transducer protein that undergoes conformational transitions in response to fluctuations in calcium concentration or membrane voltage. The conformation sensitivity of the GFP-based reporter is achieved either via changes in the efficiency of fluorescence resonance energy transfer (FRET) between two spectral variants of fluorescent proteins or via direct effects of single spectrally instable FP variants

Materials and Methods

Fluorescence is the modality of choice for in vivo and most in vitro optical imaging approaches because: (i) in fluorescence measurements most of the detected light is emitted by the probe with little background signal; (ii) each dye molecule can generate many photons per time unit, and (iii) fluorescence generally is much more sensitive to the environment of the chromophore

For single wavelength measurement, when only relative changes in free Ca^{2+} are required, a simple formula provides an estimate:

$$\text{Free } Ca^{2+} \text{ change} = \Delta Ca^{2+} = \Delta F / F_{rest} = (F_{max} - F_{rest}) / F_{rest} \quad (4.2)$$

where F_{rest} is the dye fluorescence in the resting cell and F_{max} is the dye fluorescence at any time t after addition of the cell stimulus. For absolute free Ca^{2+} the equation changes to:

$$[Ca^{2+}] = K_d \cdot (F - F_{min}) / (F_{max} - F) \quad (4.3)$$

where K_d is the dissociation constant for Ca^{2+}, F_{max} is the fluorescence when the dye is saturated by Ca^{2+} and F_{min} is the fluorescence when there is no Ca^{2+} bound to the dye.

Ratiometric dyes can be calibrated in pure solution (Figure 4.5). But, in the cell, at the end of the experiment, the maximum fluorescence is measured when the dye is saturated with Ca^{2+}, and then the minimum when all the Ca^{2+} is removed, using a Ca^{2+} chelator such as EGTA. When ratiometric dyes such as fura-2 or indo-1 are used the calibration equation is:

$$[Ca^{2+}] = R_f \cdot K_d \cdot (R - R_{min}) / (R_{max} - R) \quad (4.4)$$

where R is the ratio of fluorescence at the two wavelengths, typically excitation around [illegible] nm for fura-2 or emission around [illegible] nm for indo-1, and R_f is the ratio of fluorescence at the [illegible] nm excitation or [illegible] nm emission, respectively, with the dye in zero or saturating Ca^{2+}. However, because there are problems of the precise conditions inside the cell, it is common practice to plot simply the ratio of fluorescence with time for fura-2 and indo-1, and a fluorescence ratio of [illegible] for single-wavelength dyes.

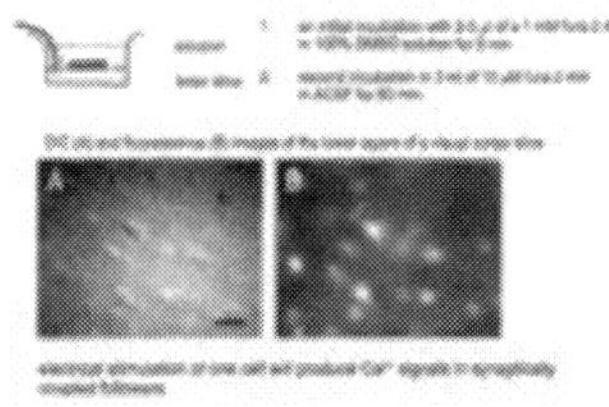

We produced numerous additional GCaMP variants and tested them in automated neuronal assays. With the aim of improving sensitivity, we focused mutagenesis on the interface between cpGFP and CaM at 16 amino acid positions, some mutagenized to near completion. Mutations were made at 18 additional sites, notably at the M13/CaM interface which can affect calcium affinity (A317) and in CaM (R392).

Dissociated rat hippocampal neurons in 24-well plates were transduced with GCaMP variants together with nuclear GFP using lentivirus-mediated gene transfer. Electrodes triggered trains of action potentials in all neurons within each well (Methods). Time-lapse images (35 Hz) of ~800 µm fields of view containing 10-100 neurons were acquired, while delivering a series of action potential trains. Fluorescence changes extracted from single neurons were used to compare the sensitivity, dynamic range, and kinetics of individual GCaMP variants and OGB1-AM. We monitored the resting brightness of the sensor by measuring green fluorescence relative to GFP fluorescence.

Results

Tangential optical sections of the cerebellar cortex were obtained by two-photon fluorescence imaging (900-nm excitation wavelength) of an in vitro cerebellar macroslice preparation at different depths below the pia surface

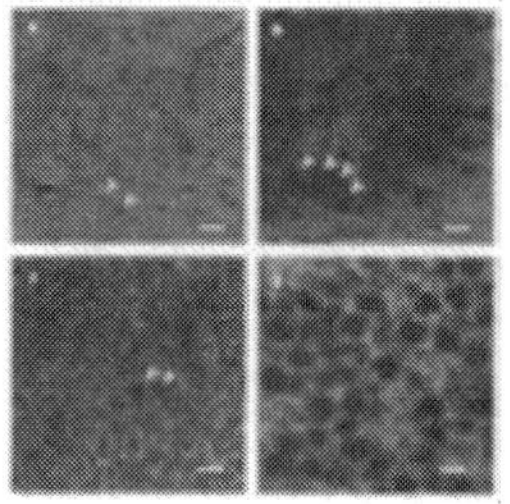

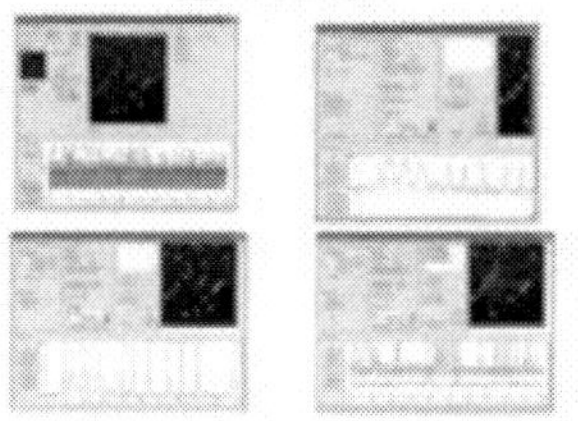

Fluorescence imaging on lobule V/VI of mouse cerebellar cortex stained with a voltage-sensitive dye. Stimulation electrode was placed at the point indicated by asterisk.

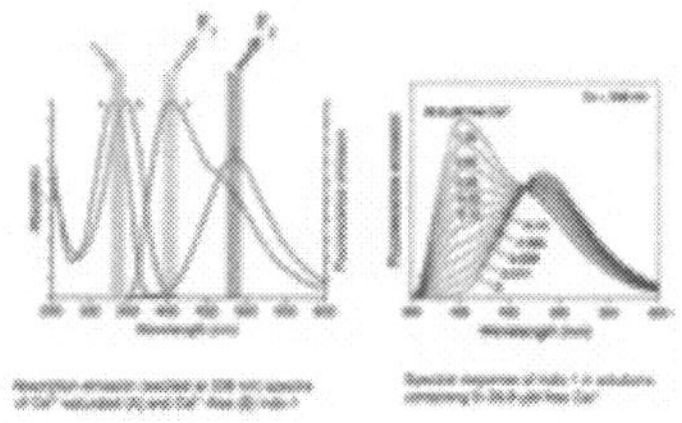

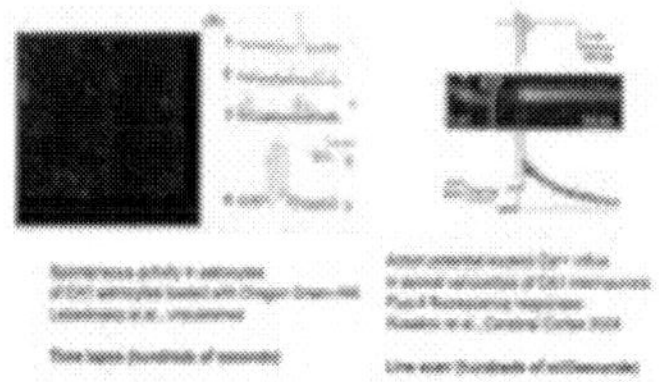

Conclusions

The development of optical methods to monitor neuronal activities began nearly 35 years ago and enormously progressed in a variety of directions. Further technical advance is required in two main fields: (i) methods for fluorescence imaging at very high frame rates in three dimensions [47] , and (ii) further improvements of available probes, in particular genetically encoded voltage sensors, suitable for deep tissue imaging.

References

Dimitrov D, He Y, Mutoh H, Baker BJ, CohenL, Akemann W, Knöpfel T: Engineering and characterization of an enhanced fluorescent protein voltage sensor. PLoS ONE 2007;2:e440.

Romoser VA, Hinkle PM, Persechini A: Detection in living cells of Ca 2+ -dependent changes in the fluorescence emission of an indicator composed of two green fluorescent protein variants linked by a calmodulin binding sequence. A new class of fluorescent indicators. J Biol Chem 1997; 272: 13270–13274.

Svoboda K, Yasuda R: Principles of two-photon excitation microscopy and its applications to neuroscience. Neuron 2006;50:823– 839

Diez-Garcia J, Matsushita S, Mutoh H, Nakai J, Ohkura M, Yokoyama J, Dimitrov D, Knöpfel T: Activation of cerebellar parallel fibers monitored in transgenic mice expressing a fluorescent Ca 2+ indicator protein. Eur J Neurosci 2005;22:627–635.

Acknowledgements

School for Engineering of Matter Transport and Energy
Barrow Neurological Institute
Ira A Fulton Schools of Engineering

509678 Targeting and Killing Melanoma Cancer Using Photoactivated Nanoparticles

Abstract:

Photodynamic therapy (PDT) uses photoactivated nanoparticles that are functionalized to target cancer cells and applies light to excite the nanoparticles which subsequently release reactive oxygen species (ROS) that set off a process resulting in cell-suicide of the cancer cells only. This study tested the polymer Poly[2,1,3-benzothiadiazole-4,7-diyl[4,4-bis(2-ethylhexyl)-4H-cyclopenta[2,1-b:3,4-b']dithiophene-2,6-diyl]] (PCPDTBT) for use as the photoactivated nanoparticle while also examining the Endothelin 3 ligand (ET3) for use as the functionalizing agent for targeting melanoma. This study was inconclusive. Although PCPDTBT was likely an effective photosensitizer for PDT, the results suggested that the functionalized nanoparticles did not target A375 melanoma cancer cells, likely due to the aggregation of the nanoparticles which exceeded the largest size that can be transported into cells at sizes of 100 nm and larger.

Copyright © American Institute of Chemical Engineers. All rights reserved.

509696 Impact of Superparamagnetic Iron-Oxide Nanoparticles and Magnetic Fields on Mixed-Species Oral Biofilm

Abstract:

Impact of Superparamagnetic Iron-oxide Nanoparticles and Magnetic Fields on Oral Biofilm

Jane Nguyen

University of New Mexico, Albuquerque, New Mexico

Data from the Center of Disease Control states that approximately 80% of the adult population in America are victims of periodontitis, an irreversible gingival disease that stems from the onset of gingivitis. The most common cause of gingivitis is poor oral hygiene which produces plaque and bacteria propagation in the human mouth through the establishment of primary and secondary colonizing bacteria. Without the proper oral care, plaque will harden under the gum line to produce tartar which serves as a protective shield for the bacteria and induces irritation along the gums. The continual gum agitation and growth of bacteria results in gingivitis. Therefore, it is imperative to prevent the aggregation of primary and secondary colonizers on the teeth and gingiva. Many attempts have been made to incorporate metal-oxide nanoparticles into dental care due to their promising bactericidal abilities, but a limitation is found in the need to develop non-toxic nanoparticles that still display high levels of antimicrobial potency.

To address this issue, saliva containing mixed bacterial species was collected from adult individuals and propagated to form biofilm. The biofilm was separated to form two samples. The control samples contained the biofilm in phosphate-buffered saline (PBS). The experimental samples subjected the biofilm to a two-fold concentration gradient of FDA-approved superparamagnetic iron-oxide nanoparticles suspended in PBS. The control and experimental samples were exposed to two different magnetic field treatments, static and alternating, induced by molybdenum magnets. A static magnetic field treatment induced magnetic fields with a strength of 4.44 kGs from the bottom of the samples for 6 hours. The alternating field treatment exposed the samples to two magnets. A magnet with a field strength of 4.44 kGs was placed below the samples and a magnet with a field strength of 0.12 kGs was placed above. Both magnets were alternated for 30 minutes over a total span of 6 hours. Analysis utilizing an XTT reduction assay indicated that both static and alternating magnetic field treatments significantly suppressed biofilm metabolic activity. Fluorescent confocal laser scanning microscopy was used to analyze the bacterial viability. Results displayed that the static magnetic treatment was more effective in disrupting the biofilm than the alternating treatment. This difference was more prominent when the treatments were used with nanoparticles. SPSS software was used to conduct a Mann-Whitney U-test to compare differences between the control and the experimental samples. The test yielded a p-value of less than 0.05, which concludes that the nanoparticles did disrupt the bacterial biofilm significantly. Overall, the nanoparticles were proven to effectively disrupt the biofilm, but their ability to kill any bacteria was minimal.

Future studies will apply the nanoparticles to an antimicrobial mouth rinse to penetrate the oral biofilm and increase the potency of the mouth rinse on oral bacteria. Two strains of bacteria will be used to simulate the formation of oral biofilm: *Streptococcus gordonii*, a primary colonizer, and *Fusobacterium nucleatum*, a secondary colonizer. The biofilm samples will be subject to a two-fold concentration gradient of iron oxide nanoparticles and a time-dependent matrix study will be conducted to observe the nanoparticles' effect on the oral bacteria after sixty seconds with ten seconds increments. The predominance of poor oral hygiene and the biomedical limitations of nanoparticles have led to the need for a potent substance that is safe enough to be used inhousehold dental care products. Iron-oxide nanoparticles have been used in a wide variety of *in vivo* biomedical applications such as MRI contrasting, therefore they are a promising candidate to use in conjunction with an antimicrobial mouth rinse to prevent gingivitis thus reduce the magnitude of the population suffering from periodontitis.

Copyright © American Institute of Chemical Engineers. All rights reserved.

509728 Ammonium-Functionalized Carbon Nanotubes for Nucleic Acid Delivery in a Hypoxia-Induced Model of Acute Kidney Injury

Abstract:

Acute kidney injury (AKI) is a detrimental side effect of pharmaceutical treatment, particularly chemotherapy, and abdominal surgery. Ammonium-functionalized carbon nanotubes (fCNT) are a renal small interfering RNA (siRNA) delivery platform that have been used to prevent cisplatin-induced AKI in mice. To explore other causes of AKI, a precision-cut hypoxia-induced murine model of AKI is described here. RNA interference (RNAi) therapy using siRNA against *Trp53* and *Mep1b*, two genes responsible for AKI progression, bound to fCNT (fCNT/si*Trp53*/*Mep1b*) decreases renal expression of pro-inflammatory cytokines in prophylactically treated mice. A robust hypoxia-induced model of AKI is developed and used to probe transcriptome changes and the therapeutic effect of fCNT/si*Trp53*/*Mep1b*.

Copyright © American Institute of Chemical Engineers. All rights reserved.

510810 Effect of Curcumin on aÃŽÂ²-42 Interaction with and Disruption of Lipid Membrane

Abstract:

Effect of Curcumin on Aß-42 Interaction with and Disruption of Lipid Membrane

Alzheimer's disease is a progressive, degenerative disorder most common in individuals 65 and older, affecting approximately 48 million individuals as of 2015. With a life expectancy of 3-9 years post-diagnosis and the high cost of caring for these individuals, viable treatments and prevention medications are necessary.

The main pathological hallmark of AD is the presence of extracellular senile plaques made of amyloid beta (Aß) protein. Two forms of Aß are found in AD: Aß-40 the most abundant, Aß-42 the most toxic. The toxicity of Aß protein is related to its capability to bind to the cell membrane which leads to the production of pores and cell death.

The Chi lab is investigating the protection of lipid membranes via the use of curcumin. In this study, the neuroprotective mechanism of curcumin on the highly toxic Aß-42 protein was investigated by using a combination of technics: Langmuir Blodgett trough, TEM and fluorescence. First, the capability of curcumin to protect the lipid monolayer against different aggregation forms of Aß-42 was tested. We showed that the insertion of Aß-42 into the membrane was reduced in presence of curcumin. Secondly, the effect of curcumin on Aß-42 aggregation was evaluated. Curcumin didn't stop the aggregation of Aß-42, but rather modified the morphology of the fibrils which were more granular. Finally, the surface activity of Aß-42 co-incubated with curcumin was characterized by an increase of the lag phase.

This study facilitated a better understanding of the molecular mechanism of curcumin's neuroprotective effect, which could contribute to the development of treatments for AD.

Copyright © American Institute of Chemical Engineers. All rights reserved.

510998 The Effects of Light Sources on Nutrient Production in the Growth Phase of *Chlamydomonas Reinhardtii*

Abstract:

The development of microalgae as a source of energy, medicine, and food is widely recognized as valuable, but the process is very complex and the effects of each individual variable are not yet fully understood. Light source is one important variable that greatly affects the growth rate and nutrient composition of algae and could be a simple method to control them in scaled operations requiring unique outcomes. The purpose of this research is to compare the growth rates of *Chlamydomonas reinhardtii*, a model alga strain, under differing light sources: Cool White and Gro-Lux. Cool White lights have a more evenly distributed spectrum whereas Gro-Lux lights have very sharp peaks around 440 and 660 nm, which are substantially closer to the peak absorbances of chlorophylls *a* and *b*. The initial cell concentration, growth media composition, and electric power input to lights were controlled. The dry weight, as well as the concentrations of protein, carbohydrates, and chlorophylls *a* and *b* were measured during the growth phase. The dry weights did not differ significantly between the different light sources. This might give the impression that lights focused on the peak absorbances of chlorophylls *a* and *b* do not yield significantly greater growth; however, power input to the light sources does not necessarily correspond to real photon flux, which was 120 $\mu mol\ m^{-2}\ s^{-1}$ for Cool White and 80 $\mu mol\ m^{-2}\ s^{-1}$ for Gro-Lux. Even at roughly equal dry weights, there was significant variation in the composition of nutrients. Cool White lights yielded 288% more carbohydrates whereas Gro-Lux produced 188% more protein and about 108% more chlorophyll. These results show that nutrient composition is affected by changing light exposure wavelengths, and could lead to greater control or optimization in the future.

Copyright © American Institute of Chemical Engineers. All rights reserved.

509924 A Microfluidic Approach to Examine How the Hemodynamic Environment Alters Circulating Tumor Cell Phenotype

Abstract:

An important step in the metastatic progression of cancer is the migration of cancer cells from a primary tumor to distal location through intravasation, in which cancer cells travel through the bloodstream as circulating tumor cells (CTCs). While in the blood vessels, CTCs experience hemodynamic shear stresses at high levels (5-60 dyn/cm^2). Previous studies identified that exposing cancer cells to fluid shear stress (FSS) induced a phenotypic change causing the cells to be more metastatic. While a novel finding, this study was limited in its ability to control the FSS used to challenge the cells and was limited to only final measurements at the end of the experiment. The goal of this project is to address this limitation by developing a microfluidic device to isolate and trap single CTCs and then interrogate them with varying amounts of FSS. The microfluidic traps were designed with a semi-circular geometry with a space in the middle to allow for optimal trapping and testing of FSS-induced cell deformability. COMSOL simulations were performed to optimize the fluid velocity profile of the device to ensure a high trapping efficiency. The simulations also identified the flow profiles needed to produce FSS on the cells that they typically encounter in the bloodstream. These flow profiles were used to test the phenotypic change in MDA-MB-231 cells trapped in the microfluidic device. Preliminary findings demonstrate the ability to efficiently trap single cells in the microfluidic device to explore how increasing the values in fluid shear stress magnitude and duration altered size and shape of the cells allowing for a more accurate determination of cell phenotype. These results will provide new insight into the mechanisms of how CTCs survive in the bloodstream and if changes in FSS results in a more aggressive cancer phenotype.

Copyright © American Institute of Chemical Engineers. All rights reserved.

510216 Single Cell Analysis on the Effects of Fluid Shear Stress on Cancer Cell Deformability and Migration

Abstract:

Metastasis, the leading cause of cancer-related fatalities, is a complex biological process involving the transit of circulating tumor cells (CTCs) through the bloodstream. While transporting through the circulatory system, these cells are exposed to interstitial flow and chemical stimuli, potentially triggering significant phenotypic changes. Recently, it has been discovered that model CTC cell lines express significantly higher levels of EpCAM (an established CTC biomarker) after exposure to fluid shear stress (FSS). However, these prior studies have been limited to population-based studies with restricted ranges of FSS duration and magnitude. Additionally, changes in cell morphology and motility have not been quantified after exposure to FSS. This project presents a microfluidic approach for assessing the phenotypic response of cancer cells to the shear forces present in the circulatory system. A trapping device comprised of two inlets (cell suspension and media), a sheath-flow focusing region, and semicircular trap array was constructed for on-chip morphological studies. Within the device, isolated cells were challenged with hemodynamic shear stress at defined magnitudes (5, 10, 15, 30, 60 dyne/cm^2) and durations (1, 5, 15, 30 minutes). Cell morphology data was acquired before, during, and after shear stress exposure. It was discovered that cellular deformability parameters (e.g., circularity, horizontal length, area) and trap retention rates responded heterogeneously with increasing values of FSS magnitude and duration. This heterogeneous response suggests the presence of distinct cell subpopulations with different gene expression, cytoskeleton integrity, or deformability attributes. Additionally, it was observed that exposure to FSS significantly increased cell motility. Further studies are underway involving both biomarker staining and the transfection of fluorescent vectors for CTC phenotype to further identify and characterize the various subpopulations observed during testing. This work demonstrates the utility of a microfluidic platform for high-throughput, single-cell phenotype analysis and provides valuable insight into the biophysics of cancer metastasis.

Copyright © American Institute of Chemical Engineers. All rights reserved.

510610 Using a Magnetic Field to Control Fiber Alignment in Electrospun Scaffolds for Tissue Engineering

Abstract:

Title: Utilizing Magnetic Electrospinning to Create Gradients in Fiber Alignment for Interfacial Tissue Engineering

Authors: Alessandra Gualtieri, Raymond K. Tindell, Julianne L. Holloway

Affiliation: Chemical Engineering; School for Engineering of Matter, Transport and Energy; Arizona State University

Tissue engineering is an ever-growing field, which aims to advance current treatments for injuries and disease, through a combination of biomaterials, cells, and signaling molecules to restore tissue function. Of particular interest, musculoskeletal impediments are the second most frequent disability and affect over 1.7 billion people worldwide. Musculoskeletal tissues, such as bone, tendon, and cartilage tissue, are important for load bearing and everyday movement. Additionally, musculoskeletal tissues are heterogeneous, where these tissues connect and form chemical and physical gradients responsible for transferring load from one tissue type to another (i.e. tendon-bone interface). Electrospinning is a common technique for fabricating nanofibrous scaffolds that can mimic the structure of the native extracellular matrix (ECM). However, current electrospinning techniques do not easily allow for the replication of the chemical and physical gradients present in musculoskeletal interfacial tissues. In this work, magnetic electrospinning was utilized to produce a tunable fiber alignment gradient that mimics the physical gradient in the tendon-bone interface, where the tendon ECM is highly aligned and the bone ECM is not aligned. Magnetic electrospinning uses permanent magnets to create a magnetic field, where the electrospun fibers will align in the direction of the magnetic field and transition to unaligned fibers as the field weakens. As a proof of concept, electrospinning was performed using poly(caprolactone). Permanent magnet strips were selectively adhered onto a rotating collection mandrel to spatially control the size of the aligned fiber compartment of the scaffold. Fiber alignment percent was quantified using scanning electron microscopy and bright field images of the electrospun fibers as a function of scaffold length and with different magnet configurations. Results successfully show a high degree of fiber alignment on the surface and very close to the magnets with decreasing alignment at distances away from the magnet. Furthermore, by placing multiple magnets next to each other in â€œseriesâ€ the length of the highly aligned compartment increases. Future research will combine this technique with offset electrospinning to independently control gradients in fiber alignment (via magnetic field) and chemistry (via degree of offset between electrospinning solutions) to better mimic the tendon-bone interfacial tissue.

Copyright © American Institute of Chemical Engineers. All rights reserved.

510938 Microfluidics for Environmental Control and Quantitative Analysis of Mouse Stem Cell Aggregate Differentiation to Motor Neurons

Abstract:

Pluripotent stem cells (PSCs) can be differentiated as three-dimensional aggregates into specific cell and tissue types for therapeutic purposes and to study mechanisms of embryonic development. Current culture methods for culturing cell aggregates include microwells, hanging drops, multiwell plates, petri dishes, and spinning bioreactors; however, these methods are often unable to reproducibly control the culture environment, track individual samples longitudinally, and image samples *in situ*. To address these challenges, we have developed a microfluidic platform that provides a controlled environment for culture and differentiation of mouse embryonic stem cell (mESCs) aggregates. We use this platform to study how cell culture environmental parameters affect differentiation of mESCs to motor neurons.

Our microfluidic devices are fabricated out of polydimethylsiloxane (PDMS) replica molded from a silicon master mold and plasma bonded to glass coverslips. Devices are loaded with Olig2-GFP mESC aggregates, which are formed by forced centrifugation of cells in microwells in a neural induction media overnight. Devices are connected to a syringe pump in a cell culture incubator, and media is perfused continuously at a set flow rate for up to nine days. Aggregates are cultured in batch stirred suspension culture as a control. Live imaging of aggregates on-chip and in batch on days four through nine indicates that aggregates on-chip are more uniform in size, and we hypothesize that the microfluidic platform can better modulate the cell microenvironment. Controlling the cell culture environment impacts differentiation and can help us study differentiation mechanisms. Initial evidence suggests that microfluidic culture conditions have significant effects on generation of Olig2+ progenitor motor neurons. We will continue to screen additional media perfusion flow rates to investigate whether culturing on-chip has an advantage over culturing in batch such as producing purer populations of motor neurons or reducing heterogeneity in differentiation.

Copyright © American Institute of Chemical Engineers. All rights reserved.

510940 A Microfluidic Device to Characterize the Effect of Orthogonal Chemical Gradients on 3D Cancer Cell Migration

Abstract:

Once a primary tumor has metastasized to another location in a cancer patient, there is approximately a 20% chance of survival. In order to prevent cancer metastasis in the future, it is necessary to understand how cancer cells migrate from a primary tumor to the circulatory system and from the circulatory system to a secondary metastatic site. Within the tumor environment, cancer cells are subjected to competing gradients including chemicals (chemotaxis), extracellular matrix proteins (hapotaxis), and variations in matrix stiffness (durotaxis) which results in the directed migration of cells. Highly metastatic strains of breast cancer, such as the triple negative cell line MDA-MB-231, will chemotax towards certain chemoattractants such as SDF-1Î± and EGF. To explore this phenomenon, we have fabricated a microfluidic device that creates a â€œflow-freeâ€ chemical gradient within a chamber of cells seeded in a 3D collagen matrix. The chemical gradient is formed by the diffusion of chemoattractant within an agarose hydrogel using a three-channel microfluidic device composed of PDMS imprinted with three parallel channels including a source channel, a middle flow-free cell channel, and a sink channel. While the chemotactic response of MDA-MB-231 cells to gradients of either SDF-1Î± or EGF have been characterized, there is little known about how these cells respond to competing gradients or combinations of these chemoattractants. To create an environment with multiple gradients that better replicates *in vivo* conditions, we designed a second microfluidic device capable of inducing parallel or orthogonal chemical gradients by modifying the initial design to include two source channels, two sink channels, and a middle flow-free cell channel in the shape of a square. This device has the capability to study how cancer cells respond to multiple gradients as opposed to singular gradients. The gradients produced by the device have been simulated using COMSOL, and preliminary results will show MDA-MB-231 cells migrating within the orthogonal device.

Copyright © American Institute of Chemical Engineers. All rights reserved.

510791 Optimization of Sonication Conditions for Nanoparticle Samples

Abstract:

Nanoparticles, although invisible to the naked eye, opens doors to colossal opportunities for improving various aspect of our lives ranging from agriculture to smart products. In biomedical research, the study of nanoparticles is commonly used for drug delivery of pharmaceuticals, proteins and therapeutics. We are developing a new protocol for surface modification of nanoparticles, and this study seeks to identify microcentrifuge tubes that are best suited for sonication, a key step in the process to disperse the nanoparticles. Plastic microcentrifuge tubes that feature a coating to provide ultra-low binding characteristics are used to prevent nanoparticles from sticking to the sides of the tube. However, we have observed particulate formation when samples are sonicated in these tubes. This could lead to a wide set of data to be misinterpreted due to contamination by these particulates. For instance, nanoparticle size and concentration measurements could be skewed because of the presence of these particulates. Therefore, our goal is to investigate the particulate concentration and size produced by tip sonication of water in several brands of microcentrifuge tubes, which are Thermo Scientific™ Low Retention Microcentrifuge Tubes, Bitoix™ Low Binding Microcentrifuge Tubes, Eppendorf™ DNA LoBind Microcentrifuge Tubes, and Eppendorf™ Protein LoBind Microcentrifuge Tubes. The temperature sensitive nature of polymeric nanoparticle samples often requires the sonication step to be performed at reduced temperature, so the effect of sonication temperature on particulate formation was also investigated. First, we sonicated ultrapure water in each of the brands of tubes, at room temperature, on ice, and on dry ice. We then examined the concentration and size of particulates formed from sonication using NanoSightâ€™s Nanoparticle Tracking Analysis (NTA). It was observed that ThermoFisher™ tubes produce less contaminant particulates compared to the other three tubes. Additionally, tubes sonicated on dry ice consistently produced less contaminant particulates compared to tubes at other temperatures. Because water sonicated in ThermoFisher™ tubes on dry ice produced results most comparable to that of fresh ultrapure water, we recommend sonication of nanoparticle samples under these conditions to produce minimal contamination.

Copyright © American Institute of Chemical Engineers. All rights reserved.

510999 Designing a Multipurpose Diagnostic Device Using Microbubbles and Microfluidics

Abstract:

Lipid coated, micron-sized bubbles, or microbubbles, combined with microfluidics shows promise in the biomedical field as tools for therapeutic and diagnostic ultrasound applications. The ability to modify their composition to selectively bind to specific targets by incorporating certain bioactive groups into the lipid monolayer shell allows the potential diagnosis of a multitude of conditions. As a proof-of-concept, a mixture of model targets and microbubbles were run through a microfluidic device after an incubation period. Microbubble-bound targets are separated from their unbound counterparts using acoustic focusing. Targets are separable because unique characteristics exhibited by the microbubbles when exposed to an acoustic environment, which stem from their density and compressibility. Previously, the binding mechanics between microbubbles composed of biotinylated lipids and streptavidin coated microspheres were explored with flow cytometry. Samples run through a microfluidic device and exposed to an acoustic environment demonstrated differential displacement of the microbubble bound microspheres. In this study, the objective is to optimize this procedure. A novel device such as this could provide an inexpensive point of care approach to early diagnostics of cancers involving circulating tumor cells which is currently difficult due to low concentrations.

Copyright © American Institute of Chemical Engineers. All rights reserved.

511058 Use of Time-Resolved Flow Cytometry to Detect Metabolic Shifts in Cancer Cells

Abstract:

Abstract

Flow cytometry is the process of passing cells individually through the path of a laser beam with the use of a fluidic stream that carries the cell suspension. With laser excitation, molecules that are labeled on the surface or interior of the cell will fluoresce. The fluorescence is detected and in turn used to make interpretations about cellular properties including phenotype, genotype, morphology, and other features. In this research we collect light scatter, fluorescence, and a parameter that is not found on conventional flow cytometers--the fluorescence lifetime. The lifetime is important because it is related to the time it takes for excited molecules to decay, which is affected by the cellular microenvironment. We use a specialty flow cytometer to measure this value (~ 1 â€“ 20 nanoseconds). This research is focused on designing and optimizing the cytometry instrumentation needed for making fluorescence lifetime measurements. To test our system, we culture immortalized cancer cells and measure endogenous fluorescence from the excitation of naturally occurring intracellular molecules. The results we obtain are values of fluorescence lifetimes that shift depending on the cellular metabolism and that change depending on the cell type and metabolic pathway taken by the cancer cell. Our overall goal is to build a tool that will measure fluorescence from the naturally occurring species in the cell (e.g. the NADH metabolite) and look for alterations in the lifetime to identify metabolic shifts in a less invasive manner. At the same time, how the metabolic pathways change once that cancer cells are treated with a drug. And determine which the best treatment to that cancer cell is.

Copyright © American Institute of Chemical Engineers. All rights reserved.

511141 Fabrication of Microfluidic Devices Utilizing Stereolithographic 3D Printing

Abstract:

While microfluidic devices provide various advantages in scaled biomedical applications, the device fabrication process requires extensive labor and skill, as well as access to appropriate clean room facilities. Recent studies have been conducted in the use of 3D printing applications as a replacement for the more tedious photolithography and soft lithography in the fabrication of these devices, accelerating and streamlining the design process. In this experiment, an Autodesk Ember stereolithographic 3D printer is used to generate molds for the fabrication of PDMS microfluidic channels. To assure accuracy of the final molds, several resolution models of various dimensions and geometries were printed using variable and non-variable exposure settings; these devices were then characterized using a DektakXT profilometer, an epifluorescence microscope, and ImageJ image processing software. From this analysis, it was determined that the height and width of the printed features was different than expected dimensions; the percent of deviation from the expected value also appeared to vary based on the magnitude of the height and width. A design of experiment was conducted and analyzed using Minitab to test for a correlation between the height and the width of the printed features. This resolution data was then compiled and used to generate calibration curves for the desired feature height and width; several prints of channels of various dimensions were then printed to test the effectiveness of the calibration curves. As a final proof of concept, two additional prints, one using the drafted calibration curves and one without, of an initial microfluidic junction design were also tested. In future studies, additional resolution models will be analyzed to improve upon the accuracy of the dimensional calibrations curves, and various microfluidic devices will be fabricated and compared to existing devices created using traditional photolithographic methods.

Copyright © American Institute of Chemical Engineers. All rights reserved.

509700 The Application of Using 3-D Printed Colorimetric Paper Sensor for the Detection of Antioxidant Activity in Southern California Native Plants

Abstract:

Free radical is highly reactive and is easy to react with molecules in biological system. Previous research have shown that free radical contributes to many chronic health problems. Provided that antioxidant reduce oxidation reaction, the purpose of this research project is to develop a low cost and fast bioassay method for detecting the presence of antioxidants in various foods/Plants using 2, 2-diphenyl-1-picrylhydrazyl (DPPH). To verify the developed method, multiple analyses were used in measuring the activity of antioxidants as it reacts with DPPH, which include electrochemical analysis, UV Spectroscopy, and the proposed paper bioassay.

Our research focused on measuring the amount of antioxidant present in stinging nettle (*Urtica dioica*) since this plant has a long history in medicine and food among native Indian tribes. Antioxidant substance is extracted from by mixing grinded stinging nettle with methanol and the resulted plant extract reacts with DPPH solution for detection. Ascorbic acid was used as antioxidant standard in assay. The reduction of DPPH absorbance from UV, oxidation current from electrochemical measurement, and HSB values from paper sensor (paper bioassay) were compared to determine the accuracy of paper sensor. Based on the growing interest in free radical and preventive methods for many of the chronic diseases, the usefulness of essential, safe nutrients in protecting against the adverse effects of oxidative damage warrants further study.

Copyright © American Institute of Chemical Engineers. All rights reserved.

510871 A High Throughput Microfluidic Device to Screen Protein Binding Ligands

Abstract:

Proteins carry numerous functions that are essential for the sustenance and growth of humans and all forms of life. However, the affinity of proteins to peptides and their interactions are not fully explored. Identifying binding peptides can be useful for protein purification, diagnostics and therapeutics. Microbeads can be used as surfaces to allow different ligands to bind to proteins. The aim of this study is to automate a platform that can sort thousands of these beads and identify which peptides have affinity to proteins such as host cell proteins. A novel, automated, high-throughput microfluidic platform is constructed to sort the beads. Fluorescent tags, such as GFPs, are used to characterize the affinity. Because the beads range in size between 100-300 µm, the result is a robust device that can address and tolerate both larger and smaller beads. Additionally, MATLAB is used to process fluorescent images taken of each bead and sort the beads into two groups. The goal is to use the platform to explore an array of different peptides and proteins.

Copyright © American Institute of Chemical Engineers. All rights reserved.

509782 Lost in Translation: Mapping the Ribosomal Active Site

Abstract:

The ribosome, a 2.5-MDa molecular machine that polymerizes Î±-amino acids into proteins, is the catalytic workhorse of the translation apparatus. The catalytic capacity of the translation machinery has attracted extensive efforts to repurpose it for novel functions. One key idea is that the natural translation machinery can be harnessed to synthesize polymers containing non-natural building blocks. Expanding the repertoire of ribosomal substrates and functions is a difficult task, however, because the requirement of cell viability severely constrains the alterations that can be made to the ribosome, a catalyst that sustains the life of a cell. In practice, these constraints have made the natural ribosome nearly unevolvable and, so far, no generalizable approach for modifying the catalytic peptidyl transferase center (PTC) of the ribosome has been advanced. We propose to address this grand challenge by using cell-free systems that harness the biosynthetic potential of cellular machines without using intact cells, thus removing cell viability constraints.

Here, we use our *in vitro* ribosome synthesis, assembly, and translation system (termed iSAT) to generate variant ribosomes with mutations in the active site, and inquire how these modifications impact protein synthesis. The active site of the ribosome is composed solely of ribosomal RNA (rRNA). This rRNA is assembled into three key loops, which facilitate translation: the peptidyl transferase loop, the A-loop, and the P-loop. The peptidyl transferase loop plays a key role in positioning substrates, is involved with peptide stalling, and is a key target for antibiotic binding. The A- and P-loops flank the peptidyl transferase loop within the ribosomeâ€™s active site. The A-loop is responsible for aminoacyl-tRNA interactions with the ribosome, and the P-loop is known to play a role in further facilitating peptide bond formation between amino acids. Because mutations to many of the nucleotides in these three active site loops confer lethal phenotypes, few to no studies exist in probing and fully characterizing them.

Understanding the effects of single-base mutations on these active site nucleotides will provide insight into the basic biochemistry of these bases in translation, but also provide the groundwork for engineering the catalytic center of the ribosome. Using iSAT, we have assembled mutant ribosomes possessing single-base substitutions of 23S rRNA nucleotides in these active site loops. By successfully quantifying full-length protein synthesis kinetics of iSAT-assembled wild type and mutant ribosomes, we unexpectedly found many key PTC mutations, which were expected to abolish ribosomal activity, still permitted full-length protein synthesis. We also assessed translation fidelity and ribosome assembly. Our work provides a comprehensive mapping of the impacts of multiple ribosomal mutations on protein synthesis. The understanding gained facilitates efforts to engineer and evolve ribosomes for synthetic biology.

Copyright © American Institute of Chemical Engineers. All rights reserved.

509426 Mithramycin Revisited: Exploring the Mechanism of Toxicity in Failure of a Phase I Clinical Trial

Abstract:

Hepatotoxicity is one of the major hurdles that leads to failure during drug development. In most cases, hepatotoxicity is detected in the late stages of drug development either in animal toxicity studies or clinical trials. However, considerable time, effort and resources are spent by the time a compound that may not be viable as a therapeutic agent has undergone animal studies or clinical trials and then is demonstrated to have hepatotoxic side effects. Moreover, patients may suffer when toxicity is first uncovered in first-in-human trials. Mithramycin A (MTH), which we previously found to be active against Ewing Sarcoma, faced this obstacle and therefore failed phase I clinical trials. We thus sought to characterize the mechanism of hepatotoxicity of MTH so compounds with similar hepatotoxic pathways could be avoided in the future.

Herein, we used HepG2 spheroid cultures to assess the hepatotoxicity of MTH. MTH treated spheroids declined in both size and viability at treatment concentrations 1,000 fold lower than those treated with an FDA approved drug, amiloride, which has low hepatotoxic risk and thus served as a control. We then used chemogenomic screening in a pooled *Saccharomyces cerevisiae* gene deletion collection to generate a hypothesis on the mechanism of hepatotoxicity of MTH. We found 48 non-essential deletion strains. Next, we tested the hypothesis that MTH exerts its hepatotoxicity by disrupting calcium ion homeostasis of the liver. By using florescent calcium labeling in a 2-D HepG2 culture model, we could detect a noticeable increase in free cytosolic calcium levels just 30 minutes after treatment which continued 24 hours post-treatment. While MTH has previously been reported to bind to DNA as its target mechanism of action, our results suggest that the compound modulates its off-target toxicity via the disruption of calcium homeostasis in the liver.

Copyright © American Institute of Chemical Engineers. All rights reserved.

509950 Immunogenic Lipid Encapsulated Mesoporous Silica Nanoparticles for Ovarian Cancer Immunotherapy

Abstract:

2017 NSMS Research Experience for Undergraduates

UNM| May 30th-August 4th

Immunogenic lipid encapsulated mesoporous silica nanoparticles for Ovarian Cancer Immunotherapy

Karen Sanchez[1], C. Jeffrey Brinker[2],3, and Rita E. Serda*[3]

Department of Chemical and Materials Engineering, New Mexico State University, Las Cruces, New Mexico, 88003, NSF/UNM Center for Micro-Engineered Materials, Department of Chemical and Biological Engineering, The University of New Mexico, Albuquerque, New Mexico 87131

Email: ritaserda@unm.edu

Introduction

Traditional cancer treatments expose the entire body to high levels of toxic agents causing adverse side effects. The goal of immunotherapy is to increase specificity to cancer cells by eliciting an immune response against the abnormal cells. This study evaluated mesoporous silica nanoparticles (MSNs) and lipid-coated mesoporous silica nanoparticles (LC-MSN) as delivery vehicles for vaccine development. The nanoparticles were loaded with the model antigen ovalbumin and monophosphoryl lipid A (MPL-A) to both direct the immune response and activate dendritic cells (DC). The nanoparticles were compared based on cellular uptake, activation, and antigen processing and presentation by DC. DC, the most potent antigen presenting cells, were derived from bone marrow extracted from C57BL/6 mice. Flow cytometry was used to confirm DC phenotype (CD11c+), measure activation (CD86 expression), and evaluate presentation of the ovalbumin peptide SIINFEKL in association with H-2K^{b}. Confocal and scanning electron microscopy were used to visualize association of DC and nanoparticles.

Results and Discussion

This study determined the effectiveness of ovalbumin loading in MSN and LC-MSN and the ability of DC to process and present the delivered antigen. PLC-MSNs and MSNs were loaded with a mass ratios of silica cores to ovalbumin of 4:1 and 10:1. The CBQCA assay was used to quantitate protein loading and the resulting LC-MSNs and MSNs were added to plated DCs. 72 hours later, the cells were collected and analyzed using a flow cytometer. LC-

MSNs proved to be more effective than the pure MSN with respect to both loading of the ovalbumin and for processing and presentation of the antigen.

DCs treated with LC-MSNs for either 30 minutes or 120 minutes were visualized using confocal microscopy. Images acquired on the Zeiss confocal microscope supported internalization of the nanoparticles by the DC, with intracellular trafficking of the nanoparticles along the microtubules to the perinuclear region of the cell.

Conclusion

LC-MSNs were more effective than MSN for both loading the ovalbumin antigen and for processing and presentation of the antigen by DC.

References

(1.) Chiang, Cheryl Lai-Lai., Kandalaft, Lana E., (September 1, 2013). A Dendritic Cell Vaccine Pulsed with Autologous Hypochlorous Acid-Oxidized Ovarian Cancer Lysate Primes Effective Broad Antitumor Immunity: From Bench to Bedside. *NIH Public Access*.

(2.) Dunfee, Paul N., Lin, Yu-Shen., Dunphy, Darren R., (July 15, 2016). Mesoporous Silica Nanoparticle-Supported Lipid Bilayers (Protocells) for Active Target and Delivery to Individual Leukemia Cells., *ACS NANO*. 10, 8325- 8345

(3.) Slutter, Bram., Zhi, Ding., Bouwstra, Joke A., Bal, Suzzane., Jiskoot, Wim., Adjuvant effect of cationic liposomes and CpG depends on administration route.

(4.) Wang, Min., Zander, Thomas., Liu, Xiaoyan., Liu, Chao.,(2015). The effect of temperature on supported dipalmitoylphosphatidylcholine (DPPC) bilayers: Structure and lubrication performance. *Journal of Colloid and Interface Science*, 445, 84-92

---

[1] Department of Chemical and Materials Engineering, New Mexico State University

[2] Center for Micro-Engineered Materials, The University of New Mexico

[3] Department of Chemical and Biological Engineering, The University of New Mexico

Copyright © American Institute of Chemical Engineers. All rights reserved.

509833 Enhancing Ethanol Production with RNA-Engineering of Zymomonas Mobilis

Abstract:

A key challenge facing the biofuel industry is the toxic nature of ethanol and other pretreatment chemicals to the microbes that catalyze the conversion of biomass to ethanol. *Zymomonas mobilis* has been recognized as a useful bacterium for this catalysis because it naturally produces and tolerates ethanol in high quantities. Recently discovered small regulatory RNAs (sRNAs) in *Z. mobilis* that change expression under ethanol stress indicate the possibility to engineer increased ethanol production and tolerance. In this study, we evaluate which sRNAs most positively affect ethanol production by overexpression or deletion from the genome. This way, we can engineer a more robust strain of *Z. mobilis* that is apt for ethanol production by combining different overexpressions and deletions of these RNA regulators. In doing so, we serve the biofuel production community seeking to develop the most efficient microorganisms to enable chemical production in a cost-effective and sustainable way. By increasing ethanol yields directly and allowing bacteria to live longer in their own product, we demonstrate that multiple sRNAs can be manipulated to enhance ethanol tolerance and ultimately that sRNAs are powerful tools for metabolic engineering in this organism.

Copyright © American Institute of Chemical Engineers. All rights reserved.

AMGEN Foundation

Amgen Scholars Program

Biosynthesis of a Metallophore in Actinobacteria

Brianna Lax[1], Nick Harris[2], Wenjun Zhang[3]

[1]Department of Chemical Engineering, University of Michigan, Ann Arbor
[2]Department of Plant and Microbial Biology, University of California, Berkeley
[3]Deparment of Chemical and Biomolecular Engineering, University of California, Berkeley

Introduction

The Biosynthetic Pathway

The biosynthetic pathway of interest is present in *Mycobacterium tuberculosis*, *Mycobacterium marinum*, and *Streptomyces coeruleorubidus*. It consists of five genes which produce a small molecule believed to be a metallophore. The oxidase (*ScoE*) performs an unprecedented mechanism for isonitrile synthesis which uses a single substrate as both the carbon and nitrogen source.

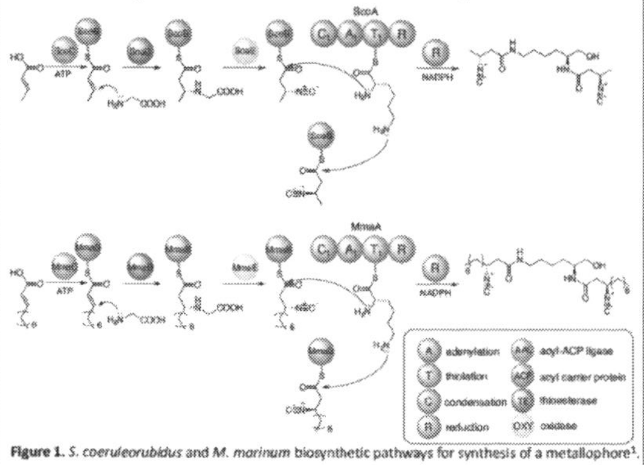

Figure 1. *S. coeruleorubidus* and *M. marinum* biosynthetic pathways for synthesis of a metallophore[1].

Metal Transport

Comparison of intracellular concentrations of various metals between wild type *M. marinum* and *ΔMmaA-E* reveals that the product of this biosynthetic pathway may be involved with metal transport either in or out of the cell.

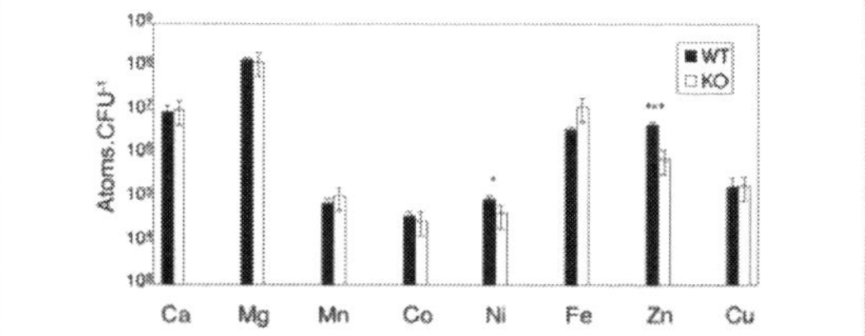

Figure 2. Difference in various intracellular metal concentrations between wild type *M. marinum* and *ΔMmaA-E*. $*P \leq 0.05$ $***P \leq 0.001$[1].

Methods

Immobilized Metal Affinity Chromatography (IMAC)

Figure 3. IMAC with a nickel column for purification of recombinant proteins with a polyhistidine tag.

Quadrupole Time-of-Flight Mass Spectrometry (Q-TOF MS) and Targeted MS-MS

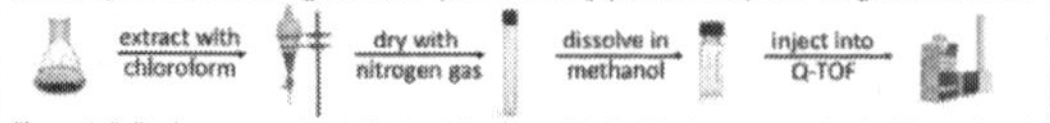

Figure 4. Cell cultures are extracted using chloroform, dried with nitrogen gas, dissolved in methanol, and analyzed using the Q-TOF.

Characterization of *Mycobacterium marinum* Product

Heterologous Expression and Comparative Metabolomics

Heterologous expression of the biosynthetic genes from the *M. marinum* pathway (*MmaA-E*) generated a product, 505.3748 m/z, in *Escherichia coli* that was not present in the empty vector control. This product was characterized using liquid chromatography-mass spectrometry (LC-MS) and targeted MS-MS. Comparative metabolomics between wild type *M. marinum* and *ΔMmaA-E* showed that this mass is further modified by other genes in *M. marinum*.

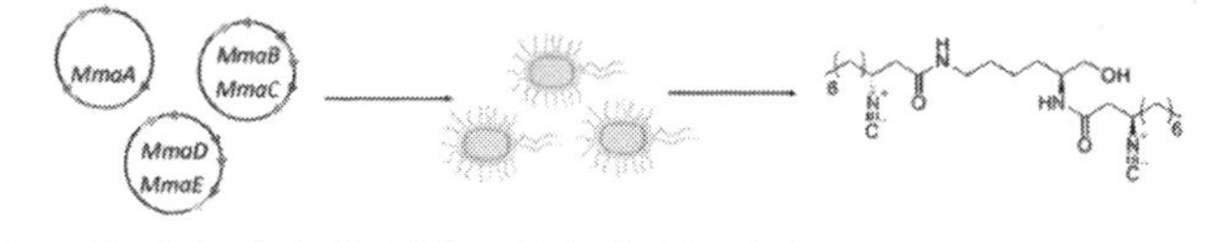

Figure 5. Biosynthetic production of isonitrile lipopeptides from *E. coli-MmaA-E* cultures.

Labeled Substrate Feedings

Labeled substrates were fed to *M. marinum* cultures and LC-MS was used to analyze shifts in the mass of interest (1090.6750 m/z) to verify the incorporation of glycine.

Figure 6. LC-MS extracted ion chromatograms of *M. marinum* product with unlabeled glycine feeding (left) and ^{15}N labeled glycine feeding (right).

Upregulation of Biosynthetic Genes

The isonitrile lipopeptide is produced in trace amounts under normal growth conditions. Purification and characterization require greater amounts of compound. Increased production by overexpression of the gene which upregulates the biosynthetic pathway, *SigC*, was performed by designing a plasmid with an anhydrotetracycline-inducible promoter to regulate *SigC*.

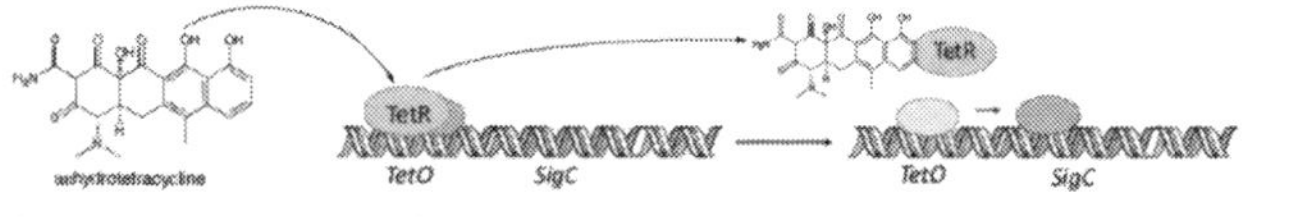

Figure 7. Repressor protein TetR is bound to the transcription factor, shown in purple, inhibiting transcription of gene *SigC*. When anhydrotetracycline is introduced into the system, it binds to TetR and allows for transcription of *SigC*.

Mycobacterium marinum Knockouts

Four knockouts from the *M. marinum* genome, consisting of genes which are close in location to the biosynthetic pathway, were performed to see the effect each of these genes has on the product of the pathway. The four knockouts include the oxidase, a major facilitator superfamily (MFS), a proline-proline-glutamate family protein (PPE), and a P-type ATPase (CtpB).

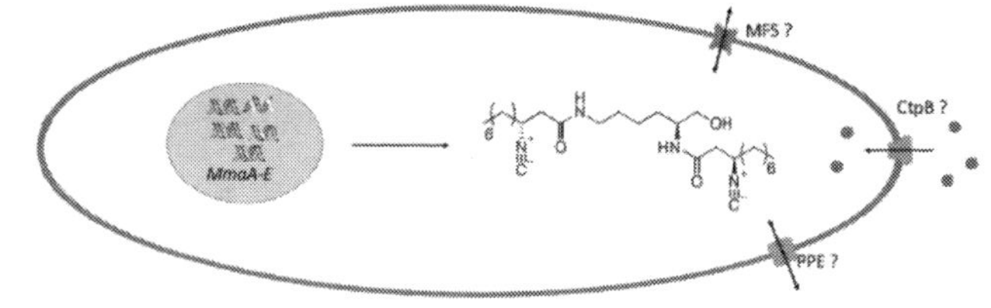

Figure 8. Hypothesized function of the proteins produced by each of the four knocked-out genes and their interactions with the putative metallophore.

Reconstitution of Oxidase *ScoE*

Substrate Limitation

The oxidase *ScoE* can only function if the substrate is bound to an ACP. The thioesterase cleaves the bond between the substrate and the ACP *in vitro*, as well as its intended function of a glycine addition to a fatty acyl ACP, inhibiting the function of the oxidase.

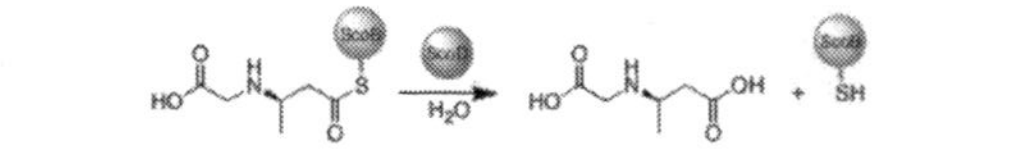

Figure 9. Hydrolyzed isonitrile lipopeptide produced *in vitro* as a result of undesired thioesterase activity.

ACP Glycine Adduct Synthesis

The *ScoE* substrate, an ACP glycine adduct, was synthesized via the elongation of (*R*)-3-aminobutanoic acid with *tert*-butyl bromoacetate, the addition of a phosphopantetheine arm via coenzyme A, and the loading of the substrate onto the ACP (*ScoB*) using a phosphopantetheinyl transferase (SFP). This substrate was used for *in vitro* reconstitution of the activity of *ScoE*.

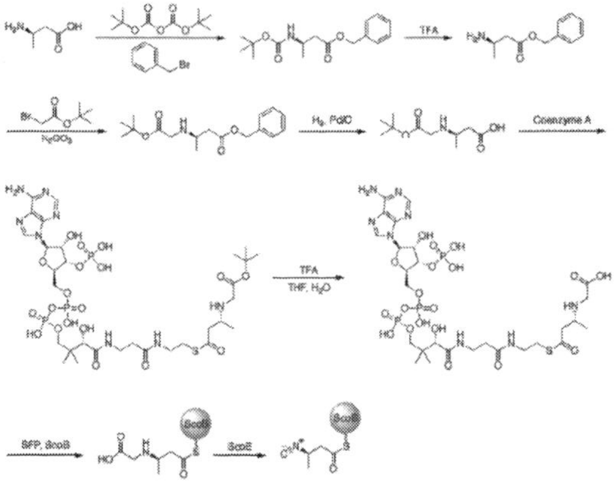

Figure 10. Synthetic reaction scheme for *ScoE* substrate and *ScoE* product.

Future Work

- Knockout *M. marinum* cultures will be used to determine the effect of the knockout gene on compound production and transport using inductively coupled plasma-optical emission spectroscopy
- *In vitro* reconstitution of *ScoE* using the synthetic substrate will be used to confirm that the function of this enzyme is isonitrile synthesis
- Mass spectrometry (MS) and targeted MS-MS will be used in conjunction with labeled substrates to determine the structure of the product

References

1. Harris, N. C., et al **2017**. *PNAS*, 114(27), 7025-7030.

Thank you to my mentor Nick Harris, my PI Wenjun Zhang, the other members of the Zhang Lab, the Berkeley Amgen Scholars Program, and the Amgen Foundation for their support and funding.

510086 Quantitative PCR Analysis of Antibiotic Resistance Mechanisms for Design Optimization of Bacteriocin Antibiotic Scaffolds

Abstract:

Overuse and lack of novel antibiotics has led to resistant phenotypes of bacteria especially in the developing world with limited access to healthcare. Unfortunately, difficulty in innovation and minimal return on investment for development of antibiotics have caused a lack of novel antibiotics for treatment of bacterial infections. (1) Antimicrobial peptides (AMPs) synthesized by bacteria called bacteriocins hold great promise as potential scaffolds for antibiotics due to their antibacterial activity. Here we tested bacterial resistance against a novel synthetic bacteriocin with cationic properties, syn-safencin. A known resistance mechanism against cationic antibiotics in gram positive bacteria is charge modification in the cell wall through the D-alanylation of lipotechoic acids (DA – LTA) to repel cationic antibiotics. We measured the effect of this resistance mechanism against syn-safencin by monitoring the *dltABCD* operon encoding the genes responsible for this resistance. We were able to quantify the relative expression using a reverse transcription – real time PCR (qRT – PCR) approach. The expression fold changes of mRNA transcripts were calculated using the CT method where CT is the cycle threshold value obtained from qPCR analysis. The results did not show any significant expression fold changes when treated with syn-safencin compared with the vehicle control of 10% DMSO. These results suggest that the DA – LTA is not a mechanism of resistance of syn-safencin, and other mechanisms involving surface charge modification should be monitored for resistance. Finding the mechanisms of resistance against synthetic bacteriocins will allow for the design optimization of peptide candidates that account for known resistance mechanisms.

(1) Fischbach, M. A. 2010, 1089 (2009).

Copyright © American Institute of Chemical Engineers. All rights reserved.

510552 Assessment of Neonatal Platelet Dense Granule Trafficking Downstream of Protease-Activated Receptors and $P2Y_1$/ $P2Y_{12}$

Abstract:

Assessment of neonatal platelet dense granule trafficking downstream of protease-activated receptors and $P2Y_1$/ $P2Y_{12}$

Kendra R. Jones,[1] Anh T. P. Ngo,[2] Annachiara Mitrugno,[2] Anne D. Rocheleau,[2] Sandra Baker-Groberg,[2] Joseph E. Aslan,[2] Stewart Worthington,[3] Alysia Cox,[3] Susan Lattimore,[3] Michael Recht,[3] Kristina M. Haley,[3] Owen J. T. McCarty[2]

[1]School of Chemical, Biological, and Environmental Engineering, Oregon State University, Corvallis, OR

[2]Department of Biomedical Engineering, Oregon Health & Science University, Portland, OR

[3]The Hemophilia Center, Oregon Health & Science University, Portland, OR

Introduction: Platelet dense granule secretion follows activation of protease-activated receptors (PARs) 1 and 4 in human platelets to result in ADP release with subsequent signaling of the ADP receptors $P2Y_1$/$P2Y_{12}$ and amplification of adult platelet activation and aggregation[2,3]. Neonatal platelets are hyporesponsive to multiple platelet agonists as compared to adult platelets. These mechanisms are unknown, yet neonates do not manifest a clinical bleeding tendency. Our group has shown that neonatal platelets (neonatelets) are hyporesponsive to the PAR1 agonist, thrombin receptor activator peptide 6 (TRAP-6), in comparison to adult platelets. Therefore, we hypothesize that the hyporesponsivity in neonates is due to either a deficiency in platelet dense granule function and thus ADP release, and/or a lack of synergy between $P2Y_1$/$P2Y_{12}$ and PARs. This study aimed to investigate: 1) dense granule trafficking and release, and 2) PAR cleavage using small volume, whole blood samples. Blood was tested using two novel assays developed by our group; a super-resolution structured illumination microscopy (SR-SIM) and chemiluminescent secretion assay. We also utilized fluorescent-activated cell sorting (FACs) to assess synergy between $P2Y_{12}$ and PARs by detecting markers of platelet activation.

Materials and Methods: In accordance with an Institutional Review Board-approved protocol, fresh adult blood was obtained from healthy donors by venipuncture into 3.8% sodium citrate, and fresh full-term neonatal blood was collected at 24 hours of life via heel stick into trisodium citrate (0.38% w/v). Glass bottom dishes were coated with Poly-L-lysine and seeded with adult or neonatal whole blood previously fixed and diluted with 4% paraformaldehyde. Samples were stained with platelet dense granule marker (CD63), PAR1 (ATAP2), PAR4 (14H6) and imaged using 100x SR-SIM. Platelet dense granule release was measured as light output generated following an ATP-luciferase reaction. Additionally, FACS was performed using whole blood treated with TRAP-6, ADP, AYPGKF (PAR4 agonist), or a combination of ADP and TRAP-6/AYPGKF, and labeled for CD62P and PAC-1. Total platelet glycoprotein (GP) expression was measured using a BioCytex kit.

Results and Discussion: Both dense granule trafficking via SR-SIM and the luminescent assay suggest that neonatal dense granule secretion was unresponsive to PAR1/PAR4 activation. Exogenous ADP appears to rescue the PARs response, but not significantly when compared to the signal for ADP response alone. Additionally, flow cytometry results demonstrate GP expression does not differ in adults and neonates, indicating that the hyposensitivity to PAR1 agonists is likely physiological.

Conclusion: Our hyposensitivity results suggest that neonatal dense granule trafficking and secretion following PARs activation is impaired, when compared to adult platelets. These results provide a potential explanation for hyposensitivity to PAR agonists, and provides insight into platelet granules and synergy between PARs and $P2Y_{12}$.

References: (1) Andrew M, et al. *Blood* 1992; **80**:1998-2005. (2) Falker K, et al. *Biochemical Journal* 2011; **436**: 469-480. (3) Nylander S, et al. *British Journal of Pharmacology* 2004; **142**: 1325-1331. (4) Baker-Groberg SM, et al. *Thrombosis and Haemostasis*; **14**: 815-27.

Copyright © American Institute of Chemical Engineers. All rights reserved.

511022 WT1 Mutations Prevent Normal Hematopoiesis in Acute Myeloid Leukemia

Abstract:

The *Wilms Tumor 1* (*WT1*) gene is a tumor suppressor involved in regulating normal cell growth and differentiation. While WT1 is overexpressed in most acute myeloid leukemia (AML), approximately 10% of both pediatric and adult patients with AML have a mutation in the *WT1* gene. *WT1*-mutated AML is associated with poor outcomes for patients. At this time, however, patients with mutated *WT1* are not treated differently from patients with wild-type *WT1*. The WT1 protein is hypothesized to interact with the proteins TET2 and IDH1/2 in the WIT pathway to initiate the early steps in demethylation. Therefore, *WT1*-mutated AML should be hypermethylated compared to AML with wild-type WT1. Our work seeks to identify the role of WT1 in normal blood cell development and in leukemogenesis. We hypothesize that WT1 is crucial for normal hematopoiesis and that mutation of WT1 contributes to leukemogenesis by affecting the methylation landscape of the DNA. We initiated this work by exogenously expressing wild-type and mutant WT1 in cell lines that have no WT1 expression. In both the human AML cell line U937 and the mouse myeloid cell line 32D, wild-type WT1 expression results in differentiation by both flow cytometric and morphologic evaluation; however, expression of mutant WT1 does not result in differentiation. In addition, in the HL60 human AML cell line, which expresses wild-type WT1, the small molecule *all-trans* retinoic acid (ATRA) results in differentiation, again by both flow cytometric and morphologic evaluation. In order to identify the target genes of WT1 in the setting of wild-type WT1 expression, we have optimized chromatin immunoprecipitation (ChIP) for HL60 cells that have been stimulated to differentiate with ATRA. These samples are currently being sequenced to identify genes for which WT1 results in demethylation. These genes, which we anticipate will encode proteins already known to be involved in hematopoietic differentiation and development, will then be validated using quantitative real-time PCR (qRT-PCR) and Western blotting. We also plan to use targeted approaches to characterize the methylation landscape, or methylome, of *WT1*-wildtype and *WT1*

Our expectation is that we will find hypermethylation of these regions in cells expressing mutant WT1. We will then investigate whether hypomethylating agents, already used as treatment for certain types of AML, can revert the DNA to levels of methylation observed in cells with wild-type WT1 expression; potentially this strategy could be used in the treatment of patients with mutations in the *WT1* gene.

Copyright © American Institute of Chemical Engineers. All rights reserved.

511051 The Lipid Phase Preference of the Adenosine A2a Receptor Depends on Its Ligand Binding State

Abstract:

We formed giant unilamellar protein vesicles (GUPs) out of phase-separating lipid mixtures and incorporated the adenosine 2a receptor ($A_{2a}R$) to observe protein portioning between liquid ordered and liquid disordered phases. When no ligand is bound, $A_{2a}R$ partitions preferentially into the liquid disordered phase of GUPs, while ligand-bound $A_{2a}R$ partitions into the liquid ordered region. This phenomenon represents a potential allosteric modulation of the proteinâ €™s membrane localization, with implications to control signaling pathways via protein segregation within the lipid bilayer.

Copyright © American Institute of Chemical Engineers. All rights reserved.

511095 Validation of a Galectin-8 Reporter As a Measure of Nanocarrier Endosomal Escape and Biologic Drug Intracellular Bioavailability

Abstract:

Introduction: Endosomal escape is a critical step in the delivery of intracellularly active biomacromolecules, such as nucleic acids, peptides, and proteins. We and others recently utilized Galectin 8 (Gal8) recruitment to detect endosomal vesicle disruption by nonviral carriers[1,2]. Here, we sought to validate a Gal8 recruitment-based assay as a direct measure of endosome escape and predictor of biologic drug intracellular bioactivity (siRNA gene silencing); in addition, this methodology was utilized to rapidly optimize polymer composition and molecular weight for siRNA knockdown activity building from our recent work on siRNA nanocarriers (si-NPs)[3].

Materials and Methods: A library of poly[(ethylene glycol)-b-[(2-(dimethylamino)ethyl methacrylate)-co-(butyl methacrylate)] [PEG-(DMAEMA-co-BMA)] polymers of equivalent MW were synthesized with 0, 25, 40, 50, 60, and 75 mol% BMA using RAFT polymerization[3]. A second polymer library was synthesized with 50 mol% BMA composition with varied molecular weight within the DMAEMA-co-BMA si-NP core-forming polymer block (S, M, L, XL block length, 19-24 kDa range). Polymers were complexed with siRNA into si-NPs , which were analyzed for pH-dependent membrane disruption (hemolysis assay), siRNA loading efficiency (ribogreen), endosome disruption in living cells (Gal8-YFP recruitment assays), lysotracker studies, cell viability, and knockdown bioactivity (luciferase assay).

Results and Discussion: The balance of cationic and hydrophobic content, as well as the core block lengths affect endosomal escape and gene knockdown characteristics of polymeric si-NPs. Multiple polymers were both nontoxic and exhibited switch-like pH-dependent hemolysis. However, only one polymer (50B) was both nontoxic and had robust knockdown. This polymer was also the only polymer found to have robust Gal8-YFP recruitment. As total block length is increased in the 50B composition, endosomal escape and model gene knockdown increases. We also find that directly measuring endosomal escape as measured by Gal8 -YFP recruitment is an accurate predictor of model gene knockdown, and correlates with standard lysotracker studies.

Conclusions: Gal8-YFP recruitment is a robust new tool to screen polymer libraries for cytosolic delivery, especially for cargo whose cytosolic delivery and bioactivity may not be easily measured. It is straight forward, rapid and does not require expensive fluorescently labelled cargo. It can also work with peptides[1]. The results confirm that Gal8-YFP recruitment is a sensitive method of assessing how effective polymers are at endosomal escape, as a predictor of in vitro bioactivity.

Refs: 1. Kilchrist et al. Cell Mol Bioeng. 2016. **2.** Wittrup et al. Nat. Biotech 2015. **3.** Nelson et al. ACS Nano 2013

Copyright © American Institute of Chemical Engineers. All rights reserved.

509512 Expansion and Optimization of DIVA DNA Sequence Validation Services

Abstract:

Advances in DNA sequence validation technology has enabled the Joint BioEnergy Institute (JBEI) to offer high-throughput plasmid sequencing to its researchers. However, the types of compatible samples have not significantly expanded. This lack of compatibility leads to time and materials wasted on purifying or transforming a plasmid into *E. coli*. Here, we sought to update the current sequencing protocol to include the most utilized species and plasmid types, including *E. coli* cultures with medium and low copy plasmids, *A. tumefaciens*, *P. putida*, *S. cerevisiae*, and *P. pastoris*. Based on previous runs, the laboratory determined that a baseline coverage greater than 60x over the whole plasmid yielded the most accurate reads. This project showed that *E. coli* was the most robust species, with sequencing succeeding for all liquid cultures, regardless of copy number, and for resuspended colonies picked directly from agar plates. The remaining species experienced varying levels of coverage failure. However, minimal modification to the existing sequencing workflow, such as longer lysing times or the use of higher copy plasmids, could increase overall coverage. Overall, the sequence validation services provided by JBEI have numerous advantages over traditional Sanger sequencing with third-party companies. By expanding the compatible sample types, we are making this efficient, cost-effective sequencing service more accessible to JBEIâ€™s researcher community.

Copyright © American Institute of Chemical Engineers. All rights reserved.

509925 Engineered Biomaterials Enable Cell Replacement Therapy to Restore Motor Function in Parkinson's Disease Model Rats

Abstract:

Parkinson's Disease (PD) is a progressive neurodegenerative disease affecting ~10M patients worldwide. One of the hallmarks of PD is the selective loss of midbrain dopaminergic neurons (mDA) residing in the substantia nigra (SN) and projecting to the striatum. The associated loss of striatal dopamine input causes the characteristic debilitating motor dysfunction seen in PD patients.

Medications such as levodopa treat PD symptoms by providing a dopamine precursor to help remaining mDA neurons compensate for dying neurons' dopamine production; however, since neurodegeneration is not halted, this compensation requires higher amounts of levodopa as degeneration continues, and is still only sufficient for 5-10 years. Deep brain stimulation, an experimental late-stage treatment, risks infection, hemorrhage, and adverse mental side effects due to its invasive, poorly-understood nature. Human pluripotent stem cell (hPSC)-based cell replacement therapy is a promising alternative to existing PD treatments, as it replaces lost mDA neurons and has potential to restore lost function. Previous work in the Schaffer lab has shown that 3D biomaterials including PEG-pNIPAAM and hyaluronic acid hydrogels can aid in overcoming CRT's key barriers of scalability, differentiation efficiency, and post-transplantation survival.

In this experiment, hPSCs were first differentiated into mature mDA neurons using 3D PEG-pNIPAAM gels, then encapsulated in HA-based gels optimized for controlled release of incorporated neurotrophic and dispersion factors, and finally transplanted into the striatum of PD model rats. Two behavioral assays showed rapid and sustained restoration of motor function in the "3DGF" group, which received mDA neurons grown in PEG-pNIPAAM and encapsulated in the HA gels with incorporated factors. Immunohistochemistry showed improved survival of transplanted neurons, higher proportion of TH+ (mDA) neurons, and increased formation of synapses with host neurons in the 3DGF group, corroborating the improvement in motor function. This sustained functional improvement was not recapitulated in groups where cells were encapsulated without factors, were not HA-encapsulated, or were not cultured in PEG-pNIPAAM. The incorporated neurotrophic/dispersion factors were particularly important, since encapsulation with the factors yielded a 5-fold higher fraction of TH+ neurons than without the factors, and markedly improved motor recovery. Indeed, the two histological measures which correlated best with motor improvement were dispersion of transplanted cells and number of surviving TH+ neurons--both of which were highest in the 3DGF group. These results prove hPSC-based cell replacement therapy is capable of restoring motor function in PD model rats, and is significantly improved by our biomaterial platforms for mDA differentiation and transplantation.

Copyright © American Institute of Chemical Engineers. All rights reserved.

511614 Protein Engineered Triblock Polymers Comprised of Two Sads for Hydrogel Fabrication

Abstract:

Protein hydrogels have many unique characteristics such as their biocompatibility, controllable pore size, and adjustable mechanical properties, making them ideal candidates for biomedical applications. Using protein hydrogels as a building block, the protein secondary structure can be manipulated to impart desired properties for particular applications such as stimuli response and drug binding. Our lab has generated a mechanically responsive triblock polymer that comprises two domains, one derived from the N-terminal coiled-coil region of the non-collagenous glycoprotein cartilage oligomeric matrix protein (C), capable of binding small hydrophobic molecules, and the other from the elastin-like polypeptide motif (E), which experiences an inverse transition temperature from a soluble random coil to a ß-spiral structure. This triblock polymer is rationally designed, in which the two domains are linked as C-E-C with the E motif flanked between two C domains, which combines the respective functions of each domain. The negative control $C_{L44A}EC_{L44A}$ was constructed bearing a previously characterized impaired C_{L44A}domain deeming the triblock inelastic, unstructured and incapable of binding to small molecules. CEC and its counterpart $C_{L44A}EC_{L44A}$ were structurally assessed via circular dichroism (CD) for secondary structure, as well as accessed for thermostability, mechanical integrity, and small molecule binding capability. Furthermore, the two constructs were assessed for micellar-like particle formation to elicit the driving factors behind the formation of hydrogels. The CEC triblock demonstrated enhanced mechanical properties, structural integrity and small molecule encapsulation over its counterpart as well as compared to previously characterized variants bearing the C domain, such as EC, CE and ECE. These capabilities are a result of the addition of a second C domain. Moreover, CEC exhibited significantly denser micellar-like particle formation compared to $C_{L44A}EC_{L44A}$. Since CEC demonstrates promising biomedical applications, further studies to fabricate a free standing hydrogel consisting of the CEC polymer with patterning to substrate capabilities are underway.

Copyright © American Institute of Chemical Engineers. All rights reserved.

510026 Tuning Gel Stiffness with Surface Chemistry to Study Cancer Cell Behavior

Abstract:

Cancer metastasis occurs when cancer cells migrate away from the source of the tumor to other tissues in the body. Treatment options become limited once the cancer has metastasized, which motivates an understanding of what factors impact cancer cell migration. Metastasis is driven by extracellular matrix (ECM) cues such as chemical gradients, orientation of ECM fibers, or stiffness of the ECM that direct cell migration. In this study, we examined how cancer cells respond to changes to the stiffness of a collagen gel in a 3-D environment. Previous studies have used gel density or cross-linking in polymer networks to control the stiffness of various gels. However, changing gel density or cross-linking can result in a system with different properties, such as number of cell-ECM attachment sites, which also impacts cell behavior. It is known that attaching a soft gel to a stiff substrate and adjusting the thickness of the soft gel can be used to control local stiffness in a material. We used this approach in addition to chemical linkage between a surface and a soft gel to control the local gel stiffness. A study showed that cells had a larger area when located closer to the interface between a soft gel and hard material. Therefore, it was predicted that cells located near a chemically-modified glass surface would have larger areas due to the larger local stiffness at the surface. To test this hypothesis, MDA-MB-231 breast cancer cells were embedded in collagen gels between two glass coverslips of a specific surface chemistry (clean glass or glutaraldehyde-treated) and gel thickness. The cells were stained for F-actin, and morphological properties of the cell were measured using ImageJ. The cells were found to have larger areas when located near a glutaraldehyde-treated glass surface compared to the cells near the surface of clean glass regardless of soft gel thickness. In addition, cells found further from the glutaraldehyde-treated surface had smaller areas than those found near the surface. On the other hand, when cells were embedded in collagen between two clean glass coverslips, the measured cell area was the same regardless of distance from the surface or collagen gel thickness. This indicates that there exists some distance from the surface in which cells can no longer sense the chemical linkage between the collagen fibers and the glutaraldehyde-treated glass surface. This result is significant because it demonstrates that cells respond differently in the ECM when located near stiffer tissue within the body. This finding could be used in tissue engineering applications where surface linkage might control cell function or in fundamental studies where researchers are trying to understand responses to stiffness.

Copyright © American Institute of Chemical Engineers. All rights reserved.

510323 Towards the Development of an in Vitro 3D Co-Culture Cancer Model for the Design of Dendrimer-Anti-Cancer Drug Conjugates Â

Abstract:

Towards the Development of an in vitro 3D Co-Culture Cancer Model for the Design of Dendrimer-Anti-cancer Drug Conjugates

Justin Moore, Elizabeth Bielski, and Sandro R.P. da Rocha*

School of Pharmacy & Chemical and Life Science Engineering

Virginia Commonwealth University Ð Richmond, VA 23298-0533

Two dimensional (2D) monolayer cells cultured on flat substrates have been for decades an important *in vitro* tool used to test and understand numerous biomedical problems. Traditionally, these cell culture assays have been a monolayer of cells cultured *in vitro*. However, these 2D cultures have recently come under scrutiny for their significant departure when compared to the *in vivo* tumor microenvironment they are designed to mimic. In 2D monolayer cell cultures cell-to-cell/cell-to-matrix interactions, biomedical gradients and signaling pathways are lost, tumor characteristics that significantly affect the drug transport. The long-range of this project is to establish a 3D in vitro model to test the ability of dendrimer-anti-cancer therapeutics to treat solid tumors, and the impact of their chemistry in terms of its ability to penetrate deep regions of the solid tumor environment. The first part of my project is establishing a more relevant model of the tumor microenvironment. This consisted of a transition from the traditional 2D cultures to a 3D model that better represents the architecture of solid tumors. Furthermore, the establishment of a co-culture system using cancer-associated stromal cells was an appropriate subsequent step in to achieve a more relevant in vitro system to test and understand the cytotoxicity profiles of the different systems. The stromal cell type used were cancer-associated fibroblast (CAF) because of their important role in creating an extracellular matrix which has been shown to impact drug penetration. Optimization of the ratios of cancer/stromal cell lines was required to see what initial ratio would result in a 70%/30% stromal/cancer cell ratio, a ratio representative of the in vivo microenvironment. Briefly, at a range of different ratios of 4T1 metastatic breast cancer cells and NIH 3T3 GFP fibroblast cells were seeded in flat-bottom 96 well plates, which were coated with agarose gel. The MCS were allowed to grow for 4 days then were dissociated. Then, using Flow Cytometry the co-cultured spheroids were characterized. It was found that the fibroblast and 4T1 cells are growing with similar growth kinetics within the co-culture. Previously, using an MTT assay, we tested the cytotoxicity of doxorubicin (DOX) on mono-culture MCS, and then compared these results to a 2D model system assay. The MTT assay showed that much higher concentration of DOX are required to achieve 50% of cell death (IC50%) in the 3D spheroid (ca. 13 fold). Additionally, Dendrimer-Dox-conjugate had an increased toxicity profiles in both system which may suggest that the dendrimer conjugation has more ability to penetrate the spheroids and each cell in general. Next steps include establishing a cytotoxicity protocol for the 3D co-culture system and how manipulating (size, charge, and functional groups) impacts the Dendrimer-DOX penetration.

Corresponding author: srdarocha@vcu.edu

Acknowledgements: VCU for a Start-up Grant; NSF-DMR Grant # 1643770 and VCU IMSD - NIHR25GM090084 for a scholarship for JM.

Copyright © American Institute of Chemical Engineers. All rights reserved.

510422 Femtosecond Laser Patterning of Cell Scaffold Materials for 3D Vascularization Modelling

Abstract:

Femtosecond Laser Patterning of Cell Scaffold Materials

for 3D Vascularization Modeling

Sabrina McCarthy[1], Nannette Hernandez[1], Alessandro Enrico[2],

Anna Herland[3], Frank Niklaus[2] , Gšran Stemme[2]

1 Department of Chemical Engineering, City College of New York

2 Department of Micro and Nanosystems, School of Electrical Engineering, Royal Institute of Technology

3 Department of Physiology and Pharmacology, Karolinska Institutet

Creating functional, biological tissue constructs is essential for drug screening. Current approaches consist in cell-based vascularization, or scaffold molding and subsequent channel perfusion. However, the first is difficult to achieve and to control, while the second often break sterile conditions because of the molding procedure, and is still limited in terms of 3D geometry of the possible vascular network.

Here we propose 3D patterning of channels by using a femtosecond laser. As non-contact technique, sterility of the sample is maintained, and design freedom in terms of patterning is provided by the nonlinear nature of multiphoton absorption that leads to point-like modification in correspondence of the focal spot of the laser beam.

Recent interests have been devoted to achieving vascularization in ECM based scaffolds such as collagen. When creating these constructs, providing sterility to cells while also creating a mechanically stable ECM based scaffold in a time and cost efficient manner is crucial. Femtosecond laser technology, as opposed to UV molding and soft lithography, allows for a non-contact technique and the freedom to pattern in 3D geometry.

Experiments using femtosecond laser technology to pattern vessel like structures in collagen can provide more insight into what parameters are needed to ablate through a liquid medium. By adjusting power, repetition rate, and feed rate ablation in collagen immersed in liquid is possible without modifying the layers above.

Copyright © American Institute of Chemical Engineers. All rights reserved.

510751 Synthesis of Temperature Responsive Core-Shell Magnetic Nanoparticles Via Atom Transfer Radical Polymerization

Abstract:

Magnetic nanoparticles have been widely studied for various applications in medicine, environmental remediation, etc. The controlled functionalization of these particles is key for successful application. In this work, research was completed to investigate the effect of coating Fe_3O_4 magnetic nanoparticles with a hydrogel shell consisting of the temperature responsive material N-isopropylacrylamide (NIPAAm) with a novel hydrophobic crosslinker curcumin multiacrylate (CMA). NIPAAm-co-CMA core-shell nanoparticles were synthesized through surface-initiated atom transfer radical polymerization (ATRP). A kinetic study was conducted to test the effect of reaction time on coating thickness. The resulting particles were characterized using dynamic light scattering (DLS), thermogravimetric analysis (TGA) and fourier transform infrared spectroscopy (FTIR). DLS was used to analyze the thermo-modulated changes in particle size, TGA was used to analyze the extent of polymer loading, and FTIR was used to identify functional groups.

Copyright © American Institute of Chemical Engineers. All rights reserved.

510468 Exploration of a Modular Microcapsule Scaffold System for Cartilage Tissue Engineering

Abstract:

Shallow defects in human articular cartilage defects do not heal, and defects that reach the subchondral bone are replaced with fibrous tissue that lacks desired mechanical properties. Autologous cartilage grafts cause donor site morbidity, and joint replacements require complex surgeries. This study investigated a modular tissue engineering approach for cartilage repair. Rat bone marrow mesenchymal stem cells (MSCs) were encapsulated in microcapsules made of complexes of chitosan with blends of hyaluronan (HA) and chondroitin-4-sulfate (CSA), and an optional interior of collagen type I gel. Chitosan is biocompatible and biodegradable, and HA and CSA are glycosaminoglycans (GAGs) found in cartilage extracellular matrix (ECM) that have been shown to enhance the chondrogenic differentiation of MSCs in terms of the ECM the chondrocytes produce. The capsules are formed by dripping a suspension of MSCs in GAG solution into a stirring chitosan solution. An ionic complex of the cationic chitosan and the anionic GAGs forms at the droplet boundaries, trapping the cells inside spherical membranes with diameters on the order of hundreds of micrometers. These capsules can be fused together in molds to potentially form large tissue constructs. Capsules made with a blend of HA and CSA (HA/CSA) had spherical walls, and encapsulated MSCs formed multiple aggregates within 24 hours. Capsules made of HA had thick, wrinkled walls, and the encapsulated MSCs remained dispersed. MSCs in collagen-containing capsules formed single, large aggregates. By day 7, the HA-collagen capsules exhibited a 38% reduction in diameter, likely due to contraction of the collagen gel by MSCs, which may be desirable for creating dense tissue cultures. When cultured in chondrogenic medium, the MSC aggregates in HA/CSA capsules became smooth but remained as distinct aggregates, while the HA capsules filled in with opaque material, and underwent a 39% diameter reduction over 28 days. At 21 days, the capsules contained 82±4, 81±3, 69±6, and 106±16 pg collagen/cell seeded in the HA growth, HA chondrogenic, HA/CSA growth, and the HA/CSA chondrogenic cultures, respectively. Histological staining with safranin-O and toluidine blue showed the production of proteoglycans in all culture conditions. The contraction of the HA capsules and the high collagen production in HA/CSA capsules make both capsule types promising for further study.

Copyright © American Institute of Chemical Engineers. All rights reserved.

510759 Functionalized Gold Nanoparticles As Novel Alzheimer's Therapeutics

Abstract:

Aggregation of amyloid-β protein (Aβ) into toxic structures in the brain has been indicated as a major etiological factor in Alzheimer's disease. Much of contemporary Alzheimer's research seeks to obtain a greater understanding of the Aβ aggregation process, and evaluate compounds that inhibit it. Previous research has explored the effects of inorganic nanoparticles with functional coatings on Aβ aggregation. One hypothesis suggests that such systems influence this process by acting as point sources of local solution destabilization. This project describes the modeling and synthesis of gold nanoparticles (AuNPs) functionalized with poly(acrylic acid) (PAA) chains as a novel aggregation inhibitor. PAA samples were synthesized and grafted onto AuNPs in a ligand exchange process. These PAA-AuNPs were characterized through various chemical analysis methods. The previous hypothesis was confirmed in part by study of such NP systems with a statistical thermodynamics model for molecular dynamics. Equilibrium calculations reveal that the polyelectrolyte brush "pulls" species in toward the surface to relieve negative charge density, affecting solution characteristics such as local pH. Therapeutic potential of these PAA-AuNPs was evaluated with direct introduction into an Aβ monomer solution; aggregation was subsequently measured with fluorescence spectroscopy. The combined theory and experiment will yield a better understanding of amyloid aggregation mechanics, and a process for optimizing the design of an aggregation inhibitor.

Copyright © American Institute of Chemical Engineers. All rights reserved.

510486 New Amphiphilic Polyanhydride Copolymers for Vaccine Delivery

Abstract:

Polyanhydride nanoparticle-based vaccines (i.e. nanovaccines) have been explored as next generation vaccines due to their adjuvant characteristics and ability to release encapsulated proteins over a long duration, allowing for the possibility of single-dose administration of poorly immunogenic proteins. These nanovaccines have been shown to enhance immune responses as evidenced by sustained antibody production, robust germinal center formation and the ability to enhance cytotoxic T cell responses.

Polyanhydride copolymers comprise of 1,6-bis(*p*-carboxyphenoxy) hexane (CPH), 1,8-bis (*p*-carboxyphenoxy)-3,6-dioxaoctane (CPTEG), and sebacic acid (SA), with each of these monomers offering its own beneficial properties. In particular, nanoparticles with high CPTEG molar composition have demonstrated improved stability of secondary and tertiary structures of encapsulated proteins, and enhanced cellular internalization and persistence. SA-rich chemistries have also demonstrated enhanced cell internalization, and improved shelf stability of encapsulated proteins, likely due to their higher glass transition temperatures. However, copolymer chemistries consisting of CPTEG and SA have not been studied to date.

In this study, several different chemistries of CPTEG and SA copolymers were synthesized by melt condensation and their physical properties were evaluated. Polyanhydrides with glass transition temperatures higher than body temperature were synthesized into nanoparticles encapsulating ovalbumin, a model antigen. Particle size was determined via scanning electron microscopy and the stability of released ovalbumin was evaluated using circular dichroism and fluorescence spectroscopy. The final part of this study explored nanoparticle internalization in RAW 264.7 macrophages to compare how well the CPTEG:SA chemistriesâ€™ were internalized by the cells in comparison to the CPH:SA and CPTEG:CPH chemistries. The studies performed indicate that the CPTEG:SA copolymers are promising candidates for vaccine delivery.

Copyright © American Institute of Chemical Engineers. All rights reserved.

510494 Gel Encapsulation of Mel28 Cells Using DEX/E3 Water-Water System

Abstract:

Mariah Gallegos, University of New Mexico

Mentor: Dr. Andrew Shreve, Department of Biomedical Engineering; Dr. Nick Carroll, Department of Chemical Engineering

Gel encapsulation of Mel28 cells using DEX/E3 water-water system

Aqueous two-phase systems using protein and glucan have potential for droplet generations which in turn can be used to encapsulate spherical cells with further implications for 3D biological printing. The phase-separation being investigated includes a mixture of Dextran and E3. This water-soluble solution phase separates once a critical concentration of Dextran and E3 has been obtained. This process is Thermodynamically driving which is temperature dependent. The fluid can then be put into a capillary device in which droplets are formed. E3 acts as a gel pursuer for the microdroplet. The outer microdroplet layer is protein-rich and the inside contents is Dextran-rich. Using this type of water-water interactions will allow for a cell to be encapsulated within the protein gel with the use of crosslinking. This environment is cyto-friendly and due to the cross-linkage creates more attractive mechanical properties. Using E3 as the precursor, within the water-water system, will also create a great platform for 3D printing biological due to the stability and degradation of the protein. Thus, this research will be important towards mass printing biological material.

Copyright © American Institute of Chemical Engineers. All rights reserved.

510622 Formulation of Novel Dry Powder Antibacterial Aerosols Using Central Composite Design of Experiments

Abstract:

Pulmonary bacterial infections are often difficult to eradicate through intravenous or oral administration of antibiotics due to the presence of bacterial biofilms and the difficulty in achieving the desired drug concentration at the site of infection. Bacterial biofilms consist of bacterial colonies surrounded by a polysaccharide matrix that protects the embedded bacteria from external threats such as antibiotics. As a result, high doses of antibiotics are typically administered, leading to adverse effects. Our lab is working on developing combination therapies consisting of an antibiotic and a nutrient dispersion compound to increase the susceptibility of the biofilm bacteria to antibiotics. The nutrient dispersion compounds act as bacterial nutrients and entice the bacteria out of biofilms, resulting in greater exposure to antibiotics. This allows for administration of lower antibiotic doses, decreasing the likelihood of adverse effects.

The objective of this study is to develop a dry powder aerosol containing colistin sulfate (antibiotic) and sodium citrate (nutrient dispersion compound), that has high yield, good aerodynamic properties, and high drug loading. Dry powder aerosols allow for direct administration of drugs to the lungs, enabling delivery of lower doses of the drug and faster onset of therapeutic activity. A central composite design of experiments, generated using Minitab® software, was used to study the effect of formulation and spray drying process parameters on the yield and aerodynamic properties of the powder. A two-level central composite design was used to generate 31 formulations across the sample space, varying solution flow rate, atomizer flow rate, leucine concentration and solution concentration. Dry powder aerosols were generated using a Büchi 190 spray dryer, keeping temperature and pressure drop constant. Leucine was used to increase the flowability of the powders. Water content of the powders was measured using a coulometric Karl Fischer titration. Pareto chart analysis of spray drying data showed that increasing atomizer flow rate and solution concentration significantly increased powder yield. Yields ranged from 10% to 70% with a majority of yields between 40% to 60%. Water content studies showed water contents between 1 to 7%.

Future studies will examine contents of colistin sulfate and sodium citrate in the powders using the high-performance liquid chromatography (HPLC) and citrate assay technique respectively. Aerodynamic properties such as mass mean aerodynamic diameter (MMAD), fine particle fraction (FPF), and emitted fraction (EF) will be measured using the next generation impactor (NGI).

Copyright © American Institute of Chemical Engineers. All rights reserved.

510813 Rational Design of Genetically Engineered, Gold and Cell-Binding Polypeptides for Fabricating Thermoresponsive Cell Culture Substrates

Abstract:

Chronic wounds do not adequately recover through the natural healing process and have become a significant challenge for healthcare systems worldwide. In the US, chronic wounds affect an estimated 6 million people per year, costing more than $25 billion annually due to complications and over $18.5 billion in associated care. Peripheral arterial disease, neuropathy, and immunodeficiency disorders are conditions that can lead to chronic open wounds and lesions. These conditions are especially prevalent in diabetic patients due to immobility and lack of blood flow to the extremities. Current technologies for chronic wound healing are aimed at dressings which consist of various hydrogels and collagen. Modern tissue engineering methods aim to expand the understanding of cell sheets and cell hydrogels which could form a â€˜cellular scaffoldâ€™ to improve recovery time. However, these methods often rely on proteolytic enzymes such as trypsin, which damages the extracellular matrix (ECM), cell-cell junctions, and surface cell receptors. Other methods use polymers such as poly(N-isopropyl acrylamide) (pNIPAM), which relies on ECM molecules or chemical conjugations of cells for binding and release with no specific cell binding motifs. For this study, elastin-like-polypeptides (ELPs) were designed with a thermally responsive motif, allowing us to control and target cell binding and detachment in culture without damaging the ECM. These proteins were deposited onto gold surfaces via specific cell-binding cysteine domains. The location of the cell binding motifs within the ELP proved to be an important factor for detachment of the cells. ELP layers with cell binding domains near the surfaces allowed for thermally controlled cell adhesion, similar to those found in pNIPAM. Decreasing temperature below the lower critical solution temperature (LCST) causes the ELPs to adopt a random coil formation, which leads to retraction of the cell binding motifs, thus detaching the adherent cells. The cell lines tested for detachment included bovine aortic endothelial cells (BAECs) and human umbilical vein endothelial cells (HUVECs). This study indicates that varying cell lines can be consistently recovered from the surfaces as cell sheets for tissue engineering purposes.

Copyright © American Institute of Chemical Engineers. All rights reserved.

510948 Analyzing Chemical Compound Modulators of the Vibrio Cholerae PTS and Biofilm Formation

Abstract:

Phosphoenolpyruvate-carbohydrate phosphotransferase system (PTS) is a multistep chemical process which regulates the intake and use of carbohydrates by bacteria. It is used for the transfer of sugars in *Vibrio cholerae* and requires phosphate transfer. In the absence of transported sugars, phosphate will accumulate on the protein components of the system and will indirectly analyze the PTS sugars present and metabolic state of the bacteria cell, allowing the bacteria to form biofilms. Biofilms are multi-layered communities of bacterial cells that grow attached to one another. In addition to controlling sugar uptake by bacteria, the PTS regulates several cellular functions such as chemotaxis, glycogen metabolism, catabolite repression and the aforementioned biofilm formation.

In this study, *V. cholerae* MO10 served as the model organism. It was grown in the absence of chemical compounds as a negative control, while an altered strain lacking PTS activity served as a positive control. A chemical screen was used to identify chemical compounds that promote the production of *Vibrio cholerae* biofilms. Growth kinetics was evaluated in absorbance-based microplate assays, and quantitative biofilm assays were performed in borosilicate glass tubes. All assays were monitored by spectrophotometry.

Biofilm induction was confirmed by monitoring the growth of *Vibrio cholerae* in the presence or absence of chemical compounds. Significant biofilm growth was seen in LB with three specific compounds, and one of the three compounds was further analyzed in Minimal Media Pyruvate. Biofilm growth was seen to be concentration dependent.

Further assays are underway to see the effect of this candidate compound on *Staphylococcus aureus* biofilms. Growth kinetics and quantitative biofilm assays are being performed in absorbance based microplate assays, and monitored by spectrophotometry.

These assays will determine whether or not biofilm induction occurs through interactions with the bacterial PTS. By modulating the PTS, chemical compounds that regulate biofilm formation in gram negative and gram positive bacterium will be identified and characterized. This may lead to the development of new microbial control strategies with applications in biological, medical, and pharmaceutical sciences.

Funder Acknowledgement(s): This study was supported by a grant from NSF (Research Initiation Award #1505301) awarded to Patrick Ymele-Leki PhD, Assistant Professor of Chemical Engineering, Howard University, Washington, DC 20059.

Copyright © American Institute of Chemical Engineers. All rights reserved.

510829 Designing Size-Controllable Bicelles through Addition of Polyethylene Glycol Conjugated Lipid and Cholesterol

Abstract:

Our study focuses on effectively controlling the morphology of lipid-based nanodiscs (also known as "bicelles"). We have found that polyethylene glycol conjugated (PEGylated) lipid and cholesterol can enhance the stability, control the size of nanodiscs, and prevent a nanodisc to vesicle transition. Low incubation temperature can also improve the structural stability of the bicelles. Size-uniform bicelles were self-assembled in lipid mixtures composed of zwitterionic dipalmitoyl (di-C 16) phosphatidylcholine (DPPC), dihexanoyl (di-C6) phosphatidylcholine (DHPC), negatively charged dipalmitoyl (di-C 16) phosphatidylglycerol (DPPG) and sodium chloride. Small angle X-ray scattering, dynamic light scattering and transmission electron microscopy were used to characterize the nanostructure. We varied PEGylated lipid and cholesterol concentrations in the ranges of 0 - 5 mol% and 0 - 10 mol% of the total lipid composition, respectively. The samples with lower concentrations of PEGylated lipid (0%, 1%, 2.5%), and cholesterol (0%, 2%) showed substantial growth compared to the samples with higher concentrations. A structural transition from bicelles to vesicles was observed after the low PEGylated lipid and low cholesterol samples were incubated at 25 for several days, but was not present in samples containing higher amounts of cholesterol (5%, 10%) or PEGylated lipids (4%, 5%). Low temperature (4) significantly stabilizes the size of bicelles. It is possible to increase the size of bicelles at 25 until the desired size, then stabilize them by incubation at low temperature. These results have significant relevancy for pharmaceutical applications as it has been reported that nanodiscs have higher cellular uptake than vesicles. The ability to control the size of the nanodiscs would make it possible to optimize the system to allow for drug delivery to tumors with high efficacy.

Copyright © American Institute of Chemical Engineers. All rights reserved.

510954 Designing Next Generation Polyanhydride Copolymers for Nanoparticle-Based Vaccines

Abstract:

Polyanhydride nanoparticle-based vaccines (*i.e.* nanovaccines) offer many benefits over traditional vaccines, such as possessing controllable release profiles, surface erosion kinetics, vaccine dose-sparing capacity, chemistry-dependent immunomodulation, and protein shelf-life stability. These nanovaccines are comprised of copolymers of 1,6-bis(*p*-carboxyphenoxy) hexane (CPH), 1,8-bis(*p*-carboxyphenoxy)-3,6-dioxaoctane (CPTEG), and sebacic acid (SA), with each of these monomers offering its own beneficial properties. In particular, nanoparticles that are high in CPTEG content have been shown to improve cellular internalization and persistence as well as facilitate the retention of encapsulated protein secondary and tertiary structure. Nanoparticles high in SA content have demonstrated improved cell internalization and shelf-life stability of encapsulated proteins. Therefore, CPTEG:SA copolymers have been synthesized to maximize the properties of each individual component. These new copolymer chemistries were characterized for molecular composition and molecular weight via nuclear magnetic resonance spectroscopy (NMR) and glass transition temperature and melting point via differential scanning calorimetry (DSC). As expected, several of the CPTEG:SA copolymers have glass transition temperatures greater than body temperature and therefore could be useful for vaccination applications in the future. Flash nanoprecipitation was utilized to form nanoparticles from these novel chemistries which were examined by scanning electron microscopy (SEM) which showed very discrete particulate structures. Cellular uptake was investigated through flow cytometry which provided significant evidence that CPTEG:SA nanoparticles are readily taken up by J774 macrophages providing evidence of their potential viability as nanovaccines.

Copyright © American Institute of Chemical Engineers. All rights reserved.

511038 Chitosan Hydrogels As an Alternative Injectable Treatment for Degenerative Bone Disorders

Abstract:

Chitosan Hydrogels as an Alternative Injectable Treatment for Degenerative Bone Disorders

Trent Faulkner[1], Soheila Aliakbarighavimi[1], Allison Josselet[2], Ethan Lungren[1], Ying Wu[2], Yisheng Sun[2], Ferris Pfieffer[2,3], Christina Goldstein[3], Caixia Wan[1,2], and Bret Ulery[1,2]

[1]Department of Chemical Engineering, University of Missouri, Columbia, MO 65211

[2]Department of Bioengineering, University of Missouri, Columbia, MO 65211

[3]Department of Orthopaedic Surgery, University of Missouri, Columbia, MO 65211

Osteoporosis is one of the most common orthopedic disorders, affecting approximately 10 million people in the United States alone. Osteoporosis of the spine is of particular concern as it increases the risk of vertebral compression fractures (VCFs). Vertebroplasty and kyphoplasty are minimally invasive methods commonly used to treat these fractures and consist of injecting poly(methyl methacrylate) (PMMA), a quickly solidifying bone cement, into the porous bone tissue. However, the high compressive strength of PMMA results in undue pressure being placed on the vertebrae adjacent to the treated area, increasing the chance that those vertebrae will experience a future VCF. Additionally, PMMA injections do not serve to aid in regeneration of the degraded bone, instead only providing mechanical support.

Aiming to develop a bioactive and mechanically supportive injectable treatment for degenerative bone disorders and fractures, we created a hydrogel composed of a polymeric chitosan backbone crosslinked with genipin and phosphate. Initial tests of the hydrogel found a maximum compressive strength of 50 kPa, considerably less than what is necessary to mechanically support bone. Looking to improve the compressive strength, we suspended negative, neutral, or positive cellulose nanocrystals (CNCs) in chitosan before gel fabrication and subsequent strength testing. The results of these trials showed minimal changes in compressive strength upon addition of positive or negative CNCs, but over a two-fold increase with the addition of neutral CNC. While still far from the strength of bone, we are currently investigating additional additives that may eventually lead to a mechanically competent and bioactive injectable alternative to PMMA.

Copyright © American Institute of Chemical Engineers. All rights reserved.

511161 A Stimulus-Responsive, in Situ Forming, Nanoparticle-Laden Hydrogel for Ocular Drug Delivery

Abstract:

Glaucoma and other chronic optic neuropathies are the leading causes of blindness worldwide. These diseases are characterized by damage to the optic nerve at the base of the retina, and most medications targeting optic neuropathies are administered as eye drops. However, since corneal penetration efficiencies are typically <5% for such dosage forms, not only are these formulations ineffective, but the dosage of drug is much higher than required, which incurs higher costs. There is a clear, unmet need for novel transcorneal drug delivery vehicles that achieve higher drug penetration at significantly lower dosages. To this end, we have developed a stimulus-responsive, in situ forming, nanoparticle-laden hydrogel for spatiotemporal and dosage-controlled release of poorly bioavailable drugs into the aqueous humor of the eye. The hydrogel is formulated as a composite of hyaluronic acid (HA) and methylcellulose (MC). Both polymers are biocompatible, highly mucoadhesive and approved by the FDA as GRAS. The amphiphilic nanoparticles are composed of poly(ethylene oxide) (PEO) and poly(lactic acid) (PLA). Experimental design aided the identification of hydrogel composition and nanoparticle content in the formulation for stimulus-responsive switching between thixotropy and temperature-dependent rheopexy. The liquid formulation has a sol-gel transition temperature of 32Â°C, which is the temperature of the ocular surface, and its gelation characteristics permit the formulation to form a thin, uniform coating over the cornea through blinking of the eyelid. These properties facilitate application of the medication as an eye drop immediately prior to the patientâ€™s bedtime, and also permit easy and scalable manufacturing, which greatly simplifies the path towards development of a regulated industrial-scale manufacturing process. We subsequently tested the efficacy of our formulation in whole-eye experiments using porcine eyeballs by loading the nanoparticles with cannabigerolic acid (CBGA), a molecule that is a close mimic of an approved glaucoma medication. Our formulation exhibits over a 300% increase in transcorneal penetration over control formulations. Our work paves the way for the introduction of novel products targeting ocular diseases to the market.

Copyright © American Institute of Chemical Engineers. All rights reserved.

509755 Dynamic Ex Vivo Model of Lung Micromechanics

Abstract:

Cystic Fibrosis is caused by a mutation in CFTR (Cystic Fibrosis Transmembrane Conductance Regulator), which is mechan-osensitive. Creating an environment which closely follows the mechanical strain experienced by the alveolar-capillary membrane will enable accuracy in further studies into Cystic Fibrosis, CFTR behavior, and therapeutic delivery to cure symptoms of as well as the disease itself. Research has previously explored the effects of uniaxial stretching on such cells, but has yet to thoroughly investigate a biaxial, dynamic environment on the CFTR. A "Lung on a Chip" (LoC) device has been fully constructed which employs electromagnetic actuation of a PDMS membrane combined with a hydraulic force transmission layer to generate biaxial stress and strain. The device has been partially optimized and the efficiency of fluidic volume transfer—and therefore fluidic force transmission—as well as strain on the membrane in wells have been analyzed. While desired fluid force transmission and strain were not achieved with the current embodiment, a strong foundation exists for further optimization.

Copyright © American Institute of Chemical Engineers. All rights reserved.

510909 Assessing Cytocompabitility of Novel High Ductility Magnesium Alloys

Abstract:

Magnesium-based alloys have attracted considerable attention in the last decade due to its biocompatibility, ability to degrade *in vivo*, and suitable mechanical properties. New application possibilities, such as orthopedic implant, sutures and vascular stents, have the potential to reduce the number of second surgeries as well as significant long-term adverse effects due to perennial existence of the device. Magnesium alloys do not produce any toxic corrosion product during degradation; however, its rapid, uncontrolled degradation rates raise concerns, as the surrounding tissues get affected. Currently, there are no studies to date reported to evaluate the cytotoxicity of magnesium-based alloys on airway epithelium cells. The aim of this study is to assess the *in vitro* cytotoxicity of our proprietary patent pending ultra-high ductility (UHD) magnesium alloys specifically aimed at tracheal stent application. In this study, pure Mg, AZ31, and three alloys of different compositions containing lithium, aluminum, and zinc are tested. MTT testing and DAPI & F-action staining were performed to determine the effects of the degradation products of these novel alloys with different dilution ratios on the human bronchial epithelial cell line (BEAS-2B). These results demonstrated comparable cytotoxicity of our proprietary magnesium alloys to commercial pure magnesium and AZ31 alloys. ICP-OES and the MTT analyses show a potential impact of lithium ion concentration on the cell viability. Through DAPI & F-Actin staining, it can be seen that the alloys do not have an adverse impact on the morphology of the cells. The effect of magnesium ions on the aggregation of BEAS-2B cells can also be seen in the DAPI & F-Actin staining. The *in vitro* studies, combined with the *in vivo* experiments that are currently proceeding, will elucidate the biocompatibility of each alloy, and will also help ascertain the suitability of these novel ultra-high ductility (UHD) alloys for tracheal stent applications.

Copyright © American Institute of Chemical Engineers. All rights reserved.

510608 Molecular Dynamics Simulations of a Laminin and Elastin-Based Triblock Fusion Polypeptide

Abstract:

Computer simulations offer unparalleled insight into microscopic phenomena, such as protein assembly. Simulations allow us to model complex molecular systems and provide a link between theory and experiment, as well as predictions to guide further experiments. Using molecular dynamics (MD) simulations, we are characterizing the structural and dynamical properties of an elastin-like fusion protein. Elastin-like polypeptide (ELP)-based biomaterials hold promise in the treatment of neurodegenerative diseases; for instance, the ability of ELP chains to self-assemble allows them to serve as scaffolds for tissue growth and regeneration[1]. This assembly behavior results from a reversible inverse temperature phase transition, wherein ELPs coacervate with increasing temperature[2]. ELP-based fusion proteins preserve this useful thermal response, and the phase transition temperature can be modified by adjusting protein sequence and amino acid composition[3]. We have designed a triblock system consisting of two ELP chains fused to the N- and C-termini, using a GVG linker, of the fifth globular domain of a key extracellular matrix protein known as laminin. Our ELP chains have a repeating $(VPGXG)_n$ sequence, where X denotes a guest residue and n denotes the number of pentapeptide repeats. We have begun simulating the triblock fusion protein using the ELP sequence $(VPGKG)_2(VPGLG)_2(VPGIG)_2(VPGKG)_2$. Our system was equilibrated for 10 ns (to 310 K) and underwent free dynamics for 200 ns. Our previous simulations of a diblock LG-ELP suggest that a temperature of 310 K promotes Î²-rich structure within the ELP chains[4]; our current results indicate formation of Î²-rich structure at 310 K in this triblock fusion protein, which is consistent with. The radius of gyration (R_g) also differs between the N-terminal and C-terminal ELP chains by 1 Ã…, and the R_g value for the N-terminal chain slightly increases following 150ns. Moreover, the number of inter-chain backbone contacts between the ELP chains remains constant at an average of 198 contacts per ns. Our data further indicates that the ELP chains interact intra-molecularly, with an average of nine hydrogen bonds forming in each of the N-terminal and C-terminal ELP regions. The information obtained from our analyses will guide ongoing computational and experimental studies of this protein design. Simulations of this ELP-LG-ELP system will both aid our understanding of ELP-based materials and elucidate the thermodynamic principles that govern the desired properties of various ELP-based materials.

[1] Zhang Y, Avery RK, Vallamjo-Martin Q, Assmann A, Vegh A, Memic A, Olsen BD, Annabi N, Khademhosseini A (2015) A Highly Elastic and Rapidly Crosslinkable Elastin-Like Polypeptide-Based Hydrogel for Biomedical Applications. Adv Funct Mater. 25:4814-4826.

[2] Tarakanova A, Huang W, Weiss AS, Kaplan DL, Buehler MJ (2017) Computational Smart Polymer Design based on Elastin Protein Mutability. Biomaterials 127:49-60.

[3] MacEwan SR, Chilkoti A (2010) Elastin-Like Polypeptides: Biomedical Applications of Tunable Biopolymers. Biopolymers 94:60â€"77.

[4] Tang J, McAnany C, Mura C, Lampe KJ (2016) Toward a Designable Extracellular Matrix: Molecular Dynamics Simulations of an Engineered Laminin-Mimetic, Elastin-Like Fusion Protein. Biomacromolecules 17:3222-3233.

Copyright © American Institute of Chemical Engineers. All rights reserved.

509770 Bissap in Cosmetics

Abstract:

Bissap or also known as hibiscus subdariffa is a plant is used for teas and other medicines in Senegal. The goal of this research is to see if the seeds can be used in oils for cosmetics. This is important toUniversity of Cheikh Anta Diop of Dakar, Ecole Supeĺrieure Polytechnique (ESP), because this will be the first time that they have looked at the applications of the oil that comes from their native plant, bissap. The method that is used to gather the oil is extraction using a solvent. The reaction process used is called the Soxhlet extraction. Through this extraction with hexane as the solvent the oil is able to be collected from the ground up seeds. The final yield for each of the extractions are as follows: 1st extraction yield 6.6g of product, 2nd extraction yield 200g of product, 3rd extraction yield 244g of product. The oils where then used to make the cosmetic products that is the goal of the research. Soap and lotion were made from the oils. The next step is to finish characterizing the oil to see how beneficial it is for the hair, body, and nails.

Copyright © American Institute of Chemical Engineers. All rights reserved.

510800 Elucidating the Metabolism of Sulfolobus Solfataricus Using 13C Metabolic Flux Analysis

Abstract:

Sulfolobus solfataricus is a highly thermophilic archaeon with optimal growth temperature of about 80-85°C. A key characteristic of *Sulfolobus solfataricus* is its ability to grow in low pH environments (pH<3). This presents new opportunities for industrial application in lignocellulosic biomass processing. In this study, the cell physiology and metabolism of *Sulfolobus solfataricus* was investigated. First, an optimized minimal growth medium was developed for growing the cells in aerobic batch culture. Next, growth experiments were performed with different C5 and C6 sugars. Based on these experments, we determined that *Sulfolobus solfataricus* could grow efficiently on multiple sugars (glucose, xylose, cellobiose). The metabolism of *Sulfolobus solfataricus* was then investigated in detail using ^{13}C isotope tracing. Specifically, parallel labeling experiments were performed using ^{13}C-glucose and ^{13}C-xylose tracers, followed by mass spec analysis and ^{13}C-metabolic flux analysis to quantify intracellular metabolic fluxes. The ^{13}C-flux results provided a detailed picture of the activity of metabolic pathways in this organisms during aerobic growth on glucose and xylose. Overall, the insights obtained from this study will enable future metabolic engineering studies of *Sulfolobus solfataricus* with potential biotechnological applications.

Copyright © American Institute of Chemical Engineers. All rights reserved.

510868 Metabolic Engineering of S. Cerevisiae for the Production of Propane

Abstract:

The goal of this experiment was to produce propane from *Saccharomyces cerevisiae* through a novel biosynthetic genetic pathway using genes KivD and ADO from *Lactococcus lactis and Prochlorococcus marinus* respectively. These genes branch off of *S. cerevisiae*â€™s metabolism from ketoisovalerate in the valine pathway. With increasing fuel prices worldwide and the possibility of a future energy crisis, there has been a strong shift towards developing renewable fuel systems from microorganisms. Propane, in particular, is an advantageous substance due to its diverse uses in motor vehicles, furnace heaters, and cooking fuels. Introduction of propane synthesis into *S. cerevisiae* helps these efforts by developing a high-producing fuel strain that can easily be grown to industrial sizes. A metabolic model of this organism was examined to determine any gene knockouts that would direct more of *S. cerevisiae*â€™s metabolism towards the production of propane. Further, production of propane in *S. cerevisiae* was compared for strains with and without the additional insertion of various valine pathway genes isolated from *Penicillium chrysogenum*. The high producing nature of the *P. chrysogenum* valine pathway was exploited by the introduction of eighteen different gene sequence iterations of the pathway into *Saccharomyces cerevisiae* cells. The pathway of *P. chrysogenum* was suggested to be high producing as L-valine is a key precursor to penicillin, an evolutionarily significant molecule to this organism. Though *P. chyrsogenum* is predicted to have a higher production of propane, it is much more difficult to grow at industrially relevant levels because it is multi-cellular and aggregates in media. *S. cerevisiae* was selected as the cellular host because it has a well characterized genome and metabolism, and it is much more easily grown to industrial standards. Yeast assembly methods were used to introduce the different iterations of the *P. chrysogenum* valine pathway genes into *S. cerevisiae* each including a Nourseothricin selection gene as well. This included a high and low producing strain for pathways using one of three genes involved in the expression of AHAS and ilvD proteins (AHAS I, AHAS II, AHAS III; ilvD I, ilvD II, ilvD III). Additional analyses were performed to determine which of these genes was responsible for regulation of each enzymeâ€™s activity. Propane production of individual strains was quantified and compared to determine the efficiency of each strain. Once the highest producing strain has been selected, metabolic engineering techniques, such as increasing strain tolerance through serial subculturing and enhancing culture conditions, will be performed to optimize efficiency in this pathway while it is grown up to industrially relevant levels. Optimization of propane production in *S. cerevisiae* for use in industrial processes is an advantageous way to obtain valuable biofuels as the world faces an energy crisis.

Copyright © American Institute of Chemical Engineers. All rights reserved.

509878 Modulating the Heat Shock Response in E. coli to Regulate Membrane Protein Expression

Abstract:

Membrane proteins are essential for cell survival and show potential as pharmacological and therapeutic targets in the field of nanobiotechnology.[1,2] In spite of their promise in these fields, research surrounding membrane proteins lags since their over-expression often leads to cell toxicity and death.[3,4] Â
expression could be regulated and optimized by modifying the heat shock response of *Escherichia coli* (*E. coli*). To test this hypothesis, the membrane-protein expression pathway was reprogrammed using gene-blocks that were antisense to vital membrane protein DNA and RNA binding-site sequences. aP_{BAD} and aHtdR gene-blocks were designed to be antisense to the DNA of the arabinose P_{BAD} promotor and H. turkmenica deltarhodopsin (HtdR) transmembrane protein respectively. These sequences were then employed to be cloned into a pMM102 vector and grown in NEB-5Î± *E. coli* cells. Stable glycerol stocks of the aP_{BAD}-PMM102 in BW25113 cells were generated for further use and testing and an aHtdR-PMM102 cell stock is still in development. After inducing the aP_{BAD}-PMM102 with L-arabinose to allow for membrane protein expression, spectrophotometry was used to test the optical density and growth of the cells at 600nm. The controls, with no HtdR protein coded, grew about 40% more than the aP_{BAD}-PMM102 cells and other test cells, as expected, and no significant difference was seen in growth between the aP_{BAD}-PMM102 and the other, unaltered, membrane protein-expressing cells. Continuing, the cells with membrane protein expression present as a purple pellet when spun down and as expected, only the PMM102 cells presented purple coloring with no noticeable difference in shade while the non-membrane protein-expressing cells were a yellow pellet. Further work will include cloning a GFP gene-block instead of aP_{BAD} or aHtdR to determine if the insignificance found in the results of this experiment is due to a flaw in the experimental design or not.

References

1. Wallin E, von Heijne G: Genome-wide analysis of integral membrane proteins from eubacterial, archaean, and eukaryotic organisms. Protein Science. 1998, 7 (4): 1029-1038.
2. Hopkins AL,Groom CR(2002) The druggable genome. *Nat Rev Drug Discov* 1: 727â€"730
3. Nannenga B, Baneyx F: Enhanced expression of membrane proteins in *E. coli* with a P_{BAD} promoter mutant: synergies with chaperone pathway engineering strategies. Microbial Cell Factories. 2011, 10(105): 1475-2859-10-105
4. Baneyx F: Recombinant protein expression in Escherichia coli. Curr Opin Biotechnol. 1999, 10 (5): 411-421. 10.1016/S0958-1669(99)00003-8

Copyright © American Institute of Chemical Engineers. All rights reserved.

510885 Heterologous Reconstitutition of the Quorum Sensing System of Clostridium Difficile in Non-Pathogenic Hosts

Abstract:

It has recently been discovered that toxin secretion in hypervirulent strains of *Clostridium difficile*, a highly antibiotic-resistant opportunistic pathogen, is upregulated by a 2-component quorum sensing apparatus, leading to increased mortality in patients infected with these strains. In this work we reconstitute the quorum sensing system of *C. difficile* in non-pathogenic hosts. We discuss synthetic strategies to systematically isolate responsive regulators and development of high-throughput screens for detection of quorum sensing signaling molecules. Finally, we reconstitute a homologous system from the food-borne pathogen *Listeria monocytogenes* and examine implications of orthogonality and cross-talk between gram-positive QS in pathogens.

Copyright © American Institute of Chemical Engineers. All rights reserved.

510128 Understanding Preferential Consumption of Aromatic Compounds in Acinetobacter Baylyi ADP1

Abstract:

Rapid depletion of fossil resources and an increasing threat of anthropogenic climate change are driving a need for renewable alternatives to petrochemicals. Lignin, a common waste product from paper and agriculture industries, offers a rich diversity of aromatic molecules that can be converted into biorenewable alternatives to petrochemicals. *Acinetobacter baylyi* ADP1 (referred to as ADP1) is an attractive candidate for lignin valorization due to its ability to metabolize a diverse array of lignin monomers and its natural propensity to incorporate foreign DNA into its genome. However, ADP1 exhibits strict substrate preference among these carbon sources through mechanisms that are incompletely understood. The presented work investigates possible mechanisms and motivations underlying the native carbon source preference of ADP1 primarily using metabolomics, an analytical method that enables the simultaneous profiling of concentrations of hundreds of metabolites within the cell. As the closest predictor of cellular phenotype, metabolomics is a powerful tool for understanding changes in cellular behavior. Combined with analytical methods for measuring metabolite uptake/secretion and protein expression, these tools provide a robust framework for elucidating the mechanisms causing substrate preference in ADP1. Once the native substrate preferences of ADP1 are fully understood, this knowledge can be applied to engineer ADP1 for production of valuable products such as advanced biofuels.

Copyright © American Institute of Chemical Engineers. All rights reserved.

510464 Essential Oils and the Ability to Inhibit Biofilm Formation of Pseudomonas Aeruginosa

Abstract:

Despite the constant innovation of science and technology, bacteria are outpacing the market for new and innovative antibiotics and antibiotic resistance is becoming a looming threat. With a decreasing number of antibiotic companies due to government legislation and the expensive Research and Development process. The first case of a superbug arose in Washoe County, Nevada in August, 2016. As a result, the present study explored the use of alternate substances to eradicate the potential superbug cases. The study focused on *Pseudomonas aeruginosa* and its notorious biofilm formation. The present research examined the ability of Peppermint, Roman Chamomile, and Lemongrass essential oils to inhibit biofilms. To evaluate the effectiveness of those test substances, a biofilm assay was prepared in a 96-microtiter plate. The plate was later run through spectrophotometer to indirectly assess bacterial growth and biofilm formation. The results show that Lemongrass has the most prominent reduction of both bacteria growth and biofilm formation. The same trend was observed for Peppermint, however, the correlation was weaker than the Lemongrass. Roman Chamomile essential oil did not show any effect on antimicrobial activity as both bacterial growth increased and biofilm formation stayed stagnant. Further studies are needed to justify bacterial growth patterns that were observed in this research.

Copyright © American Institute of Chemical Engineers. All rights reserved.

511018 Quantification of Antibody Specificity Using Yeast Surface Display

Abstract:

The human microtubule associated protein tau is expressed in healthy neurons and is essential to normal function. It is known that tau undergoes phosphorylation, which is a chemical modification at threonine, serine, and tyrosine residues. Hyperphosphorylation of tau â€" phosphorylation at abnormally high numbers of residues - is involved in a number of diseases, including Alzheimerâ€™s disease and chronic traumatic encephalopathy, collectively labeled tauopathies. Finding antibodies that can bind specifically to disease-associated phosphorylated residues in tau is essential to further research and, eventually, to the determination of a clinical solution to these diseases. The human tau protein is very long and contains many phosphorylation sites, and while there are many antibodies that bind to tau their specificity to phosphorylated tau is not well characterized and is challenging to quantify. To address this problem, we displayed the tau protein on the surface of yeast cells. Using flow cytometry, the binding of commercially available antibodies was quantified. This allows for the testing of large quantities of individual yeast cells expressing the protein, and expression level for each cell can be quantified using an epitope tag inserted at the C-terminus of tau. Specific serine and threonine residues were mutated with overlap extension polymerase chain reaction (OE-PCR) to alanine, glutamic acid, or aspartic acid in order to mimic a dephosphorylated or phosphorylated state. These phosphomimetic mutations were used to ensure pseudo-dephosphorylation or pseudo-phosphorylation at specific sites, and flow cytometry binding data for each was compared to determine epitope and phospho-specificity. This platform can be used to determine the binding and specificity of commercial antibodies, as well as newly created and affinity matured antibodies.

Copyright © American Institute of Chemical Engineers. All rights reserved.

511154 The Involvement of CaBP1 in Hippocampus-Dependent Spatial Learning and Memory

Abstract:

Ca_v1 channels are L-type voltage-gated pathways that mediate the influx of Ca^{2+} into the cell upon membrane polarization. Ca_v1 channels, mostly $Ca_v1.2$ and $Ca_v1.3$, are widely expressed in the brain and are associated with a variety of psychiatric disorders, including depression and schizophrenia. Because of this, the modulation of these channels is of great research interest due to their potential uses as therapeutic targets. CaBP1 is a Ca^{2+} binding protein that is highly expressed in the retina, inner ear, and brain, and though its functions in the nervous system are still primarily unknown, it has been previously established that CaBP1 suppresses Ca^{2+}-dependent inactivation (CDI) of Ca_v1 in heterologous expression systems. CaBP1 is thought to inhibit CDI to restrict excessive calcium influx of the cells, which could be pathological. To analyze if CaBP1 plays a role in psychiatric disorders by modulation of Ca_v1 channels, mice lacking expression of CaBP1 (C-KO) were used to investigate the role of CaBP1 as a regulator of voltageâ€“gated $Ca_{vÂ}$
behavioral experiments were conducted showing that CaBP1 KO mice have mild learning defects compared to wild-type (WT) mice. In electrophysiological analyses, Ca_v currents underwent greater Ca^{2+}-dependent inactivation in C-KO than in WT neurons form the hippocampus of mice. Our findings establish CaBP1 as a regulator for spatial learning and memory, possibly through inhibiting CDI of CaBP1.

Copyright © American Institute of Chemical Engineers. All rights reserved.

510861 Investigation of Novel Cellulose-Binding TÃPirin Proteins in Caldicellulosiruptor Species

Abstract:

The genus *Caldicellulosiruptor* represents a group of extremely thermophilic bacteria with potential for production of biofuels and chemicals. Traditional methods of producing biofuels incorporate chemical and thermal pretreatment of plant biomass, followed by enzymatic degradation, to produce fermentable sugars. The use of *Caldicellulosiruptor* could consolidate these processes into a single step, requiring no biomass pretreatment. The ability of *Caldicellulosiruptor* to bind to cellulosic biomass is due in part to novel proteins known as tāpirins. Once bound, *Caldicellulosiruptor* solubilizes the recalcitrant cellulosic biomass, an important step in the degradation of sugars needed for processing. In order to study these unique proteins, tāpirins from several different species were expressed in *Escherichia coli*. These proteins were then purified via immobilized metal ion affinity chromatography (IMAC), tested for their ability to bind different substrates, and prepared for crystallography. The C-terminal domains of two tāpirins, Calkro_0844 and Calhy_0908, have been successfully crystallized. These two proteins, despite their low homology, bear significant structural resemblance. To determine the specific role of tāpirins and pili in binding to cellulose and complex biomasses, a series of genetic knockouts are also being performed in *Caldicellulosiruptor bescii*. Overall, tāpirins and pili likely play an important part in the ability of *Caldicellulosiruptor* to bind and degrade cellulosic biomass, which is crucial in the consolidated bioprocessing approach to biofuel production and chemical synthesis.

Copyright © American Institute of Chemical Engineers. All rights reserved.

509838 Quantification of Phospholipid Vesicle Degradation Kinetics Catalyzed By a Model Phospholipase

Abstract:

In this study, kinetics of giant unilamellar vesicles (GUVs) degradation upon addition of secretory phospholipase A2 ($sPLA_2$) were monitored under inverted microscopy. The effects of PEGylation and different combinations of saturated and unsaturated phospholipids were quantified. Secretory phospholipase A2 ($sPLA_2$) is an enzyme present in the human body and can be found in excess in many pro-inflammatory conditions. The rationale directing the design of switchable liposomal drug carriers is based on the hypothesis that liposomes remain stable at low concentrations of $sPLA_2$ and degrade spontaneously at a concentration higher than the threshold concentration of $sPLA_2$.

The morphologies of the GUVs were observed before and after adding enzyme and degradation kinetics were quantified by calculating the change in surface area of the GUVs. A threshold concentration of $sPLA_2$ was discovered, beyond which the GUVs shrink significantly fast. More specifically, combinations of 1, 2-dipalmitoyl-sn-glycero-3-phosphocholine (DPPC), 1, 2-dilinoleoyl-sn-glycero-3-phosphocholine (DLPC, 18:2) were used in this study.

It was found that above 1Î¼g/ml of $sPLA_2$, DPPC GUV degradation became obvious within the observation time (30 min). With the presence of the unsaturated DLPC, the threshold concentration of $sPLA_2$ is reduced. The structures formed after degradation were also different for saturated and unsaturated lipids. While saturated lipids formed crystal-like structures, unsaturated lipids were seen momentarily shrinking and then relaxing back to a spherical shape. These results will help design a system for targeted drug delivery.

Copyright © American Institute of Chemical Engineers. All rights reserved.

510677 Thiol Modified Fe@Sio 2 As an Effective Recycling Magnetic Platform for Mesp

Abstract:

Thiol modified Fe@SiO_2 as an effective recycling magnetic platform for mESP

Zakary Ford[1], Tun Yun Hsueh[2], Ruben Michael Ceballos[2], Lauren F. Greenlee[1]

Ralph E. Martin Department of Chemical Engineering, University of Arkansas, Fayetteville,

AR 72701, USA

Department of Biological Sciences, University of Arkansas, Fayetteville,

AR 72701, USA

Cellulosic ethanol, a "second-generation" biofuel, has yet to become cost competitive with fossil fuels on a commercial scale. To lower the cost of biofuel, two steps within the ethanol production process have been studied thoroughly: pretreatment and enzymatic hydrolysis. In general, pretreatment aims to prime the cellulose network for sugar reducing enzymes by disrupting cellulose, hemicellulose, and lignin interactions. The second area of research focus comprises of the discovery of novel enzymes, the use of enzymes "cocktails," engineering of enzymes, and the use of platforms. Through prior work, results have shown that the utilization of a mobile enzyme sequestration platform (mESP) developed from a NASA prototype scaffold known as a "rosettazyme" mESPs are useful in enhancing sugar reduction efficiency on acid- and alkaline- pretreated cellulosic feedstock. However, without an efficient recycling process, the technology will still not result in lower costs and thus will not contribute to a cost-competitive biofuel. Therefore, the focus of this study is to develop a suitable nanoparticle technology to efficiently recover mESPs.

In this presentation, we will show recent results on continued work to develop thiol functionalized Fe@SiO_2 core\shell nanoparticles obtained through co-precipitation and sol-gel methods. In particular, we will discuss the progress of obtaining characterization data from transmission electron microscopy (TEM), energy dispersive x-ray spectroscopy (EDS), and x-ray photoelectron spectroscopy (XPS) showing successful attachment of SiO_2 and thiol to the surface of the Fe nanoparticle. Furthermore, we will discuss our efforts to determine the efficiency of attachment of our nanoparticle to mESP subunits.

Copyright © American Institute of Chemical Engineers. All rights reserved.

510213 Uncovering the Effects of Oxidative RNA Modifications on the Formation of Protein-RNA Complexes

Abstract:

Cells exposed to oxidative stress result in RNA oxidations that are linked to many diseases, including Alzheimer and Parkinsonâ€™s diseases. Because RNA oxidations inherently change the chemistry and therefore the configuration of a folded RNA strand, these altered properties can result in changes in RNA-protein interactions. These alterations and their biochemical effects on the fate of RNA molecules are not well understood. To study this gap, we have used as a model *E. coli* Polynucleotide phosphorylase (PNPase), a RNA-binding protein that interacts with high affinity to 8-oxo-guanosine and is possibly linked to a repair mechanism of oxidized RNA. We have used biochemical methods to study the binding between PNPase and a selected set of RNA oxidations chemistries. We have analyzed the specificity of PNPase for oxidized nucleosides using electrophoretic mobility shift assays (EMSAs). For this, we have cloned the gene encoding *E. coli* PNPase in an inducible plasmid for recombinant protein expression, transformed it into an *E. coli* BL21 strain, and induced protein expression by addition of IPTG. After protein purification using the hexahistidine tag in PNPase, EMSAs were performed for determination and quantification of the protein-RNA affinity. We particularly tested the specificity of PNPase binding for the most common types of oxidation occurring in RNA, identified upon oxidative stress of cells exposed to UV, ionizing radiation, and hydrogen peroxide. This approach has resulted in a better understanding of the natural protein recognition mechanisms of PNPase.

Copyright © American Institute of Chemical Engineers. All rights reserved.

510373 Light-Driven Nitrogen Fixation Via a Mediated CdS Nitrogenase System

Abstract:

We have been working with the enzyme MoFe nitrogenase which is capable of reducing N_2 to produce NH_3, a commodity chemical that is produced on the scale of millions of tons every year. The goal of our project is to further develop the method of using photoreactive nanoparticles to drive this reduction process. To do so, we are using CdS nanoparticles that are able to transfer excited electrons to the nitrogenase enzyme when exposed to light, ascorbic acid as an electron donor to replace these electrons, MoFe nitrogenase, and viologen as an electron transfer mediator. The kinetics of electron transfer from CdS to viologen using different ligands was investigated, and it determined electron transfer occurs more quickly with TGA ligands than ME ligands. During these reactions, the amount of ascorbic acid was also varied, and it was found that at lower levels of ascorbic acid, the rate of electron transfer increased significantly. Finally, reactions to determine how using viologen affected the yields of NH_3 and H_2 were performed by placing a solution of the following conditions under a solar simulator for 2 hours: 119 nM CdS, 112 mM Ascorbic Acid, 1.7 mM Viologen in 0.1M PB. The preliminary results showed that using viologen increased the amount of NH_3 produced but decreased the amount of H_2 produced. This likely means that viologen is altering the pathway by which NH_3 is being produced.

Copyright © American Institute of Chemical Engineers. All rights reserved.

510955 Probing H3K9me3 Using Engineered Recombinant Protein Probes

Abstract:

Trimethylation of histone 3 lysine 9 (H3K9me3) is an important epigenetic modification that is associated with gene repression and silencing. Detection and quantification of H3K9me3 is critically important for understanding its role in epigenetic regulation and potential applications of H3K9me3-based diagnostic or prognostic assays. The current detection approaches, however, depend on the use of commercial antibodies, which are costly and exhibit significant quality variations among suppliers. Here, we have designed a sensor based on recombinant proteins expressed in E. coli. The sensor can recognize H3K9me3 with high accuracy and comparable sensitivity to commercial antibodies. The production cost and batch-to-batch variations of our sensor is significantly lower compared to antibodies. Our sensor thus offers a useful alternative tool for the detection of H3K9me3.

Copyright © American Institute of Chemical Engineers. All rights reserved.

510958 Development of a Peptide Mass Fingerprinting Method for Green Fluorescent Protein

Abstract:

Peptide mass fingerprinting (PMF) is a protein identification method in which a protein of interest is cleaved into small peptides using a digestive enzyme that cleaves at very specific peptide bonds. The peptides are then separated by high performance liquid chromatography (HPLC) and identified either through UV or mass detection. This study was conducted to create a PMF method for green fluorescent protein (GFP), and then to evaluate variations in the peptide map when the protein was subjected to forced degradation. A GFP digestion method was used involving heat denaturation, a buffer at an elevated pH, and the digestive enzyme, trypsin. The PMF method was developed and validated using HPLC and HPLC-MS. Following validation, degradation studies involving oxidation, hydrolysis, heat, and light were used to determine which peptides were targeted. For example, amino acids such as methionine, cysteine, histidine, tryptophan, and tyrosine are susceptible to oxidation. Theoretically, peptides incorporating these amino acid residues should be oxidized resulting in chromatographic peak shifts. These degradation results helped ensure the stability of GFP at the Biomanufacturing Training and Education Center (BTEC) at North Carolina State University. GFP serves as BTECâ€™s model therapeutic protein in their biomanufacturing process, and these results will help with the understanding of the degradation pathways of the protein. More detailed results will be presented.

Copyright © American Institute of Chemical Engineers. All rights reserved.

510975 Antibody Purification Via Affinity Membrane Chromatography Utilizing Nucleotide Binding Site Targeting

Abstract:

Current antibody use in therapeutics and research is increasing rapidly in both complexity and demand. Therefore, there is a constant, growing need for more efficient laboratory tools and methods to facilitate antibody engineering, production, and, in particular, purification. As of now, the predominant method of antibody purification is affinity chromatography, wherein the affinity molecule is Protein A or Protein G, proteins derived from bacteria that bind to IgGs via the Fc region. This process delivers a high yield and purity, but has several drawbacks, including high cost, difficulty in isolating the affinity molecules, harsh elution steps, leaching of the affinity molecule, and limited reusability.

Our lab has developed an alternative method. The Nucleotide Binding Site (NBS) is a conserved, underutilized binding site in the Fab variable region of the antibody. We have identified a number of ligands that have moderate affinity for the NBS. For the chromatographic matrix, regenerated cellulose (RC) membranes were selected for their mechanical stability, inexpensive price, and ease of use. Membrane chromatography allows the solute and matrix to interact in the open pores of the membrane, rather than the dead pores of a resin molecule, thereby decreasing the pressure drop and allowing rapid flow-induced mass transport. By conjugating our NBS-ligand to the RC membranes, it is possible to selectively capture IgGs from solutions containing contaminants such as cell culture media of hybridomas or ascites fluids. Furthermore, the elution step is extremely gentle, only requiring 0.5 M KCl salt buffer at neutral pH, as opposed to the highly acidic (pH 2.5-3.0) conditions of the rivaling Protein A method.

After functionalizing RC membranes, they can be packed into columns in a variety of formulations. Previously, our lab has demonstrated their effectivity in HPLC columns. More recently, we have scaled them down and specialized them for use by laboratory researchers. Packing 5-20 membranes into a centrifugal spin column allows unrivaled ease and efficiency in many research settings, with each small-scale purification taking only 5 minutes or so. Such spin columns have proven very effective in our laboratory, with antibody recovery of >90% in samples spiked with contaminants. They also compare very favorably with competing products which utilize Protein A. Not only is their antibody recovery comparable or higher, but also they are highly reusable, cheaper, and require much gentler elution conditions.

Copyright © American Institute of Chemical Engineers. All rights reserved.

509233 Steps Toward Heterologous Biosynthesis of Coenzyme M

Abstract:

Coenzyme M (CoM) is a key cofactor in the biological production of methane and in the breakdown of alkenes and epoxides. CoM-dependent metabolic pathways are of interest because they could be used for the bioconversion of unconventional and harmful substrates like alkenes to biofuels and other useful chemicals. However, these useful metabolic pathways cannot currently be implemented in typical metabolic workhorses such as *Escherichia coli* since these organisms do not naturally produce CoM. This work demonstrates production of the first two intermediates in the bacterial CoM biosynthesis pathway in *E. coli*. Predicted CoM biosynthesis enzymes from *Xanthobacter autotrophicus* Py2 and their homologs from *Bacillus subtilis* or *E. coli* were expressed in *E. coli*. Pathway intermediates were detected in cellular extracts from strains containing enzymes from *B. subtilis* and *X. autotrophicus* Py2. Additionally, the first enzyme of the pathway, which has never been previously characterized, was characterized in vitro to obtain kinetic parameters. This work is significant because it achieves steps in CoM biosynthesis in *E. coli*, which is important for enabling use of a broader range of substrates in this heterologous system.

Copyright © American Institute of Chemical Engineers. All rights reserved.

510913 TIM Protein Binding Enhancement with Increased Fluidity of Membranes

Abstract:

As a healthy cell undergoes apoptosis, a cell's self-regulation of membrane fluidity and phospholipid composition of the inner and outer leaflet begin to fail. The cell membrane can no longer keep its shape and becomes fluid, or loosely packed. In addition, negatively charged phosphatidylserine [PS] is exposed on the outer leaflet; a signal to the external environment that the cell is dying. TIM [T-cell immunoglobulin and mucin domain] proteins in the external environment selectively bind to PS in the presence of calcium. The TIM protein family all possess a tryptophan, a hydrophobic and fluorescent residue, near their binding site. Using this fluorescent property, we were able to study the differences in binding affinity for TIM3 and TIM4 when the context of PS changes – i.e. increased fluidity of the cell membrane seen in dying cells. Using previous x-ray reflectivity data, the crystalline structure of TIM3 and TIM4 bound to a tightly packed membrane can be seen. The results show that TIM4's tryptophan is satisfied in the hydrophobic core of the cell membrane when the binding event occurs, however, TIM3's tryptophan remains in the hydrophilic exterior when bound to a tightly packed membrane. Our experiments explore the hypothesis that exposing TIM3 and TIM4 to membranes of increased fluidity will increase TIM3's binding affinity and have small impact on TIM4's.

Copyright © American Institute of Chemical Engineers. All rights reserved.

510832 Structural and Energetic Analysis of Disulfide Bonds in the Protein Data Bank

Abstract:

Disulfide bonds between cysteine residues play a key role in determining structure and regulating function of biological molecules. These bonds generally stabilize native protein structures along with other secondary and tertiary structures. For example, a disulfide formed between two cysteine residues separated in sequence space can establish their proximity in three-dimensional space; thus, a disulfide bond provides biological systems with a way to branch covalent interactions that tether different regions of peptide backbone. However, unlike rotations about the single bonds of a polypeptide chain(Ï†,Ïˆ), there is a significant barrier to rotation about a disulfide bond. The energetic cost of rotating about the disulfide bond restrains available conformations further and provides a potential source of energy for allosteric transformations. For instance, if a protein fold rotates a disulfide bond up hill in energy that energy can be released via reduction of the disulfide bond. Strained disulfides have been observed and characterized in the protein data bank for X-ray and NMR structures. In this study we seek to elucidate the structural information associated with disulfide bonds, strained or unstrained, and quantify their effects on protein structure. To this end, we constructed a database of cysteine information extracted from protein data bank files containing disulfide bonds. We analyzed 6238 X-ray structures (with resolution better than 3.0 ÌŠA) and 1452 NMR structures, each with sequence identity no greater than 90%. For high resolution X-ray structures (resolution less than 1.7 ÌŠA), we identified shifts in the Ramachandran plots due disulfide strain; these shifts are muted for low resolution and NMR structures. Further analysis of the internal coordinates of disulfide bonds revealed key differences between high resolution X-ray structures and NMR structures. NMR and low resolution X-ray structures exhibited significantly higher strain across these bonds, compared to high/ultra-high resolution X-ray structures. Thus, many highly strained bonds present in these structures can be attributed to the characteristics of their experimental determination, and our analysis reveal evidence of modeling artifacts, especially in NMR structures. The results of this study can have a direct impact on increasing the accuracy of models constructed from structural biology experiments.

Copyright © American Institute of Chemical Engineers. All rights reserved.

510846 Efficient CHO Platform Development for Anti-HER2 Monoclonal Antibody Production

Abstract:

Chinese hamster ovary (CHO) has become a nearly ubiquitous platform for biopharmaceutical production over the last few decades. Efficient antibody production platforms with high quality and improved efficacy are increasingly necessary due to emerging antibody-based therapies. In this study, CHO cells were genetically engineered for high yield, stable anti-HER2 antibody expression. Subsequent processes including basal medium screening, bioreactor parameter optimization, and antibody purification with next-generation chromatography (NGC) were developed and optimized to obtain a robust and scalable platform for anti-HER2 antibody production. Additionally, proteomics analysis was used to understand the relationship between the host cell protein profiles and the bioprocesses associated with recombinant protein expression. The achievement of this study can benefit the development of both biosimilar and innovative biopharmaceutical production platforms.

Copyright © American Institute of Chemical Engineers. All rights reserved.

509785 Mass Transfer Effects of Particle Size on Brewing Espresso

Abstract:

The extraction process for coffee is complicated and serves as a great model to study mass transfer, fluid mechanics, and thermodynamics. In this paper, we studied the brewing of espresso both experimentally and through validation with a mathematical model. First, the particle size distribution of finely ground coffee grinds was determined using a particle size analyzer. The ground coffee was shown to have a bimodal distribution, with smaller particles (50 â€“ 75 micron range, representing cells or cell fragments) mixed in with larger particles (200 â€“ 400 microns). Researchers have posited that significantly different mass transfer mechanisms exist between the two particle size regimes, with fast dissolution occurring from the cells and a slower, diffusion limited mechanism dominating for the larger particles. To test this hypothesis, espresso was brewed using both normally ground coffee and coffee that was ground and then sieved to remove a significant portion of the larger particles. As the hydraulic permeability of this sieved coffee was too low, microspheres of inert glass (roughly 300 microns in diameter) were mixed into the fines to lower flow resistance. Different masses of espresso were then collected and converted to mass ratios by dividing by the original mass of coffee grinds. A refractometer was using to measure the total dissolved solids in the espresso and determine the degree of extraction. From these values, it was determined that the smaller coffee particles extract at a faster rate, and are capable of reaching higher levels of total dissolved solids, when compared to the normal grind sized espresso. High-Performance liquid chromatography and mass spectrometry were also performed to study whether chemical differences arose between the different brewing techniques. The results showed that the compounds extracted for both brews are mostly identical. Finally, data from these experiments was then used to test a recently developed mathematical model for espresso brewing.

Copyright © American Institute of Chemical Engineers. All rights reserved.

510939 Chameleon Molecules: Imidazole Functionalized Poly(3-hexylthiophene) (P3HT) Properties and Integration in Alginate Hydrogels

Abstract:

Poly(3-hexylthiophene-2,5-diyl) (P3HT) is a polymer that has been heavily studied for its application in photovoltaic cells. Here we seek to develop new biological uses for this polymer. Using Kumada coupling and Soxhlet extraction we have synthesized P3HT polymers functionalized with imidazole groups, deemed P3HT-Im, with varying molecular weights. The physical properties of long, medium, and short length variations of these polymers have been investigated using carboxymethyl cellulose, carboxymethyl amylose, and sodium dodecyl sulfate. The polymers in conjunction with sodium dodecyl sulfate red shift under the critical micelle concentration. Upon reaching charge neutrality a coacervate forms. Above the critical micelle concentration micelles form, shown by dynamic light scattering, and solutions blue shift. Unlike with SDS, polymers with CMA or CMC blue shift or red shift respectively up to a maximum concentration where changes in fluorescence no longer occur. Visible induced color changes of the solutions occur for each polymer and each binding unit, also apparent under UV light. The polymers ability to cause blue shifts and red shifts in solution depending on the compound they interact with, the concentration, and the number of repeat units allows for a wide variety of applications. Recently the polymers have been incorporated into alginate hydrogels to investigate timed release of the polymer and biocidal capabilities of the gels.

Copyright © American Institute of Chemical Engineers. All rights reserved.

509961 Heat and Mass Trends in Rebuildable Drip Atomizer Electronic Cigarettes

Abstract:

Electronic cigarettes and vaporizers have become one of the fastest growing industries in the United States. Even though these technologies are widely used, their risks are not well understood. With so much still unknown about these devices, research is still needed such that appropriate and reasonable regulations can be put forth. In the ever progressing market of electronic cigarettes, the evolution of the e-cigarette is trending towards rebuildable drip atomizers (RDA) in which the user wraps a heating coil, to the complexity and design of their own choosing. These devices also offer some level of power (and thus temperature) control. In this experiment, the wicking and evaporation trends in a dual coil RDA were investigated. The goal of this project is to better understand the hazards associated with these devices, which was done by investigating the wicking and evaporation in these devices. Before the wicking and evaporation trends can be studied, an understanding of the temperature variation throughout the coil and wick must be established. The evaporation trends were investigated with respect to composition of the e-liquid, the applied voltage from controller, and saturation of the wick. Likewise, the wicking rates were researched in relation to density of wick and composition of e-liquid. The wicking trends were then investigated further with respect to the evaporation rates.

Copyright © American Institute of Chemical Engineers. All rights reserved.

510732 Characterizing a Peptide Therapeutic Derived from the Cholesterol Recognition Amino Acid Consensus (CRAC) Motif of a Bacterial Toxin

Abstract:

Membrane cholesterol is a common target of pathogens, including bacteria and viruses, as well as bacterial toxins. Many of these pathogens exploit a cholesterol recognition amino acid consensus (CRAC) motif to facilitate cholesterol recognition. We have previously identified and characterized the CRAC motif used by a member of the repeats-in-toxin (RTX) family of toxins, the leukotoxin (LtxA) produced by *Aggregatibacter actinomycetemcomitans*, to bind to target cells. We hypothesized that a peptide consisting of the LtxA CRAC motif would bind strongly to cholesterol to inhibit the activity of LtxA. Using solid-phase peptide synthesis, we created a series of peptides based on the CRAC motif of LtxA and used a variety of techniques, including isothermal titration calorimetry, surface plasmon resonance and confocal microscopy, to measure the peptidesâ€™ affinity for cholesterol and ability to inhibit LtxA activity. Our wildtype CRAC peptide binds cholesterol with a comparable affinity as the toxin from which the sequence was derived (LtxA). This peptide specifically inhibits binding of LtxA to cholesterol and thus inhibits internalization and cytotoxicity of LtxA. A control peptide, in which the CRAC sequence has been scrambled, has a reduced affinity for cholesterol and does not inhibit the activity of the toxin, demonstrating the specificity of the interaction. A panel of peptides containing residue substitutions altering the peptideâ€™s net charge demonstrated that the peptideâ€™s charge significantly influences affinity for cholesterol and subsequent inhibition activity; highly charged peptides had no affinity for cholesterol, but surprisingly, a decreased charge did not enhance affinity. Previous simulations corroborated with our experimental results indicated the peptide resides on the surface of the membrane, and the peptide must maintain the proper balance of hydrophobic and hydrophilic character to retain activity. Importantly, the peptide does not affect cell viability over extended time periods, supporting its possible therapeutic use to inhibit the activity of cholesterol-binding pathogens. Although recognition of membrane cholesterol is a vital step in the mechanisms used by numerous pathogens, a method to inhibit this process without disrupting cellular function has not yet been uncovered. Here, we have designed a cholesterol-binding peptide for this purpose and demonstrated that it is effective in blocking bacterial toxin activity and is non-toxic to host cells.

Copyright © American Institute of Chemical Engineers. All rights reserved.

511026 Drug Release of paclitaxel from Ace-DEX Particles: An Experimental and Mathematical Model

Abstract:

Microparticles (MPs) can be used to enhance local drug delivery and control release while protecting the drug from degradation and reducing off-target toxicity. Utilizing acetalated dextran (Ace-DEX) as a MP drug delivery platform allows for triggered release in acidic environments, such as the endosome of phagocytic cells, tumor microenvironments, and sites of inflammation. Ace-DEX is a biodegradable, acid-sensitive polymer that is formed when water-soluble dextran is chemically modified with acetal groups. Reaction time during polymer synthesis determines the ratio of cyclic to acyclic acetal groups, where a longer reaction time yields a higher cyclic acetal coverage (CAC). Since cyclic acetal groups are more stable and have increased resistance to hydrolysis over noncyclic groups, higher CAC Ace-DEX degrades more slowly than lower CAC polymer. In this work, the effect of CAC on the release of paclitaxel (PTX), a hydrophobic small molecule chemotherapeutic, was evaluated in a pH 7.4 environment to mimic physiological pH. A mathematical model to predict drug release from Ace-DEX MPs *in vitro* was developed based on a combination of relationships modelling simple diffusion and particle degradation via surface erosion. This model was evaluated against our labâ€™s previously published data for resiquimod release from Ace-DEX particles in pH 5 and pH 7.4 environments before adjusting parameter values to predict PTX release. Our initial results indicate that our mathematical model can accurately predict drug release from Ace-DEX particles, even while varying polymer CAC. To further evaluate our model, we have planned *in vivo* experiments with our particles.

Copyright © American Institute of Chemical Engineers. All rights reserved.

510605 Sorbitol: An Additive with Different Applications

Abstract:

Sorbitol: An additive with different applications

Universidad Nacional de Colombia, facultad de ciencias agrarias, Programa Ingenieria Agronomica

Author: Germán David González Struberg, gdgonzalezs@unal.edu.co

Context

Sorbitol is a sampleable and chemically non-reactive polyalcohol, which allows it to withstand high temperatures without a change in its properties. It is a refined low-calorie sugar (containing 30% with respect to sucrose) which has a sweetness of 60% with respect to glucose. It is considered a natural sweetener of low power, with low caloric power, belongs to the family of polyols (along with mannitol, xylitol, lactitol, maltitol and isomalt), being the most used.

Sorbitol is used as an additive in the food, cosmetics, pharmaceutical, paper and leather industries. It is also a raw material for various chemicals such as vitamin C, surfactants and resins.

PRODUCTCION

Catalytic hydrogenation of glucose:
Industrially the D-sorbitol is produced by means of the catalytic hydrogenation mainly using two types of catalysts (Ni, Ru). Generally this reaction is carried out under the following conditions with constant stirring.

Table1. Average operating conditions of obtaining sorbitol.

Process	**Temperature(°C)**	**Pressure (bar)**	**reaction time(h)**
Batch	**120 -150**	**70**	**2-4**
Continuos	**150**	**180**	**1**

The main raw material is glucose which can be obtained from different crops such as cassava, through an enzymatic hydrolysis process. The enzymatic hydrolysis in the last 30 years has displaced the acid hydrolysis, due to the availability of new enzymes. Nowadays most of the starch hydrolysis is done using enzymes, since this technique has advantages such as: control of the formation of undesirable products and greater flexibility of the product.

Sorbitol: An additive with different applications

Universidad Nacional de Colombia, facultad de ciencias agrarias, Programa Ingenieria Agronomica

Author: Germán David González Struberg, gdgonzalezs@unal.edu.co

Context

Sorbitol is a sampleable and chemically non-reactive polyalcohol, which allows it to withstand high temperatures without a change in its properties. It is a refined low-calorie sugar (containing 30% with respect to sucrose) which has a sweetness of 60% with respect to glucose. It is considered a natural sweetener of low power, with low caloric power, belongs to the family of polyols (along with mannitol, xylitol, lactitol, maltitol and isomalt), being the most used.

Sorbitol is used as an additive in the food, cosmetics, pharmaceutical, paper and leather industries. It is also a raw material for various chemicals such as vitamin C, surfactants and resins.

PRODUCTCION

Catalytic hydrogenation of glucose:
Industrially the D-sorbitol is produced by means of the catalytic hydrogenation mainly using two types of catalysts (Ni, Ru). Generally this reaction is carried out under the following conditions with constant stirring.

Table1. Average operating conditions of obtaining sorbitol.

Process	**Temperature(°C)**	**Pressure (bar)**	**reaction time(h)**
Batch	**120 -150**	**70**	**2-4**
Continuos	**150**	**180**	**1**

The main raw material is glucose which can be obtained from different crops such as cassava, through an enzymatic hydrolysis process. The enzymatic hydrolysis in the last 30 years has displaced the acid hydrolysis, due to the availability of new enzymes. Nowadays most of the starch hydrolysis is done using enzymes, since this technique has advantages such as: control of the formation of undesirable products and greater flexibility of the product.

Figura 2. Blocks diagram of sorbitol production.

Figura 2. Blocks diagram of sorbitol production.

Copyright © American Institute of Chemical Engineers. All rights reserved.

509697 Synthesis and Combustion Analysis of Biofuels from Waste Dates

Abstract:

At present, about 85% of the worldâ€™s energy demand is being met by non-renewable sources of energy. Due to the limited resources and huge environmental impact from their combustion, there is a need to gradually shift from non-renewable to renewable and carbon-neutral sources of energy such as biomass-derived fuels (biofuels). For this reason, the production of liquid biofuels from carbohydrate-rich energy crops is the focus of intensive research. This paper presents a novel pathway for the synthesis of a liquid biofuel, 2,5-dimethylfuran (DMF), from waste dates. DMF is an oxygenated fuel with high energy content (similar to gasoline), and can be produced from glucose and fructose, which are major components in dates, through a catalytic process. A two-step process is suggested in which the sugars are initially converted to 5-Hydroxymethylfurfural (HMF), and then, the HMF is converted to DMF. The process requires the usage of a biphasic system in which the date extract is firstly heated with alumina catalyst ($AlCl_3.6H_2O$). The organic layer containing HMF is then refluxed over a carbon supported palladium catalyst to produce DMF. A combustion analysis of DMF/gasoline blends is presented, where the effect of DMF on reducing the pollutant emission tendency of gasoline is seen.

Copyright © American Institute of Chemical Engineers. All rights reserved.

510733 Surfactant Interaction with Ulcerative Colitis

Abstract:

Approximately 907,000 Americans currently suffer from ulcerative colitis, a condition characterized by inflammation of the large intestine or rectum that can lead to hospitalization or surgery. Treatment of this disease often includes anti-inflammatory medication or immunosuppressants. Here foams are an attractive delivery platform, offering relatively high bioavailability, low systemic exposure, and improved patient comfort. However, the surfactants used to generate these foams may adversely affect the diseased mucosa. Initial testing evaluated two alternative surfactants for use in topical drug delivery platforms: sodium caseinate and L-Î±-phosphatidylcholine. Both were compared to the biocompatible surfactant Pluronic Â® F-127 using stability and density tests, and biocompatibility tests performed on mini-guts. Unexpected results from these experiments point to a complex interplay between surfactants and diseased tissue. This led to the current project investigating surfactant-ulcerative colitis interaction using organoid-bacteria co-cultures. Emphasis is placed not only on understanding the possible effects of increased permeability on bacterial translocation, but also on possible implications for current and future delivery platforms.

Copyright © American Institute of Chemical Engineers. All rights reserved.

510830 Nanosuspensions and Solutions for Preparation of Nanocomposites and Amorphous Solid Dispersions: Comparative Assessment of Drug Release

Abstract:

The number of poorly water-soluble drug candidates coming out of drug discovery has increased over the past decades. The low water solubility of these drugs poses a great challenge since it is frequently the rate-limiting step in the intestinal drug absorption which negatively affect bioavailability. Two major platform approaches which improve the dissolution rate and bioavailability of poorly water-soluble drugs are the nanomilling of a drug suspension followed by drying (nanocomposites) for higher surface area or formation of an amorphous solid dispersion (ASD) for higher saturation solubility. Here, we aim to assess the dissolution enhancement of griseofulvin (GF, a poorly water-soluble drug) from nanocomposites vs. ASDs, both having identical formulations. GF nanocomposite was prepared by wet media milling (nanomilling) of an aqueous GF suspension followed by spray drying. Two different polymers, hydroxypropyl cellulose (HPC) and Soluplus®, with an anionic surfactant, sodium dodecyl sulfate (SDS), were used for the stabilization of drug nanoparticles. ASDs were prepared by spray-drying a solution of the respective drugâ €“polymer ratio in acetone-DI water. The suspensions were characterized via laser diffraction, while the spray-dried powders were characterized using laser diffraction, X-ray diffraction (XRD), Scanning electron microscopy (SEM), and dissolution testing. XRD diffractograms confirmed that for both polymers, spray-drying of the aqueous nanosuspension led to drug nanocrystals in the polymeric matrix (nanocomposite), whereas that of the drug solution led to ASD. Dissolution testing at supersaturating conditions showed faster drug release and higher extent of supersaturation from ASD than from nanocomposites in case of SoluplusÂ® but showed similar performance in case of HPC.

Copyright © American Institute of Chemical Engineers. All rights reserved.

510616 Assessment of Experimental Solubility Determination Methods for Cooling Crystallization Process Design of Carbamazepine in Ethanol

Abstract:

The determination of an accurate solubility curve is critical for efficient design of the crystallization process. Inaccurate determination can lead to use of a higher or lower supersaturation level that would result in undesired extra nucleation or crystal growth. Here several solubility measurement techniques are evaluated for determining solubility data for crystallization process design. In this study, carbamazepine (CBZ) was chosen as the model drug due to its various polymorphic forms. The goals were to evaluate the effect of (a) dissolving methods, (b) sampling separation methods, and (c) quantification techniques on the accuracy and precision of thermodynamic solubility data of CBZ form III. The mechanic stirring dissolving method demonstrated similar mean values but with less variance compared to the shake-flask dissolving method. Through analysis of means, there was no significant difference between using centrifugation or filtration as the sampling separation method. But it is noted that a minimum of 30 minutes is required for settling time to ensure the measurement accuracy of the equilibrium concentration. Between the two off-line concentration quantification methods (gravimetric and HPLC), the HPLC method always showed lower mean values, but its difference from the gravimetric values are $< 4.6\%$. Also, the solubility curve developed from on-line measurement was compared to off-line concentration quantification methods, and the difference for the mean values is below 5%. In addition, the dissolution profile of the metastable form (form II) was analyzed, and the results suggest that this is a mono-tropic system between 5 °C and 50 °C for form II and form III. This work presents a comprehensive study on the solubility data of CBZ and the solubility determination methods that are important during the process development and design of crystallization processes.

Copyright © American Institute of Chemical Engineers. All rights reserved.

510625 Characterization of Pilot Scale Single-Use Bioreactors

Abstract:

Traditional stainless steel bioreactors are relatively well characterized in terms of mixing efficiency and oxygen mass transfer. Disposable bioreactors, however, are relatively new, and understanding of key operational parameters is constantly evolving. These insights are also critical for determining optimal sparging configurations and impeller speeds, since single-use units cannot be scaled up using the traditional power per unit volume criterion. Mixing times were determined for a 200 L disposable bioreactor based on pH shift following acid addition. In general, high volume, high agitation rate sub-surface additions were found to produce the shortest mixing times. Oxygen mass transfer rates (k_La) were determined for 200 L and 1000 L units, based on different sparging configurations and air flow rates. The 2 Î¼m and 20 Î¼m spargers were generally found to be superior to 2 mm T-spargers in terms of overall oxygen mass transfer. Future work can build upon these insights by examining how operational parameters affect the viability of cell culture, elucidating the influence that various agitation and sparging strategies have on cell shear rate, antifoam requirements, and CO_2 stripping in single-use bioreactors.

Copyright © American Institute of Chemical Engineers. All rights reserved.

510709 Utilizing Michael Addition Chemistry for the Synthesis of Drug Eluting Gauzes for Oral Wound Applications

Abstract:

Oral wounds are a debilitating side effect from many conditions including oral infections, dental and periodontal surgery, and oral mucositis caused by chemo and radiation therapy. These patients are often treated with corticosteroids to reduce inflammation. When applied as creams, topical steroids possess limited retention time, reducing therapeutic efficacy. This work investigates the potential for using Michael addition chemistry to create an oral drug eluting wound dressing. Michael addition of corticosteroid drugs to chitosan containing gauzes has the potential to offer tunable extended release of drug directly to the wound. The hydroxyl groups of the anesthetic eugenol and the steroid triamcinolone acetonide were each modified with an acrylate group via base catalyzed esterification with acryloyl chloride. Complete conversion was determined via FTIR and HPLC. The modified drug was conjugated with chitosan via surface modification in solution. Drug release profiles were generated for both synthesized gauze with known chitosan capacity and HemCon ChitoGauze Pro. Preliminary results indicate successful conjugation and release of eugenol and triamcinolone acetonide.

Copyright © American Institute of Chemical Engineers. All rights reserved.

510746 Single Use Bioreactor Equipment: Justification and Sustainability?

Abstract:

To whom it may concern,

Please find below the abstract for the report titled:

Single Use Bioreactor Systems: Justification and Sustainability?

A draft poster is available to accompany this abstract upon request.

Thank you for your consideration

Yours Sincerely

Jordan Harrison

Abstract

This report provides an assessment of the justification, and sustainability of single use bioreactor systems (SUBS). Bioprocessing in general is addressed in this report, however focus is concentrated on the production of various forms of Biopharmaceuticals. Justification is assessed on the grounds of financial viability, and operational practicality. Sustainability is evaluated though the calculation of the total carbon emissions from comparable SUBS and traditional re-usable systems.

The production of Biopharmaceuticals encompasses a range of processes with considerably differing operational requirements; This therefore presents an inherent challenge to provide a general justification for the use of SUBS. To address this issue, this report begins with a comprehensive examination of the frequently cited advantages and limitations of SUBS. This is then analysed to determine the most significant and common factors of consideration when justifying the use of SUBS. The primary parameters that were identified include: Compatibility, scale, product value, and throughput.

The validity of these findings is then tested by considering their applicability in the context of a variety of bioprocess applications (including: gene therapies, antibody treatments, and Vaccine production). This testing process validates the proposed considerations, and illustrates to the reader the significant difference in challenges and requirements for SUBS within bioprocessing. Additional consideration is given to the business model that is applied to a bioprocess. This analysis identified small scale, flexible, multi-product facilities as being ideally suited to utilising SUBS. Conversely, Industrial scale productions, and microbial fermentation processes are highlighted as best suited to reusable facilities. The business model most ideally suited to benefit from the integration of SUBS was found to be contract development and manufacturing organisations (CDMOâ€™s).

This detailed assessment is then summarised to produce a generalised justification flowchart which presents the key criteria required to assess the suitability of a process to the use of SUBS. This flowchart can serve as a starting point to determine if further process specific assessments are warranted for the integration of SUBS.

Economic assessment concludes that where the use of SUBS is feasible, an economic benefit can be achieved. This is largely due to increased flexibility in operations, as well as in facility design. Reductions in cost of goods manufactured (COGM) can be achieved by enabling process streamlining and de-bottlenecking. Sustainability is assessed by comparing the energy consumption, materials consumption, and waste production of SUBS in comparison of traditional multi use (MU) facilities. The high expenditure of energy required in the cleaning and sterilisation of MU systems makes traditional MU facilities less sustainable than SUBS processes.

The production, supply chain, and future of SUBS industry is also examined in this report. It is concluded that the SUBS market share will increase significantly as the industry matures. This is a result of increasing applications, and improvements in component reliability; as experience in usage increases and the necessary supply chain control is established.

Copyright © American Institute of Chemical Engineers. All rights reserved.

510731 Application of Computational Fluid Dynamics (CFD) in Biotransportation of Complex Fluid in the Human System,

Abstract:

Application of Computational Fluid Dynamics (CFD) in Biotransportation of Complex Fluid in the Human System,

Liem Thai and Kazeem B. Olanrewaju,

Department of Chemical Engineering, Prairie View A& M University

Abstracts:

Most of the pathological conditions are connected with anomalies in the flow of complex fluid (blood) and its interaction with various organs in the human system. A clear and detail understanding of the flow process and mechanism of interaction between the complex fluid (blood) and surrounding organs can help prescribes a proper diagnostic procedure in addressing pathological conditions in the human body. Computational fluid dynamics coupled with other multiphysics relating structure and chemical reaction will be adopted to model this fluid-structural interaction within the biological context. This study intends to use Comsol Multiphysics modeling platform because of its robustness and the ability to coupled fluid-solid interaction with module for transport for concentrated species and reacting flow in normal duct and porous media.

Copyright © American Institute of Chemical Engineers. All rights reserved.

511108 Development of a Biological Seed Treatment for Bacterial Diseases

Abstract:

Bacteria can cause important plant and seedling diseases, which are challenging to control due to a lack of effective bactericides. One potential biological bactericide are bacteriophages, which have been shown to control plant pathogenic bacteria in foliar applications and when directly coated on seeds. In order for bacteriophages to be used as a seed treatment they must have compatibility with current seed coating polymers. Seed coating polymers are not only used for adhesion to the seed, but they can also help stabilize the biologic seed treatment. For this study the objectives were to test the compatibility of *Clavibacter michiganense subsp. nebraskense* bacteriophages with common seed treatment polymers in liquid suspension and when treated on maize (*Zea mays subsp. Mays*) seeds, assess shelf-life of seed treatment mixtures containing bacteriophages for multiple weeks, and determine if the treatments had an effect on germination. Bacteriophages were mixed with polymers and stored in a liquid form. From the results of the liquid storage study, compatible polymers were selected for on seed storage in which common stabilizers, skim milk and sucrose, were added to the polymer-bacteriophage mixture before drying on the seed. The mixtures dried in two hours after coated on seeds and the number of active bacteriophage was unchanged. The bacteriophages remained active on coated dry seeds and had a shelf-life for at least two weeks at the lower storage temperature. The addition of stabilizers increased the stability of active bacteriophages and solubility of the polymer coating, making the bacteriophages more available. These bacteriophage seed coatings have the potential to neutralize *Clavibacter michiganense subsp. nebraskense*, before they cause Gossâ€™s wilt.

Copyright © American Institute of Chemical Engineers. All rights reserved.

509613 Directing the Self-Assembly of Multiple DNA Origami Nanostructures in a Single Reaction

Abstract:

DNA origami is a programmable self-assembly technique that combines a single-stranded DNA template, called a "scaffold," with hundreds of custom DNA strands, called "staples," to fabricate nanostructures in a bottom-up assembly process. These nanostructures are designed with unprecedented geometric precision and have promise for applications in cancer therapeutics, biosensors, nanorobotics and more. Despite these promising applications, the DNA origami self-assembly process is not well-understood, and only a few studies have explored detailed folding pathways and mechanisms. In addition, many applications require the higher order assembly of multiple structures, which is currently carried out in a multi-step assembly processes that leads to inefficient formation of the multi-structure assembly. The aim of this work is to study the competitive self-assembly of simultaneously folding two or more unique structures as a means to study detailed folding mechanisms and pathways and to establish methods for the efficient assembly of complex higher order assemblies. We tested the possibility of folding two structures with similar but distinguishable geometry, in the same folding reaction where the staples for both structures compete for the same scaffold, and we explored the effect of adjusting the staple concentration of each structure during folding. Knowing one of the two structures is more energetically favorable, we chose to keep its concentration constant while varying the concentration of the least favorable structure. By adjusting the staple concentrations we can tip the balance toward formation of either structure and adjust relative yields. Interestingly, we find that chimeras that form over a short timescale appear to revert to well folded structures over long time periods. These findings establish a foundation to assemble complex multi-structure assemblies in one step, and this work advances the field of nanomanufacturing by establishing thermodynamic and kinetic principles for the controllable and scalable self-assembly of a entire fleet of nanostructures simultaneously.

Copyright © American Institute of Chemical Engineers. All rights reserved.

509725 Towards Modeling of Methane Recycling Lake Washington Microbial Community

Abstract:

The ability to understand and manipulate biological communities is increasingly becoming important as these communities have the capability to address global problems in regards to energy, food, and disease. However, there is still a necessity to adopt a systems-level approach in this regard. The long-term goal of this project is to first better understand the functionality of a microbial community and then design a synthetic community through constructing a multi-faceted approach similar to the â€˜biobricksâ€™ method of standardizing biological parts. This approach will characterize the functionality of a microbial strain, its interaction with other members in the community, and ultimately, a specific goal such as over producing a desired product can be achieved by adding, removing, or changing the population of specific strains. The community that this project is interested in is Lake Washington (a fresh lake) community characterized by methane cycling. This community is of interest due to its ability to use sedimentary methane as an energy and carbon source. To achieve this goal, genome-scale metabolic models (GSM) need be reconstructed to understand the role of individual strains in the community and also characterize the interactions between individual strains. Thus far, the draft model for methane-oxidizer Methylobacter Tundripaladum (mbT702), has been reconstructed which has 702 genes, 1298 metabolites, 1167 reactions. Then, flux balance analysis (FBA) and flux variability analysis (FVA) are utilized to identify gaps, errors, and inconsistencies within the draft model as well as optimizing the biomass for the model. These will then be manually fixed through the utilization of literature and gap filling techniques. In addition, a model of another major-player in the Lake Washington community, Methylomonas, will be reconstructed. Our next goal is to use the reconstructed models to construct and analyze a two-member microbial community.

Copyright © American Institute of Chemical Engineers. All rights reserved.

510824 Design and Characteristics of Biodegradable and Implantable Batteriesâ€‹

Abstract:

Degradable, transient electronics is an emerging technology with applications including zero-waste environmental sensors, hardware secure memory modules, and temporary biomedical implants. Temporary biomedical implants with biodegradability and biocompatibility characteristics provide the multifunctional operation to assist or monitor a transient biological event, such as wound healing and cardiac patch. The biodegradable power supply is an essential component of many such systems. This poster will outline the fabrication of a biocompatible and biodegradable battery for use in implantable medical devices. The ionic liquid based electrolyte is synthesized from a choline functionalized gelatin polymer. The anode and cathode is made of metal-infused lignin fibers. The fibers are electrospun from a solution of lignin, polyethylene oxide, and choice metal salts followed by carbonization. The carbonization process is carried out at 800°C to bond the metals to the lignin, increase conductivity, and improve biocompatibility. Future testing will reveal the levels of conductivity, biocompatibility, and biodegradability achieved.

Copyright © American Institute of Chemical Engineers. All rights reserved.

510900 Migration of DNA Due to a Combination of Parallel Pressure Gradient and External Electric Field

Abstract:

A combination of a parallel electric field and pressure gradient can cause DNA molecules to migrate transverse to the field lines. We hypothesize that the phenomenon is due to an electrically induced velocity disturbance. A Brownian Dynamics method for a bead-spring model of a polymer, that incorporates this electrically induced velocity disturbance, was constructed and used to predict the dynamics of DNA molecules transported through a microfluidic channel due to the pressure driven flow and electric field. To test the predictions, corresponding measurements of the probability distribution of the centers-of-mass of the DNA molecules were made in multiple channels having different cross-sectional dimensions. Preliminary simulation results match, at least qualitatively, the migration velocity, extent of migration, and entry length measured in the experiments.

Copyright © American Institute of Chemical Engineers. All rights reserved.

510044 Atmospheric Pressure Raman Investigation of Binary Clathrate Hydrates in Their Role in New Energy Applications

Abstract:

Clathrate hydrates are hydrogen-bonded water molecules that form cage-like structures at high pressures and low temperatures. These clathrate hydrates enclathrate guest molecules and have applications in flow assurance, hydrogen storage, desalination, carbon capture and storage, and natural gas resources. Cyclopentane (CP) and tetrahydrofuran (THF) are both structure II clathrate formers that occupy the large cages within the structure, and are of interest because they promote clathrate hydrate formation in addition to reducing the required formation pressure. Other studies have observed a synergistic effect between THF and CP clathrate hydrates when small gases like carbon dioxide occupy the small cages, however, no studies have been done to analyze the relationship between CP and THF clathrate hydrates on their own. In this study, Raman spectroscopy is used to analyze these binary clathrate hydrate mixtures at varying molar ratios of CP/THF/water to determine the dissociation temperature. Additionally, the role of promoters in increasing the speed of clathrate hydrate formation is examined to determine the impact and optimum composition for clathrate hydrate formation. Our findings include that 0.5CP:0.5THF:17H2O and 0.75CP:0.25THF:17H2O (molar ratio) mixtures form most quickly with the greatest clathrate hydrate conversion, and that the latter mixture has a higher dissociation temperature than CP clathrate hydrates alone. This optimum mixture is favorable both thermodynamically and kinetically. Because it is more stable and can be formed/maintained at atmospheric pressure and higher temperatures, it has major implications for new energy technologies. Preliminary findings also reveal the mechanism in which the binary clathrate hydrate dissociation occurs, leading us to believe that THF actually remains within the large clathrate cages longer than CP. Further investigation is required to understand this phenomenon.

Copyright © American Institute of Chemical Engineers. All rights reserved.

510181 The Effects of Mesitylene on the Cold Flow Properties of Model Crude Oils

Abstract:

-
Introduction
Theory and Background
The Effect of Mesitylene on the Cold Flow Properties of Model Crude Oils
Ugochukwu Okeibunor and Michael Senra
Department of Chemical and Biomolecular Engineering, Lafayette College, Easton, PA 18042
Experimental Methods
Results and Analysis
Future Work
References
[1] Senra, M., Scholand, T., Maxey, C., & Fogler, H. S. Role of polydispersity and cocrystallization on the gelation of long-chained n-Alkanes in solution. Energy & Fuels, 235947-5957. Nov. 2009.
[2] Nur Fatmala. Wax in Subsea Pipelines, February 2016. Web August 7
th
2017.
The cold flow properties of a fluid represent its movement and characteristics at low temperatures. These properties are often dictated by how the precipitated solutes form crystals.
Conclusions

v Microscopy can be used to analyze the crystal structures of the gels formed in this work to better validate the results and hypotheses.
v Increasingly complex n-alkane systems can be studied to determine if the effects seen for model crude

oils can be extended to actual crudes.

Figure 2. Cross-polarized microscopy images for slowly cooled n-alkane solutions. Panel a) for 4% C36 solutions at 33°C, panel b) for 4% C36/5% C28 solution at 25 °C and panel c) for 4% C36/6% C32 solution [1]

Acknowledgements

Thanks to the Lafayette College EXCEL program and the Department of Chemical Engineering for facilitating the completion of this project.

The deposition of wax in subsea oil pipelines has been a major problem in flow assurance. These formed gel deposits can reduce oil production and once they occur, are costly to eliminate.

Figure 1. Wax formed in pipeline [2]

Figure 4. TA Instrument Discovery HR-2 rheometer

Figure 5. Storage and loss moduli of a fluid under stress

q Small aromatic molecules can influence the thermodynamic and gelation characteristics of model crude oils.

q Their effects can be a function of the amount of aromatic present and the complexity of the crude oil.

q The mechanisms by which the aromatic influences the system are widely varied.

q Proper prediction of crude oil gelation should require information about the aromatics content.

Figure 6 shows that the gel point of the C36 system increased from 33.5°C to 40°C, while the gel point of the C32 system had a slight increase from 28°C to 30°C. A hypothesis for this behavior is that increasing mesitylene concentrations decrease the solubility of C32 and C36 in dodecane, thus causing gelation to occur at higher temperatures.

Figure 8 shows that the cloud point range was about 2°C for all systems. However, monodisperse systems showed a slight upward trend while polydisperse systems showed a slight downward trend. For each mesitylene concentration, the gel point obtained by differential scanning calorimetry was generally about 2°C lower, but similar trends were observed.

Figure 6. Gel Point Trend for Monodisperse Systems.

Figure 8. Cloud Points for the n-alkane Systems Obtained by Calorimetry

Figure 7 shows that for both systems, an initial increase in mesitylene concentration by 1% - 2%

caused a sharp decrease in the gel points. At higher concentrations, the gel points rose sharply and remained relatively constant. It is hypothesized that low amounts of mesitylene disperses the formed crystals, while high amounts of mesitylene can self-aggregate, limiting their influence on the crystal structure.

a)

b)

c)

Figure 7. Gel Point Trend for Polydisperse Systems

In this work, dodecane (C12), octacosane (C28), dotriacontane (C32) and hexatriacontane (C36) served as model crude oils. The aim of this work was to investigate the effects of mesitylene, a small organic, on the cold flow characteristics of these model crude oils.

Previous work has shown that C36 (Figure 2a) forms large crystals, producing a volume-spanning network that is conducive to gelation. However, adding another n-alkane can influence how these crystals form. C28 disperses C36 from forming large and uniform crystals (Figure 2b), while C32 reduces the gelation of C36 by forming cocrystals with C36 (Figure 2c).

Studies have shown that n-alkanes, the primary components of crude oil, are largely responsible for this deposit [1]. This phenomenon occurs because of their low solubility.

Figure 3. Chemical Structure of Mesitylene

This project aims to analyze the effects of a small aromatic molecule on the gelation characteristics of different n-alkane systems. Specifically, the effect of adding mesitylene on the gel formation of these model oil systems was studied by rheology.

v Cold flow properties studied in this work were the gel point and the cloud point.

v The gel point represents the temperature where initial gel formation occurs and the cloud point represents the temperature where crystals can first be seen.

v Rheology and differential scanning calorimetry were used to analyze the pour points and cloud points respectively.

v Experimentation methods ensured elimination of thermal and shear history by holding sample quiescently at a constant temperature for a substantial period of time.

Copyright © American Institute of Chemical Engineers. All rights reserved.

509241 Design of a Novel Electrochemical Membrane Reactor for Hydrogen Production Via the Sulfur-Ammonia Water-Splitting Cycle

Abstract:

The solar-driven hybrid sulfur-ammonia (S-NH_3) water splitting cycle is a novel approach to produce hydrogen through a large-scale design. Originally proposed by T-Raissi *et al.* (2006), this cycle involves the photochemical oxidation of ammonium sulfite as a solar-photocatalytic hydrogen production step. Later, the resulting ammonium sulfate is decomposed into ammonia, sulfur dioxide and oxygen through a high-temperature solar thermochemical oxygen evolution step. Finally, ammonium sulfite is regenerated by means of chemical absorption of ammonia and sulfur dioxide in aqueous phase. Thereby, only water is needed to release hydrogen and oxygen in this cycle. This process has many advantages like high-purity hydrogen production at ambient temperature; generation of value-added by-products such as ammonium sulfate; exploitation of the entire solar spectrum and consumption of greenhouse gases (*i.e.* sulfur dioxide). However, even though the hydrogen production step has been studied, there is a lack of information on aspects which have strong impact on hydrogen production rates like pH, kinetic parameters, and/or efficient reactor design. Moreover, new approaches have been proposed to enhance the large-scale viability of the S-NH_3 cycle, including an electrochemical hydrogen production step established by Taylor *et al.* (2014). This electrochemical approach still makes use of solar energy resources, but *via* thermodynamic cycles which take advantage of the high temperature gases to produce electricity. Thereby, the electrooxidation of ammonium sulfite is carried out with the same advantages as described above.

In this work, we propose a novel electrochemical reactor for the hydrogen production step for the S-NH_3 cycle, which uses an ionic exchange membrane to achieve pH control. The reactor is designed in such a way that allows a suitable and dynamic analysis of hydrogen production at laboratory-scale.

In order to support this proposal, a preliminary pH-control analysis was carried out. Figure 1 illustrates the chronoamperograms resulting from the electrooxidation of ammonium sulfite at room temperature. The best two conditions of reaction rates were using a buffer and an Anion Exchange Membrane (AEM), as shown. Thus, the latter was selected as the most practical pH-control method for this reaction. Moreover, hydrogen production was verified by Gas Chromatography with Thermal Conductivity Detector (GC-TCD). Based on these results, a two-chamber electrochemical membrane reactor separated with an AEM was developed to allow the hydrogen production analysis, as shown in Figure 2.

The reactor performance was later analyzed using a Design of Experiments (DOE) approach to establish an optimal operating condition. Finally, hydrogen production rate at such condition was verified by GC-TCD. Currently, kinetic and mass transport studies are being carried out to discuss a better reaction design for this electrochemical approach.

Figure 1. Electrooxidation kinetics of $(NH_4)_2SO_3$.

Figure 2. Conceptual principle of operation of the proposed electrochemical membrane reactor.

Copyright © American Institute of Chemical Engineers. All rights reserved.

509256 Development of Nanofluids Based on Synergistic Effect of MgO Nanoparticles and a Quaternary Amine (CTAB) for the Inhibition of Formation Damage Caused By Fines Migration

Abstract:

Development of Nanofluids based on Synergistic Effect of MgO Nanoparticles and a Quaternary Amine (CTAB) for the Inhibition of Formation Damage Caused by Fines Migration

Rebeka Díez, Farid B. Cortés* and Camilo A. Franco*

Grupo de Investigación en Fenómenos de Superficie – Michael Polanyi, Facultad de Minas, Universidad Nacional de Colombia Sede Medellín, Kra 80 No. 65–223, Medellín, Colombia.

*Corresponding authors

e–mail: fbcortes@unal.edu.co, caafrancoar@unal.edu.co

Abstract

Fines migration is a common problem in the oil and gas industry that causes the decrease of productivity of a well due to the displacement of the mineral solids that adhere to the porous media plugging the pore throats of the formation. The primary objective of this work is to evaluate a nanofluid composed by nanoparticles of MgO and a quaternary amine (Hexadecyltrimethylammonium bromide - CTAB) to enhance the capacity of the porous media for the retention of fines nanoparticles. MgO nanoparticles were synthesized by the sol-gel method using $Mg(NO_3)_2{\cdot}6H_2O$ as a precursor. Nanoparticles were characterized by dynamic light scattering, point of zero charge, TGA and FTIR. Different sizes 11,39nm, 42,80nm and 86,20nm were obtained. Nanofluids were prepared by dispersing the nanoparticles in a CTAB-containing aqueous solution. Retention tests were performed using Ottawa sand 20/40 packed beds and fine particles suspensions at concentrations of 2 wt% in deionized water. Individual and combined effects of nanoparticles and CTAB were evaluated in different treatment dosages. Based on the experimental results, the synthesized MgO nanoparticles/CTAB nanofluids showed a higher retention of fines in comparison with the systems treated only with CTAB or with MgO nanoparticles with efficiencies up to 200% higher. Also, was developed a design of experiments aimed to find the optimal concentration for the treatment proposed using the STATGRAPHICS Centurion XVI software (StatPoint Technologies Inc.). This work proposes a new alternative for the injection of nanofluids as a treatment for the problem of fines migration to optimize the productivity of oil and gas wells.

Keywords. CTAB, Fines Migration, Formation Damage, MgO nanoparticles, Nanofluids.

Copyright © American Institute of Chemical Engineers. All rights reserved.

Impact of Coal Quality on Environmental Pollution and Available Technologies For Reducing Its Extent- An Overview on Coal Pollution in Bangladesh

Sabrina Shawreen Alam[1], Md Golam Jakaria[2]
[1,2] Department of Chemical Engineering, Bangladesh University of Engineering & Technology(BUET) (BSc.)
[1] University of Toledo, Toledo, OH 43606. [2]University of Rhode Island, Kingston RI 02881.

THE UNIVERSITY OF TOLEDO 1872

BUET

Introduction

Coal is a heterogeneous substance which is widely used as a source of energy worldwide. About 42% of global electricity supply is produced from coal-fired plants by coal combustion. Coal contains varying amounts of carbon, hydrogen, nitrogen, oxygen, and sulfur as well as trace amounts of other elements, including mineral matters in its matrix. The distribution of its constituents in coal is of significant importance, since it affects the reactivity, heating value and environmental impact of combustion. 28% of global carbon dioxide emission is accounted by coal fired power plants. SOx, NOx, COx are primary gaseous pollutants. Presence of heavy metals and chlorine intensifies its toxic behavior. Large quantity of coal residues are produced after combustion. Improper disposal of such residues can be a major threat to public health. Boropukuria is the only coal fired power plant in Bangladesh. Due to energy crisis, Bangladesh government is planning to set more coal power plants. Very few research has been done on coal in Bangladesh and now it's high time to understand the potential environmental hazards associated with coal power plants.

Clean Coal Technologies

Widely Used Technologies
- Washing
- Electrostatic precipitators
- Fabric filters
- Flue gas desulfurization
- Low-NOx burners
- Re-burning techniques coupled with NOx burners

Advanced Technologies
- Selective Catalytic Reduction (SCR)
- Pressurized Fluidized Bed Combustion
- Ultra-clean coal (UCC)
- Underground coal gasification (UCG) in situ

Significance & Goals

Bangladesh being mostly dependent on natural gas based power before, now is moving towards coal power. Inadequate research, insignificant rules & regulations are established for preventing coal pollution till now.

Therefore, the objective of this study is to-
- Study the chemical composition of Boropukuria coal.
- Identify the main toxic constituents and their contribution in coal matrix.
- Predict environmental distribution of coal burnt products and residues.
- Establish targets for making coal power more environmentally sustainable in Bangladesh context.
- Produce data for policy making.

Methods

Sample Preparation

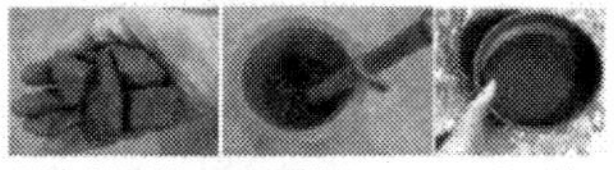

Proximate Analysis
Gravimetric method
Ultimate Analysis

Instrument: Micro CHNSO Analyzer

Ash Composition
Burnt coal ash sample pelletized into 1cm tablets by mixing with appropriate binder.
Analytical Device : Shimadzu Lab Center XRF-1800
Gross Calorific Value by Bomb Calorimeter
Quantifying Trace Elements
Hg – by acid absorption in bomb calorimeter
Cl – by reaction with Na_2CO_3 in combustion bomb
As – by acid extraction

Instrument : HACH Spectrophotometer

XRD Analysis

Instrument: Rigaku Ultima IV X-ray Diffractometer
Leaching Test
Period: 6 months
Instrument: HACH Spectrophotometer
Estimation of Gaseous Pollution Load
Simulation by FactSage 7.1 Software

Results

Proximate Analysis

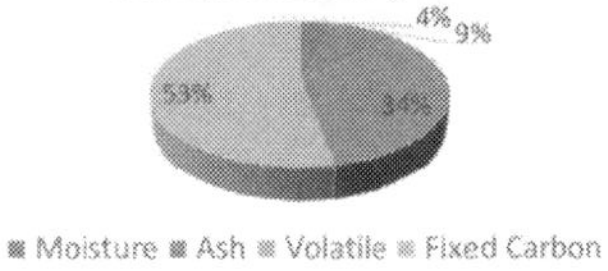

Ultimate Analysis

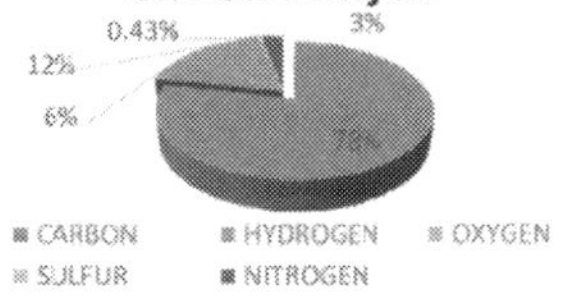

Trace Elements

Element	Amount in Coal
Mercury	0.00656 ppm
Arsenic	1.29 ppm
Chlorine	0.12 ppm

Calorific Value
Gross Calorific Value 30.61 MJ/Kg

XRD Results

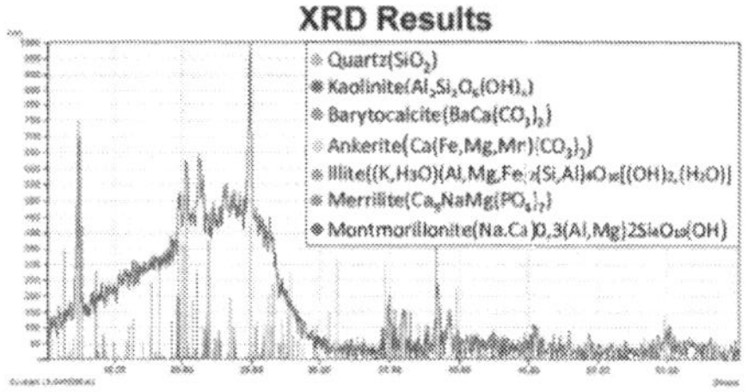

Ash Analysis Results in wt.%

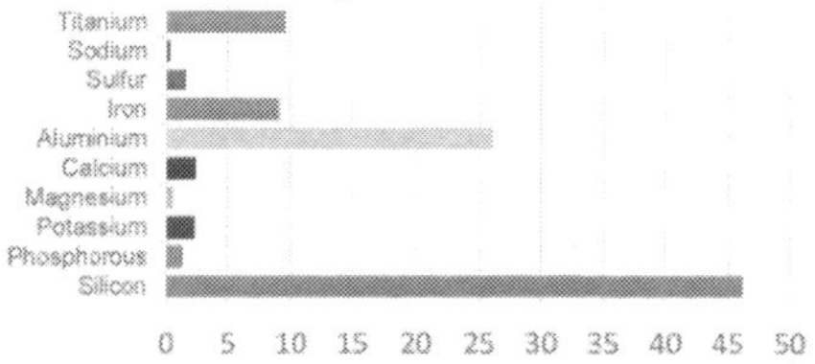

Simulation Results (1513K, 20% Excess air)

Gas	SO_2	SO_3	NOx	CO_2
Kg/GJ	0.5010	0.195	3.54	86.83

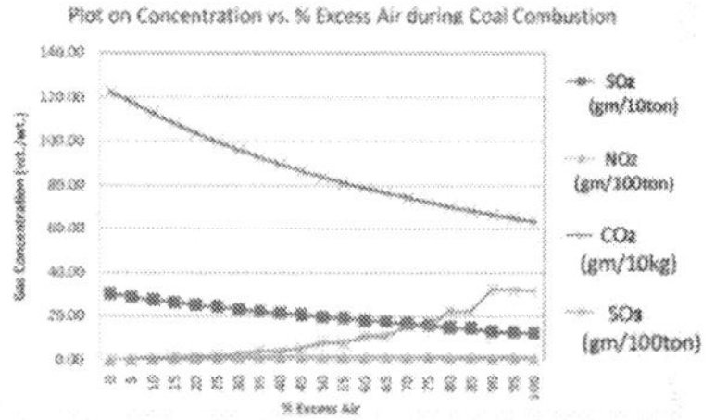

Leaching Test Results

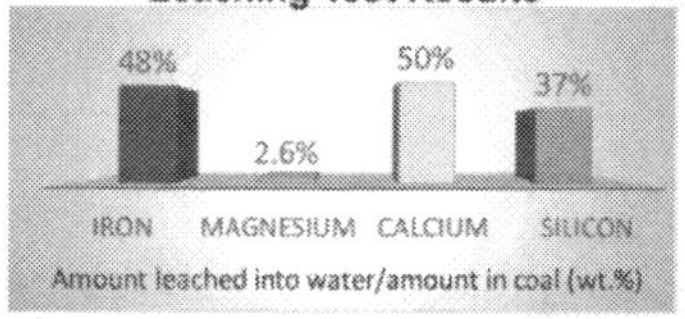

Conclusion & Future Directions

Sulfur (wt.%) lies in the allowable limit and FGD plants can efficiently remove upto 70% SOx. Presence of trace elements require further investigation.
Overall, this study can has created path for future research on-
- Quantitative and analytical analysis of gaseous effluents of Boropukuria.
- Trace metals in gaseous phase and their state.
- Interaction between Mercury and Chlorine.
- Leachability of trace elements to predict environmental impact.

Acknowledgements

Research Advisor: Dr. Kazi Bayzid Kabir, Department of Chemical Engineering, Bangladesh University of Engineering and Technology, Dhaka-1000, Bangladesh.
For Travel Support: Dr. Glenn Lipscomb, Head, Department of Chemical Engineering, University of Toledo, OH. 43606.
Lab Facilities: Dept. of Glass & Ceramic Engineering, Dept. of Civil Engineering, BUET. Material Science Research Laboratory, DU. Dhaka-1000, Bangladesh.

References

J.G. Speight,Handbook of Coal Analysis, 2nd Edition, John Wiley & Sons,2015
EIA, EPA and IAE Data

510851 Kinetic Modeling for Cellulosic Butanol Production Based on Multi-Omics Analysis

Abstract:

Butanol has a higher energy density and lower volatility than gasoline and is similarly compatible with existing energy infrastructure. Furthermore, butanol can be produced using a Clostridium cellulovorans biomass conversion platform. This study utilized a platform overexpressing heterologous acetaldehyde/alcohol dehydrogenase (adhE2) gene allowing for cellulose to butanol conversion. Using this platform, we have previously produced butanol with >3 g/L, a yield of >0.14 g/g, and a selectivity of >3 g/g from corn residues. To achieve a high butanol production, this study integrated fermentation data and the multi-Omics experimental data through two types of mathematic model. A static model based on end product yields was created to simulate flux balance analysis of carbon using fermentation data. A dynamic model was created to additionally incorporate intracellular metabolite concentrations to correct for fermentation end product yields. Resultantly, the dynamic model successfully simulated the intracellular metabolite fluctuation from pyruvate to butanol. Important results included carbon flux distribution reduction from C2 to C4 in the mutant with cellulose substrate. Key regulators for metabolic engineering were evaluated and strategies for improved butanol production were proposed.

Copyright © American Institute of Chemical Engineers. All rights reserved.

509692 Co-Pyrolysis of Potato Peels and Polystyrene : A Novel Technique to Recover the Upgraded and Valuable Bio Oil

Abstract:

The whole world is searching for novel and sustainable techniques for production of renewable fuels and valuable chemicals via a route which is clean, economical, energy saving and fast process. In this context, mixture of potato peels and polystyrene in different weight ratio were taken as feed for production of bio oil through co-pyrolysis. The product bio oil could be reformed to produce a renewable and clean source of energy from waste biomass and plastics. This novel technique undoubtedly will reduce the threat to depletion of fossil fuels. The potato peel was selected due to its high hydrocarbon contents and abundance in nature while the latter (polystyrene) has high hydrogen content, thus would add value to the oil through synergistic effects. Also, its consumption to produce a fuel, would minimize the environmental menace created by plastics.

TG-DTA analyses were carried out on raw materials to investigate their thermal degradation properties and other tests like proximate and elemental analyses were done to characterize it. The pyrolysis of pure potato peels was carried out initially at different temperatures and the optimum temperature was found to be 500°C to achieve highest liquid yield. Then co-pyrolysis of processed potato peels and polystyrene was performed at this temperature in different weight ratios and the oil obtained with 1:3 weight ratio of potato peels and polystyrene showed the bio oil yield of 67.17 weight % which is quite higher than yield of pure potato peels (38.1 weight %). The GC-FID and FTIR spectrum results confirmed that the synergistic effects between biomass and polymer during co-pyrolysis resulted in reducing the oxygenated compounds in the oil, as evident by the rise in calorific value from 24.18 MJ/kg of bio oil obtained from pure potato peels (500°C) to that of 33.42 MJ/kg through co-pyrolysis of potato peels to polystyrene (1:3) at 500°C. Also, other properties and fuel tests were performed for characterizing liquid, like pH, density, flash & fire point, water content, ASTM distillation which proved that the bio oil produced via co-pyrolysis process has a potential to be used as fuel.

The solid product Char was also analyzed for its properties through proximate analysis, fixed carbon test, calorific value and FTIR to determine the possibility of developing the char into a solid fuel or its use as an activated carbon in extraction processes of heavy metals. It was concluded that the co-pyrolysis of potato peels and polystyrene evaluates as a promising technique for recovering upgraded bio oil and valuable char.

Copyright © American Institute of Chemical Engineers. All rights reserved.

510837 Air Oxidation Kinetics of Commercial-Grade Iron Powder As an Oxygen Carrier for Chemical Looping Methane Reforming

Abstract:

Iron-based oxygen carriers are potential candidates for hydrogen production through chemical looping methane reforming (CLMR) for the low cost and high H_2 yield. The oxidation kinetics of commercial-grade iron powder is tested in a thermogravimetric analyzer (TGA) between 500 and 900 . Iron oxides prepared at various temperatures are analyzed with X-ray diffraction (XRD). The oxidation rate is generally low despite a noticeable increase at temperatures above 700 . Multiple iron oxide phases are found in the oxidized iron particles, mainly including hematite (Fe_2O_3), magnetite (Fe_3O_4), and wüstite (FeO). The composition of Fe_2O_3 increases with higher reaction temperatures and a longer reaction time. Significant particle sintering occurs at temperatures above 600 . These preliminary results suggest that the low oxidation rate, complex oxidation products, and high-temperature sintering need to be addressed in order to utilize iron as an effective oxygen carrier in CLMR.

Copyright © American Institute of Chemical Engineers. All rights reserved.

510944 Novel Extraction of Microalgae for Biofuel and Valuable Chemical Production

Abstract:

Rapid growth in the economy of the United States and its steady expansion has resulted in a concomitant demand for energy for domestic and industrial processes, 90% of which is being currently met from fossil fuel sources of coal, natural gas, and petroleum. The ever-dwindling reserves of these fossil fuel sources prompt the development of sustainable and renewable energy supplies. The development of alternative energy sources, which make for environmental sustainability while being economically competitive, is a topic of current research interest and high importance. Biofuels have a high potential as an alternative to traditional energy sources, but need continued research and optimization in order to become a viable source of energy. Photosynthetic plants such as microalgae are an ideal biomass feedstock for energy and chemical production due to their fast and robust growth in a wide range of conditions and their efficient carbon dioxide fixation. Microalgae is composed of lipids, carbohydrates, and proteins which can be extracted and converted to biofuels and other valuable chemicals. Extraction methods already exist, but literature shows that these processes fail economically due to the energy-intensive process involved in turning wet algae to dry algae. A solution to skip over this highly energy intensive step is proposed. The novel extraction method consists of the use of a switchable solvent to extract lipids, carbohydrates, and proteins all in one process while starting from wet algae. After extraction, the solvent is able to be recycled back for future use. The reusable nature of the solvent and costs saved in energy reduction make the process viable for the production of biofuels. Species selection and cultivation conditions are also examined to manipulate the algae composition during culture. The extraction method is deemed to be viable due to the successful conversion of lipids to biofuel and by fermenting carbohydrates to valuable chemicals in a more economically feasible way.

Copyright © American Institute of Chemical Engineers. All rights reserved.

510309 Thermal and Thermodynamic Properties of Ionic Liquids and Molten Salts with High Thermal Stability

Abstract:

Ionic liquids have garnered significant interest in research due to their unique beneficial properties: vanishing vapor pressures, low flammability, chemical inertness, large liquid ranges, and wide versatility. Coupled with their tenability from being organic salts, they have promising applications to replace current volatile solvents that used in reactions, separations, bioprocessing, etc. Furthermore, since one of their properties is high thermal stability, using ionic liquids for thermal energy storage, heat transfer and other high-temperature situations are other viable applications. However, this has been recently shown to be inaccurate. Previously considered stable ionic liquids have been shown to decompose at temperatures of 200-300°C when under long-term thermal stress. Therefore, ionic liquids for heat transfer fluids, high temperature solvents, or lubricants is limited.

To this end, we have synthesized several peraryl sulfonium and phosphonium cations coupled with various anions to be thermally stable at temperatures of at least 300°C with low to negligible mass loss. Along with these, we have recently studied the thermal stability of various anions previously not considered in literature coupled with a PNP^+ cation. In this work, we present the melting points for these salts along with their solid and liquid heat capacities and compare these to traditional organic heat transfer fluids. Additionally, we examine the relative effects of the enthalpy and entropy of fusion on the melting points of homologous series of these salts along with mass loss data from long-term thermal stability tests. In the perspective of heat transfer fluids, their heat capacities were determined to compare them with common heat transfer fluids.

Preliminary experiments with two perarylphosphonium salts combined with excess aromatic molecules creates an aromatic-ionic liquid biphasic system. The phase equilibria of these molten salts with five different aromatics is explored by constructing liquid-liquid phase diagrams. These initial studies along with synthesis of novel thermally stable salts can aid future work related to thermal energy storage and high-temperature processes.

Copyright © American Institute of Chemical Engineers. All rights reserved.

510538 Plant Design and Production of Fuel from Plastic Waste

Abstract:

Abstract

The demand for commodity plastics has been increasing due to the rapid growth of the world population. This has generated high quantities of waste plastic annually because of the non-biodegradable nature of plastics. Plastic waste management is therefore, one of the many challenges confronting most countries including Ghana. As calorific value of the plastics is comparable to that of fuel, so production of fuel would be a better alternative.

This project focused on the pyrolysis of plastic waste, particularly, polyethylene, to liquid fuel. The material and energy requirements for production of 50 tonnes /day liquid fuel were estimated. Some selected equipment of the pyrolysis plant was designed. Economic analysis of the plant was also performed to determine the feasibility and profitability of the project.

The liquid fuel produced may be used in furnaces, boilers, turbines and diesel engines without further refining. The by-product, which is gaseous hydrocarbons has high calorific value that can be reused to compensate the overall energy requirement of the pyrolysis plant.

The theoretical energy required to pyrolyze 1kg of polyethylene is 1.047 MJ. The estimated calorific value of the products is about 43.3 MJ/kg. The economic analysis undertaken in regards to this project revealed over 50% rate of returns which makes the project a very feasible and profitable avenue.

Copyright © American Institute of Chemical Engineers. All rights reserved.

510620 Economics-Informed Discovery of Solar Energy Conversion Systems

Abstract:

Current methods of solar energy conversion are not yet sufficiently economical to allow solar to be a competitive global energy source. This project investigates various solar energy conversion systems to identify which have the best economic potential, and how best to implement these systems. A technoeconomic analysis was conducted of available chemicals produced via solar energy conversion to identify the best candidates. Factors considered were the different materialsâ€™ market sizes, market values, and energies of formation. Several materials have high market values but too-low of market sizes, or vice versa. Additionally, some materials require too high of an energy input to make large-scale production feasible. Copper metal was shown to be one of the products with the best potential. Deposition of copper on a solar cell could allow the production of a high-value product that would increase the profitability of a solar energy-driven production process, thus making solar competitive with conventional energy sources. Investigation into producing copper via solar energy conversion and photo-electrodeposition in a continuous-flow reactor is now being conducted to optimize its economics and feasibility. Deposition of copper metal onto a gallium arsenide solar cell illustrated feasibility of the idea, the copper deposited well and was removed easily. However, gallium arsenide solar cells are expensive, so a scalable alternative is the use of silicon solar cells. Deposition of copper metal on a silicon solar cell posed challenges that will need to be overcome. Poor deposition amounts and rates and difficulty in harvesting the copper from the surface of the cell will have to be solved to implement a practical copper deposition system. Possible solutions include the use of flow options, surface pretreatment, sonication, and precipitation. Future investigation will look into these, and any other, possibilities to efficiently remove the copper and create a continuous-flow process. A more-thorough economic analysis will also be conducted to show the scalability of a copper deposition process.

Copyright © American Institute of Chemical Engineers. All rights reserved.

510739 Selective Oligomerization of Ethylene By Nitrogen Bridged Diphosphine Nickel(II) Complexes

Abstract:

Selective Oligomerization of Ethylene By Nitrogen Bridged Diphosphine Nickel(II) Complexes

Sivaramakrishna Akella[1], Lenin Kumar V[1], Varun Kaushal[2] and **Anubhav Mishra**[2]

(1) Chemistry, VIT University, Vellore, India,

(2) Chemical Engineering, VIT University, Vellore, India

Abstract

Traditional linear alpha olefin (LAO) production processes typically produce a broad and relatively inflexible distribution of linear alpha olefins. It is very important for industry to produce valuable linear alpha olefins selectively without co-producing other low-value by-products. Selective processes like production of 1-hexene and 1-octene are the targets. In this connection, numerous catalysts for ethylene trimerization to 1-hexene have been reported, and a process is now practiced commercially. Recently, the first catalysts for the selective tetramerization of ethylene to 1-octene were reported using chromium based catalysts. The active catalyst is typically generated by a combination of a chromium source with a diphosphine ligand in the presence of an aluminoxane activator. Mechanistic studies on these catalyst systems indicate that the 1-octeneis liberated from a metallcyclononane catalytic intermediate, formed by an extension of the metallacyclic mechanism previously proven for ethylene trimerization reactions. Herein, we report the catalytic ethylene oligomerization reactions to produce LAOs using phosphine based Ni complexes (Eq.1).

Eq.1: General scheme for ethylene oligomerization reaction

Results indicate that formation of 1-hexene and 1-octene with nickel complexes are quite feasible. Here the temperature can control the product distribution, while Pd(II) complexes prefer to yield more dimerization to 1-butene. This could be due to the better stability of nickelacycloalkanes and they will be allowed to insert further ethylene molecules. Palladacycloalaknes can undergo relatively faster ß-hydride elimination to yield 1-butene. The bimetallic catalyst showed a controlled distribution of C6 and C8 products based on the reaction conditions. By fine-tuning the ligands as well as reaction conditions it is possible to achieve the selective trimerization and tetramerization. The product distribution strongly depends on the nature of remote substituent 'R', pressure, temperature and solvent employed.

Copyright © American Institute of Chemical Engineers. All rights reserved.

510742 The Possibility of Using Metal-Based Ionic Liquids for Desulfurization of Diesel Fuel

Abstract:

Since the emission of sulphur oxide (SO_x) especially from automobile exhaust has been increasing rigorously, the removal of sulphur compounds from fuel oil has become a universal subject. Ionic liquids have been studied as the alternative solvent for desulfurization process with several reports on their reasonable success in removing sulphur compounds from model oil. Nevertheless, the ability of ionic liquids to remove sulphur compounds from real diesel fuel is very challenging due to multi-component nature of diesel fuel. Therefore, this work is carried out to study the desulfurization of real diesel fuel using several metal based ionic liquid $[C_4mim][MCl_n]$ with M = Fe^{3+}, Co^{2+}, Cu^{2+}, and Zn^{2+}. The ionic liquids solution are prepared by mixing an equimolar amount of $[C_4mim]Cl$ with corresponding MCl_3 salt, which are then characterized using H-NMR and Raman spectroscopies. The ionic liquid is added to model oil at equal mass before being added to diesel fuel. Total Sulphur Analyser is used to determine the sulphur concentration after mixture is at least 6 hours settled. It is observed that these ionic liquids have high ability to remove sulphur from model oil, however unexpectedly low sulphur removal from real diesel.

Copyright © American Institute of Chemical Engineers. All rights reserved.

509291 A New Process for Catalytic Fractionation of Organo-Solv Lignin to Biofuels

Abstract:

Topic: Fuels, Petrochemicals, and Energy

Title: A New Process for Catalytic Fractionation of Organo-solv Lignin to Biofuels

By: Joseph Molle and Jonathon Waterman

With guidance from: Dr. Malek Alkasrawi and Dr. Joynal Abedin

From the University of Wisconsin â€“ Stevens Point

Our current research project has developed a new technology for converting organo-solv lignin to biofuels and high value chemicals. Organo-solv lignin is extracted and recovered in mass quantities from the delignification process of lignocellulosic materials such as wheat straw and pulp used in the production of paper. The technology utilized in this research involves a novel nickel catalyst that has been synthesized specifically for lignin fractionation to simpler, more practical chemicals. The catalyst has a particular functionality as it targets specific groups in the diverse lignin polymer to yield products such as vanillin, benzene and toluene. In addition, this new process that we have developed is economical and sustainable due to several cost savings configurations. For example, the catalyst used has been recovered and recycled while still having the ability to depolymerize lignin. Furthermore, the solvent used may be recycled, which will minimize the overall production cost and waste of the process. This process is economically viable because the chemicals produced from the depolymerization of lignin can be utilized in several industries such as the fuel, pharmaceutical, food and fragrance industries. The extracted lignin that is recovered in the delignification process of products such as paper-making pulp and wheat straw currently has little to no value. Instead of paying to dispose of this recovered lignin in landfills, which is often the only method to get rid of the lignin, this unique research creates an opportunity to convert this intricate lignin by-product into several simpler chemicals that are practical in the aforementioned industries. This research explores a unique method of recycling the largest waste product in the Pulp and Paper Industry, which provides many opportunities to expand beyond the laboratory and carry out this process on a pilot and potentially a commercial scale. The capital and operating expenditures associated with this process can be offset by the high profits from the production of these valuable chemicals and fuels with their various markets in the industrial world.

Copyright © American Institute of Chemical Engineers. All rights reserved.

509510 The Production of Biodiesel in Carbon Dioxide

Abstract:

Biodiesel is a renewable alternative to replace fossil fuels. As a liquid fuel source, biodiesel can replace petroleum based fuels without requiring major modifications to existing infrastructure. Nowadays, one of the main methods for biodiesel production involves the conversion of triglycerides (TG) to form fatty acid alkyl esters (biodiesel) via transesterification with alcohol. Currently, the major catalyst for this production is the homogeneous basic catalyst. Although these basic catalysts can convert TG to biodiesel efficiently, they would also react with free fatty acid, which naturally accompanies TG in biostocks and forms soap, which would inhibit the yield of biodiesel. To avoid the negative effects from the saponification reaction, in this project, Nafion®, a heterogeneous acidic catalyst is used for producing biodiesel from TG via the transesterification reaction and using supercritical carbon dioxide to increase reaction kinetics and to decrease mass transfer resistances for both TG and catalysts between phases. In this work, we have determined that FFA is more favored in this system than TG. The presences of both water and FFA will not bring detrimental effect to the overall yield of biodiesel from TG.

Copyright © American Institute of Chemical Engineers. All rights reserved.

509689 Aqueous Phase of Biomass Pyrolysis Oil As a Raw Material for Polymer Synthesis

Abstract:

The goal of this research is to utilize the aqueous phase of bio-oil for polymer synthesis. Bio-oil is a liquid product obtained through the pyrolysis of biomass containing lignin, cellulose, and hemi-cellulose. Currently, the aqueous phase does not have much commercial value but consists of water, alcohols, aldehydes, ketones, sugar and phenols... In order to polymerize the bio-oil components, the bio-oil was altered to contain as high of a ratio of olefins as possible. In the first stage, primary and secondary alcohols in the bio-oil were converted into aldehydes and ketones, respectively. Then, through Wittig reaction, aldehydes and ketones were converted into olefins. The Wittig reaction takes advantage of a triphenylphosphine ylide attaching to an aldehyde/ketone and taking away the oxygen atom while leaving behind an olefin. Therefore, the more aldehydes/ketones available in the bio-oil, the more olefins that may result from the Wittig reaction. Wittig reactions were attempted with both a stabilized and non-stabilized ylide to yield different olefins. Fourier transform infrared spectroscopy (FTIR) and 1H and 13C nuclear magnetic resonance (NMR) were used for characterization. The results of the Wittig reaction were analyzed through chromatography to analyze the results for the desired product. Finally, the funcrionalized bio-oil olefins were polymerized by AIBN initiator at elevated temperature.

Copyright © American Institute of Chemical Engineers. All rights reserved.

509877 Thin-Film Photo-Bioreactor: Construction and Characterization

Abstract:

This project addresses Food-Energy-Water (FEW) Nexus issues in that growing cyanobacteria to produce biofuels and other bioproducts requires large amounts of water and energy. This decreases the benefits of using cyanobacteria over fossil fuel resources such as petroleum. The purpose of a Thin-Film Photo-Bioreactor (TFPB) is to decrease this water and energy consumption. This project focused on designing and constructing a TFPB to cultivate Synechocystis sp. PCC 6803 and to observe how the cyanobacterial cell density in a thin film changed over time. A film was produced using sodium alginate and calcium chloride to evenly distribute the cells, and to maintain a smooth, uniform surface area. A TFPB was made from acrylic Plexiglass and silicone gasket material that allowed the cyanobacterial film to be used for a long time, and the temperature and airflow were held constant. Measurements indicated continuous oxygen production, an increase of cell mass per unit surface area, and good activity of Photosystem II over time

Copyright © American Institute of Chemical Engineers. All rights reserved.

510445 Electrospun Nanofiber Fuel Cell Cathodes

Abstract:

Fuel cells are promising sustainable energy conversion devices. Before widespread commercialization, several key properties of fuel cells must be optimized. One method used for improving fuel cell performance is electrospinning sub-micron diameter fiber electrodes. Electrospun catalyst/binder fiber mat cathodes were found to be highly porous, with a high catalyst surface area and improved gas transport to catalyst sites. Building on previous research at Vanderbilt that used electrospinning for fiber mat fuel cell fabrication, the present study was carried out to investigate catalytically active platinum alloy particles for the cathode in a hydrogen/air fuel cell. One alloy, PtCo has been shown to be more active than Pt for oxygen reduction, based on rotating disk experiments in the literature. To confirm the advantage of this catalyst, PtCo in nanofiber mat cathode membrane-electrode-assemblies were fabricated and tested. Experimental results show a 30% improvement in maximum fuel cell power output, as compared to the use of a convention Pt/C cathode catalyst.

Copyright © American Institute of Chemical Engineers. All rights reserved.

510631 Incorporation of Plasmonic Silver Coated Gold Nanocubes into Photosystem I Films for Photocurrent Enhancement

Abstract:

Photosystem I proteins have a near unity quantum efficiency and high charge separation. While these are optimal properties of photovoltaic materials, incorporating these proteins with inorganic materials yields low photocurrents in practice. Plasmonic gold nanoparticles are used in a variety of fields for their high absorbance cross-section. Here, silver-coated gold nanocubes are employed for their complimentary absorbance with that of photosystem I. These nanocubes are incorporated into photosystem I films and tested in a three-electrode cell with a mediator solution of 20 mM sodium ascorbate, 1 mM 2,6-dichloroindophenol and 100 mM potassium chloride. A platinum counter electrode and Ag/AgCl reference electrode were used in all photocurrent measurements. Samples were illuminated for 20s by white light from a Leica KL 2500LCD lamp at an approximate light intensity of 140 mW/cm^2. A 47 nA/cm^2 enhancement of cathodic photocurrent is demonstrated after incorporation of plasmonic Au@Ag nanocubes.

Copyright © American Institute of Chemical Engineers. All rights reserved.

510786 Modeling Hydrogen Storage System with Application in Public Transportation in Mobile, AL

Abstract:

Renewable energy sources like wind, solar, hydroelectric, and wave powers have become a sustainably viable source of energy. During peak hours, the energy generated by a renewable energy source may exceed the general demand of the population, thereby causing excess energy being wasted. One way to make use of this excess energy is to convert electrical energy to another stable, an energy efficient, and storable form such as hydrogenâ€"the field of study being called power-to-gas (P2G). P2G also involves studying the various ways to store hydrogen that would minimize any loss in energy during the process. One interesting application and case study would be the use of hydrogen fuel cell as an alternative fuel for a bus transit system. This study explores different hydrogen storage mechanisms for energy harvested from various renewable energy sources unique or feasible in Alabama to be used in a bus transit system in Mobile, Alabama. We are looking at the feasibility of the process and how this system can be optimized for efficiency and profit.

Copyright © American Institute of Chemical Engineers. All rights reserved.

510896 Anion Conducting Ionomers for Electrochemical Devices

Abstract:

The design of new anion exchange ionomers for use in electrochemical devices is a critical step in improvement of anionic fuel cells and electrolyzers. Although these materials can be made from the same precursors as anion exchange membranes (AEM), they have very different property requirements. Since ionomers are used within the electrode layers, they do not act as a separator for fuel and air but rather help transport reactants and products to and from the electrocatalyst surface. In this study, a series of anionic ionomers was synthesized and tested in fuel cell and electrolyzer configurations. These ionomers are based on fluorinated poly(arylene ether) homopolymers and block copolymers with alkyl tethers that have been modified to be used as anion conductors. *In situ* testing of these materials shows that lower molecular weight materials outperform their higher molecular weight counterparts. Additionally, the use of a block copolymer with the introduction of a hydrophobic spacer further increases device performance.

Copyright © American Institute of Chemical Engineers. All rights reserved.

510951 Increased Structural Integrity of Aqueous Processed Thick Electrodes for Lithium Ion Batteries

Abstract:

Lithium-ion batteries (LIBs) have been used extensively in portable electronic devices such as cell phones, laptops, and other personal electronics for the past two decades. However, widespread use in electrical vehicles is still limited by material cost, processing cost, and energy density capability. One approach to achieving high energy density batteries is using thick electrodes (150Î¼m thickness instead of baseline 60Î¼m thickness) to increase the volume ratio of active materials in a cell of given size. Traditional electrode processing requires the use of a flammable, toxic solvent, n-methyl pyrrolidone (NMP). The recovery steps required for NMP add significant processing costs to the overall LIB production cost. Aqueous processing has been explored as a method of eliminating NMP from electrode processing; however, aqueous processed thick electrodes have a tendency to crack. Surface tensions of isopropyl alcohol / water and methyl acetate / water systems of varying concentrations were evaluated to determine the effect of the additive solvents on the overall system, and electrodes processed with the mixed solvents were examined for cracking. Both types of mixed solvents successfully reduced surface tension and allowed thick electrode processing without cracking.

Copyright © American Institute of Chemical Engineers. All rights reserved.

510965 Hydrothermal Liquefaction of Chlorella Kessleri: Effects of Algal Growth Stage on Biocrude Production and Characterization

Abstract:

According to a 2017 study by the U.S. Energy Information Administration (EIA), U.S. motor gasoline and diesel fuel consumption accounted 30% of the total U.S. energy-related CO_2 production in the year 2016. Disproportionately high emissions of CO_2 have caused adverse effects on the environment: increasing global temperatures and changing climates. Since 2007, reducing CO_2 emissions in the U.S. economy has been an overarching goal of the U.S. Environmental Protection Agency (EPA), which will require the production of 36 billion gallons of renewable fuel by the year 2022. To help cover this quota, biocrude may be implemented as a renewable replacement for petroleum crude in the synthesis of diesel fuel. The use of biocrude as a feedstock is environmentally beneficial because CO_2 is consumed while cultivating biomass, and the raw materials come from inexhaustible sources. One feasible production route of biocrude is through whole-cell conversion of algae into biocrude. Algae can be cultivated using wastewater treatment, and the bioprocessing of algae would serve the dual purpose of water purification and oil production. The purpose of this study is to demonstrate the optimization of an algae-to-biocrude operation involving the hydrothermal liquefaction of raceway pond algae into high grade biocrude. Algae were inoculated on site under nitrogen-limiting conditions. The nitrogen to phosphorous (N:P) ratio of the growth medium was studied, and algae samples were also collected at each algal growth stage (exponential, transitional, stationary). The production of biocrude was studied in terms of yield and carbon recovery in the biocrude.

Copyright © American Institute of Chemical Engineers. All rights reserved.

509237 Using a Rotating Ring Disk Electrode to Explore the Effects of Manganese on Solid Electrolyte Interphase

Abstract:

Lithium-ion batteries (LIBs) are currently being used in electric cars due to their high energy density and low self-discharge rate. Although LIBs power electric cars, they are heavy and expensive. Most cathodes in these batteries contain cobalt (such as $LiNi_xCo_yAl_zO_2$). Manganese oxide based batteries are a promising alternative that provide higher voltage than cobalt oxides. Additionally, manganese is cheap, abundant, and exhibits excellent rate capability. Unfortunately, LIBs with manganese oxide cathodes suffer from accelerated capacity fade [1] due to manganese dissolving from the cathode and disrupting the protective behavior of the solid-electrolyte interphase (SEI). We lack the understanding of the mechanisms by which manganese disrupts the SEI [2].

In order to investigate the impact of manganese on the SEI, we carried out a series of experiments using a rotating ring disk electrode (RRDE) because it allows us to differentiate between kinetic and transport effects. The disk electrode was glassy carbon coated with the cathode material ($LiNi_{.5}Mn_{1.5}O_4$) while the ring was glassy carbon. The SEI was formed on the ring electrode by cycling the ring to low voltage in LP30, a standard battery electrolyte. Meanwhile, the disk was held at high potential to induce electrolyte oxidation and manganese dissolution. The entire assembly was rotated to transport oxidation products, including dissolved manganese, from the disk to ring. After formation of the SEI, ferrocene was added to the electrolyte and oxidized at the disk. Measuring the through-film ferrocenium reduction current at the ring therefore assesses the degree of passivation of the SEI.

Results show that oxidation current on the disk increases at higher potentials. We attribute this current to electrolyte oxidation and manganese dissolution. When the disk is held at high potential, SEI formation on the ring displays a new reduction peak at 1.2V. The peak only appears when the disk is being oxidized at potentials higher than 4.7V, and the height of the peak scales with the potential of the disk electrode. We also observe that electron transfer to ferrocenium through the SEI is less kinetically hindered when the SEI is formed in the presence of electrolyte oxidation products. These observations suggest that dissolved manganese ions are incorporating into the SEI structure and reducing its ability to passivate the electrode. As the voltage on the disk during SEI formation increases, more manganese oxidation occurs. These observations open a window for future work to determine if the effects of manganese on the SEI are predominantly electronic or morphological. Determining the mechanism by which manganese causes the SEI to fail will suggest failure mitigation strategies for materials and system.

References: -

[1] Jung Hyun Kim, Nicholas P W Pieczonka, Zicheng Li, Yan Wu, Stephen Harris, and Bob R. Powell. Understanding the capacity fading mechanism in LiNi0.5Mn 1.5O4/graphite Li-ion batteries. Electrochim. Acta, 90: 556â€"562, 2013. ISSN 00134686. doi: 10.1016/j.electacta.2012.12.069.

[2] Bhandari, Arihant, and Jishnu Bhattacharya. "Reviewâ€"Manganese Dissolution from Spinel Cathode: Few Unanswered Questions." Journal of The Electrochemical Society 164.2 (2016): n. pag. Web. 07 July 2017.

Copyright © American Institute of Chemical Engineers. All rights reserved.

510078 An Investigation of a Mass Flow Rate Method for Evaluating the Filterability of Hydraulic Fluids

Abstract:

Fine filtration is used in modern systems to remove built-in contaminants, wear debris generated within the machine and dirt entering the fluid from the environment. The ability of the fluid to retain its filterability properties is critical for efficient, reliable machine performance. This is particularly true for biodegradable fluids used in environmentally sensitive areas. ISO 13357:1 and ISO 13357:2 are internationally accepted methods for evaluating the filterability and compatibility of hydraulic fluids. This method uses a manual procedure for measuring the volumetric flow rate of a fluid as it passes through a 0.8μm filter membrane. In this research, hydraulic fluids and lubricants of various base oil and additive composition were evaluated using a modified version of ISO 13357 that utilizes mass flow rate measurements to assess fluid filterability. A comparison of volumetric and mass flow rate filtration ratios reveals that the use of mass flow rate to determine filterability yields comparable results while improving test repeatability. These findings are the basis for a new ASTM standard methods proposals for the measurement of fluid filterability and compatibility.

Copyright © American Institute of Chemical Engineers. All rights reserved.

510423 Cost-Effective Polysulfide and Polyiodide Redox Flow Battery for Large-Scale Energy Storage

Abstract:

Renewable resources have the potential to compete with coal, but provide energy only intermittently. Grid-scale electrical energy storage cost-effectively smooths out these intermittencies, which is critical to enable deeper market penetration of renewable energy sources. Redox flow batteries (RFBs) are one such technology that will help renewable energy sources gain wider use. The active species in typical RFBs are expensive and relatively rare. In contrast, this work examines sulfur and iodine as the energy storage compounds, both of which are inexpensive, earth-abundant elements. The work presented herein demonstrates the feasibility of the polysulfide-polyiodide flow battery (SIFB). The system has an open circuit voltage of 1.0 volts with a peak power density of 65 mW/cm^2. SIFBs use sodium cations as the working ion to balance charges between the electrolytes; this is advantageous because the ion-selective membranes used here facilitate the transfer of positive charges while repelling negative ones, ensuring minimal crossover occurs as determined by cyclic voltammograms of the post-cycling anolyte and catholyte. Indeed, the reactor demonstrates nearly 100% coulombic efficiency for 200 charge-discharge cycles, or 530 hours, implying minimal impact of the solutions on reactor performance. The solutions were refreshed every 50 cycles. Finally, it is shown through a techno-economic analysis that SIFBs can reach the U.S. Department of Energyâ€™s price target for electrical energy storage of \$100/kWh. Not only can they reach this target, but they are currently less expensive than even the future cost of vanadium RFBs; they outperform existing batteries and can compete with other energy storage methods.

Copyright © American Institute of Chemical Engineers. All rights reserved.

510436 High Performance Nanofiber Electrodes for Li-Ion Batteries Using Particle Polymer Electrospinning

Abstract:

Particle/polymer electrospinning was used to prepare (i) fiber mat anodes containing carbon powder and pol(vinylidene fluoride) and (ii) fiber mat cathodes containing $LiCoO_2$ nano-particles, carbon powder, and poly(vinylidene fluoride) for Li-ion batteries. Unlike previous work on electrospun LiCoO2 nanofibers prepared using sol-gel chemistry and high temperature processing, the particle/polymer fiber mat cathodes reported here were made thick with a high fiber volume fraction for high areal and volumetric capacities at fast charge/discharge rates (e.g., 0.81 mAh cm^{-2} and 62 mAh cm^{-3} at 2C) which were much greater than that of a slurry cast cathode of the same composition (0.004 mAh cm^{-2} and 0.30 mAh cm^{-3} at 2C). Full cells containing a $LiCoO_2$/C/PVDF fiber mat cathode and C/PVDF fiber mat anode were also prepared and characterized. These electrospun batteries exhibited a high energy density of 144Wh kg^{-1} at 0.1C and an areal capacity of 1.03 mAh cm^{-2} at 1C. The excellent performance of the electrospun particle/polymer cathodes and anodes is attributed to electrolyte penetration throughout the 3D fiber electrode mats, a large electrode/electrolyte interfacial area, and short Li+ transport pathways between the electrolyte and active material nanoparticles in the radial fiber direction.

Copyright © American Institute of Chemical Engineers. All rights reserved.

510987 Ion Exchange Membranes for Vanadium Redox Flow Batteries

Abstract:

Vanadium redox flow batteries (VRBs) offer a scalable, long lasting, highly efficient means of energy storage. VRBs use different oxidation states of vanadium, V^{3+}/V^{2+} in the negative and V^{4+}/V^{5+} in the positive half-cell, for chemical reactions that generate electrical energy. In a VRB an ion exchange membrane separates the two half cells, and also allows the diffusion of H^{+} ions between solutions to complete the oxidation/reduction reactions. The study of ion exchange membranes is important because the membrane controls the VRBâ€™s internal resistance and effects the tendency of undesirable crossover. In this work we study the effect of different treatments on a commercial membrane (3M825EW), and how those treatments impact the membraneâ€™s conductivity, water uptake, permeability, density and porosity as they are exposed to various concentrations of sulfuric acid and vanadium solutions. The purpose of these tests is to help us further our understanding of what features of ion exchange membranes control the resistance and crossover and to determine if the treatments improve the performance of the membrane. The observations made from these characterization techniques are that the conductivity is affected by the acid and vanadium concentration, the water uptake is inversely related to the acid concentration, and the proton mobility decreases as the water uptake decreases.

Copyright © American Institute of Chemical Engineers. All rights reserved.

510720 Non-Uniform Aging of Commercial NMC/Graphite Lithium-Ion Batteries

Abstract:

Non-uniform aging of commercial NMC/Graphite lithium-ion batteries

Abdulrahim Ayoub[1], Abdilbari Mussa[2], Maria Varini[2]

[1] Department of Chemical Engineering, The City College of New York, CUNY, New York, NY 10031, USA

[2] School of Chemical Engineering, Division of Applied Electrochemistry, Royal Institute of Technology, Stockholm, Sweden

Lithium ion batteries are well-established energy storage systems with high power and energy density. In the past few years, rechargeable lithium ion batteries have played a major role in automotive applications; however, many obstacles remain in long-term performance and durability. Decrease in capacity and/or a rise in impedance leads to electrode aging, thus, dramatically degrading the batteryÕs performance.

Herein, aging of commercial NMC/Graphite battery is investigated in (i) calendar aged cells, stored at 22°C for approximately a year, (ii) cycled cells, cycled at +4C/-1C between 10-90% SOC. Further, electrochemical analysis was done on different positions within the electrode layer to study aging as a function of position, namely edge and center. In this study, electrochemical impedance and capacity measurements were the methods of choice used to characterize aging.

Calendar and cycled aged cells displayed differences in capacity fade and impedance rise. Specifically, the cycle-aged cells showed more pronounced losses, likely associated with the growth/formation of the solid-electrolyte interface(SEI) and the loss of cyclable lithium. Considering the edge and center of the same electrode layer, variations in capacity and impedance were also noticed, possibly caused by an un-even current distribution along the electrode.

Copyright © American Institute of Chemical Engineers. All rights reserved.

510792 Autogenic Synthesis of High Rate, High Energy Carbon-Sulfur Cathodes for Rechargeable Lithium Sulfur Batteries

Abstract:

There is an increasing demand for longer lasting and higher performance energy storage. Todayâ€™s rechargeable lithium-ion batteries cannot satisfy the high-energy requirements for grid energy systems, longer-lasting electronics, and long range electric vehicles. In contrast, the theoretical performance values for lithium-sulfur batteries exceed the minimum requirements for these emerging technologies. However, technical challenges including low conductivity and the polysulfide shuttle â€“ a parasitic mechanism in which the cathode decomposes into soluble chemical intermediates during cycling â€“ inhibit battery lifetime to fewer than 10 cycles.

To address these issues, this work introduces the autogenic process to synthesize high conductivity carbon-sulfur composites. Developed in 2004 by Pol et al., the autogenic process enables high temperature reactions within a solvent-free reactor. The autogenic process allows for high sulfur loading (up to 75% by TGA) with higher specific capacity compared to mechanical mixing techniques. Specifically, the high specific capacity (ca. 1227 Ah kg^{-1}) at fast charging rates (up to 1C) is hypothesized to be due to sulfur distribution: as shown by energy-dispersive X-ray spectroscopy, autogenic synthesis selectively inserts sulfur into the carbon substrate with a highly isotropic distribution. The better sulfur distribution is believed to improve conductivity and the resulting electrochemical kinetics with autogenic synthesis.

Copyright © American Institute of Chemical Engineers. All rights reserved.

510841 Towards Advanced Porous Carbon Electrodes for Redox Flow Batteries

Abstract:

The intermittency and unpredictability of renewable energy technologies has resulted in an increased need for large-scale grid storage. Redox flow batteries (RFBs) are a promising technology to satisfy these grid storage needs due to their ability to decouple power and energy, and their potentially low cost. Carbon fiber-based porous materials are typically used as electrodes and must fulfill several functionalities: high permeability and surface area, stability under operating conditions, electrochemical activity and selectivity, and stability under compressive forces. State-of-the-art RFB electrodes are re-purposed gas diffusion layers (GDLs) typically used in polymer electrolyte fuel cells and, while there is some overlap in desired properties, they have not been optimized for this specific application.

In this work, we systematically investigate various commercially available carbon substrates, specifically three carbon papers, a cloth, and a felt. In the first part, we characterize the pressure drop characteristic and electrochemically active surface area. Highly permeable electrodes, like carbon cloth, are found to experience a smaller pressure drop in a flow cell setup and would decrease pumping costs. Higher accessed surface areas, like that of SGL 39AA carbon paper, are found to be preferable for battery performance due to their increased wetted surface in contact with the electrolyte. In the second part, we present an approach for modifying porous electrodes based on attaching polymer films onto the carbon fiber surfaces. A hydrophilic monomer was grafted onto the carbon electrode leading to a large increase in the wettability of these electrodes with water and an increase in accessed surface area by a factor of more than 30 times. This might prove to be an easy and effective method to improve battery performance in aqueous vanadium flow batteries.

Copyright © American Institute of Chemical Engineers. All rights reserved.

510997 Organic Phase Change Cooled Photovoltaic Panels

Abstract:

Photo-voltaic panels lose efficiency as the temperature of the panel increases. A typical panelâ€™s output can drop 40% or more on hot, high solar irradiation days when the highest potential for generation exists. Solar panels are typically uncooled, although CoolPV from Fafco mates a HDPE tubing mat to the rear of a panel, allowing cooling water to both improve the electrical efficiency and create hot water for residential use. Taking this concept one step further, phase change cooling was investigated with an organic liquid (R-245a) introduced in the tubing behind a CoolPV panel either at atmospheric pressure or set with a pressure relief valve at 20 PSI, and allowed to boil to cool the panel. The vapor was collected and condensed. Although this system was batch based, the final implementation would be a constant flow system similar to water cooled panels. The efficiency of standard panels, water cooled panels, and organic phase change panels are compared in this poster. The eventual aim is to couple an Organic Rankine Cycle (ORC) to an array of photo-voltaic cells, converting the removed heat energy into usable electrical power.

Copyright © American Institute of Chemical Engineers. All rights reserved.

509769 Improving Protein-Polymer Block Copolymer Self-Assembly By Optimizing Depletion Force Interactions

Abstract:

Protein-based materials have become increasingly valuable due to their diverse substrate specificity, functionality, and high catalytic activity. Self-assembly of protein-polymer block copolymers enhances protein properties with stability, chemical diversity, and possibility of synthetic polymers. Self-assembled materials also offer a high degree of control over nanoscale orientation in a highly ordered and densely packed array, expanding the scope of protein-based technologies. However, the addition of depletants may promote attractive forces between collodial suspensions. This leads to an increase in depletant translational entropy that may overcome limitations in ordered particle assembly at dilute concentrations. To clarify the effect of depletion forces in order-disorder phase systems, the entropic contributions of molecular weight and relative content of large molecule depletants on the self-assembly of mCherry and carbonic anhydrase will be investigated.

Copyright © American Institute of Chemical Engineers. All rights reserved.

REINFORCEMENT OF MAGNESIUM OXYCHLORIDE CEMENT COMPOSITES USING SHORT CARBON AND ARAMID FIBERS

Hansen Mou[1] Roque F. Góchez[1], David Nasol[1], Grant MacPherson[1], Jim Wambaugh[2], Christopher L. Kitchens[1,3].

[1]Department of Chemical and Biomolecular Engineering, Clemson University. [2]JetProducts LLC, [3]Institute of Environmental Toxicology (CU-ENTOX)

Motivation

- Construction and operation of buildings account for 38% percent of carbon dioxide emissions and 39% of total energy use in the US[1]
- On average, Americans spend 90% of the time indoors; common building materials like OSB and gypsum boards contain components that may lower the indoor air quality
- In 2014, fires caused $11.6 billion in property damage; one home structure fire was reported every 86 seconds[2]

Our **research motivation** is to develop a green building material that is fire resistant, sustainable, and does not compromise indoor air quality.

[1]http://archive.epa.gov/greenbuilding/web/html/index.html
[2]http://www.nfpa.org/research/reports-and-statistics/fires-in-the-us

Introduction & Goal

When MgO is mixed with a concentrated solution of $MgCl_2$, formation of a crystalline molecule takes place, this material is known as magnesium oxychloride (MOC).

$$5MgO + MgCl_2 + 8H_2O \rightarrow 5Mg(OH)_2 \cdot MgCl_2 \cdot 8H_2O$$

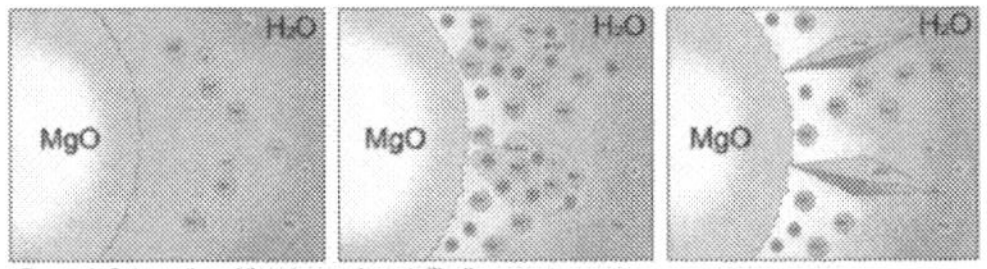

Figure 1. Schematics of the observed crystallization process. Cations and anions form a solvated phase (S) that crystallizes.

Advantages

- Low carbon footprint, as manufacturing is not energy intensive
- Structural properties similar to Portland cement, which, along with low density, makes it desirable for modular construction
- Insulation properties, which makes it suitable for structural insulated panels (SIP)
- Fire resistance, as the crystal water in the molecule delays fire propagation

Figure 2. SEM image shows MOC after crystallization reaction at 35 °C

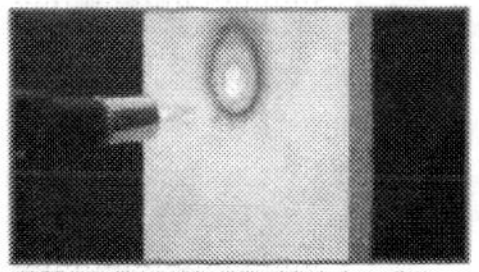

MOC can be used to make 1-hour firewalls in compliance with ASTM E119

The goal of this research is to enhance the structural properties of MOC using short, random fibers

Methods

The raw material mixture, procedure and curing conditions to make MOC composite boards were adapted from a patent previously published by JetProducts LLC [9].

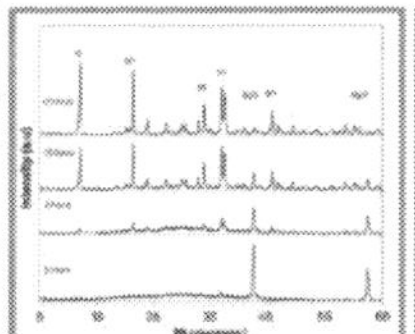

Figure 3. XRD monitoring of MOC crystallization at a 35 °C

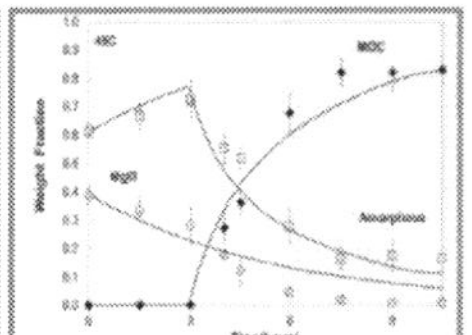

Figure 4. XRD quantitative monitoring of crystalline and amorphous phases. Continuous lines show the first-order kinetic model.

Figure 6. (Above) Fibers used to enhance MOC. Left: chopped carbon fibers (average length: 6mm). Right: chopped aramid fibers (average length: 13.6mm).

Figure 5. (Left) Uncut MOC boards reinforced with 0.5 wt.% intermediate modulus (top) and 0.5 wt.% standard modulus (bottom) carbon fibers.

- Cured for 12 hours at 45 °C, 65 %RH
- Flexural strength and modulus of rupture were assessed using a 3-point bending test geometry
- 8 samples tested for each fiber loading and type.

Figure 7. Sample cracking during the 3-point bend test

Results

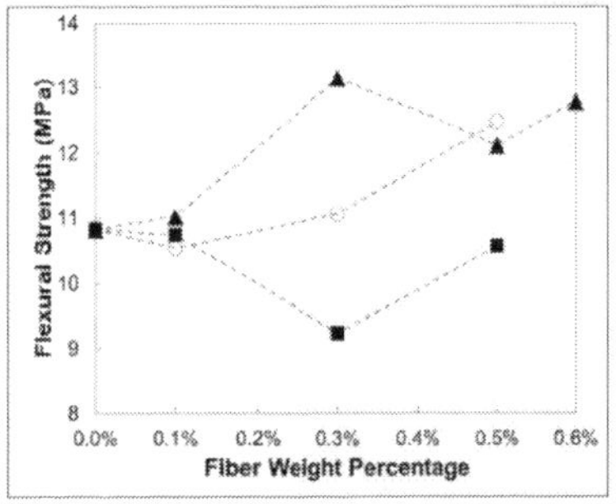

Peak flexural strength improvement of MOC observed was 21.4%, with 0.3 wt. % IMCF.

Figure 8. (Left) Average flexural strength of MOC composites reinforced with SMCF (void circle), IMCF (triangle), and aramid fiber (square). Discontinuous lines were added to guide the eye.

The short and stiff CF allowed for a homogeneous fiber distribution throughout the board, avoiding excessive clumping while mixing.

The longer and softer aramid fibers clumped together during the mixing of the mortar, creating voids and decreasing the overall flexural strength of the boards.

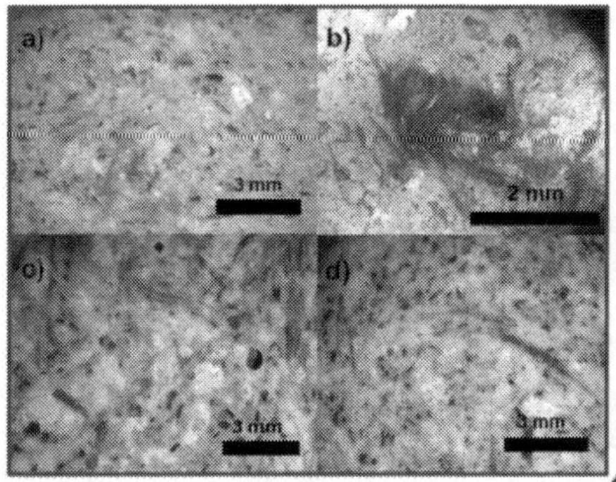

Figure 9. Optical microscope and SEM images of fiber reinforced MOC composites. Image a) shows clumped aramid fibers; b) magnified SEM image of the clumped aramid fiber; c) IMCF in the MOC matrix, not clumped; and d) SMCF in the MOC matrix, not clumped.

Conclusions and Future Work

- The flexural strength of MOC composite boards were enhanced using SMCF and IMCF, up to a 21.4% increase with IMCF at 0.3 wt. %
- Flexural strength decreased with aramid fibers due to the inhomogeneous distribution of fibers in the cement.
- These MOC composite boards were successfully prepared without excess MgOH formation
- Future work will focus on other consequences of using MOC concrete as a building material, including metal corrosion effects and water stability.
- Work will be done on the scalability of the fiber reinforcement process for MOC concrete

Acknowledgments

- JetProducts LLC for technology share and funding
- Marvin Joiner, from Leigh Fibers for providing fiber materials to this project
- Colin McMilen, for assistance with XRD measurements

For further information contact: ckitche@clemson.edu

Preparation and Characterization of Bio-Based Polyesters Derived from Xylochemicals

Rowan University

Ian C. Dunn, Kevin C. Hagan, Jord C. Yniguez, Silvio Curia, Ph.D, Prof. Joseph F. Stanzione, III
Department of Chemical Engineering, Rowan University, Glassboro, NJ 08028

Introduction

- Many polymers ubiquitous in modern life are derived from **fossil-based feedstocks**; an **unsustainable model** for the future
- Replacing **fossil-based precursors** with **bio-based building blocks** is a crucial step towards a sustainable society

Lignin

- **Lignin** is a complex **biopolymer** that gives plant cells their ridged, woody structure
- Lignin can be strategically depolymerized to yield building blocks known as **xylochemicals** [1]
- **Vanillyl alcohol** and **guaiacol** are two such building blocks that can be derived from lignin [2]

Xylochemicals → Bio-Based Polymers

- **Bisguaiacol** can be synthesized via an **electrophilic aromatic condensation** of vanillyl alcohol and guaiacol [2]
- The two hydroxyl groups in bisguaiacol can be reacted to form **polymers** [1][2]

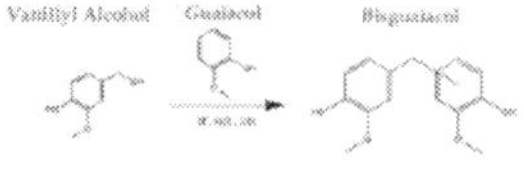

Bisphenol A (BPA)

- Widely used in polyesters
- Derived from petroleum precursors
- Linked to many adverse health effects [3]

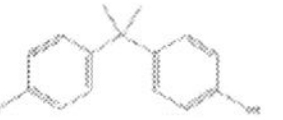

Bisguaiacol (BG)

- Similar structure to BPA
- Fully bio-based
- Less toxic than BPA [3]

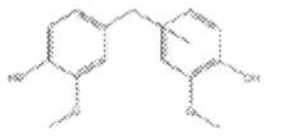

- We have focused on using BG as a replacement for BPA to produce **viable and sustainable alternatives** to fossil-based polyesters

Experimental

Preparation

- BG was synthesized via an electrophilic aromatic condensation of vanillyl alcohol and guaiacol
- BG and BPA were polymerized with various **diacid chlorides** (DACs) via an **interfacial polymerization**
- The organic phase of the reaction was extracted and dropped into hexane or methanol to precipitate polymers as amorphous solids/ powders and then dried under vacuum

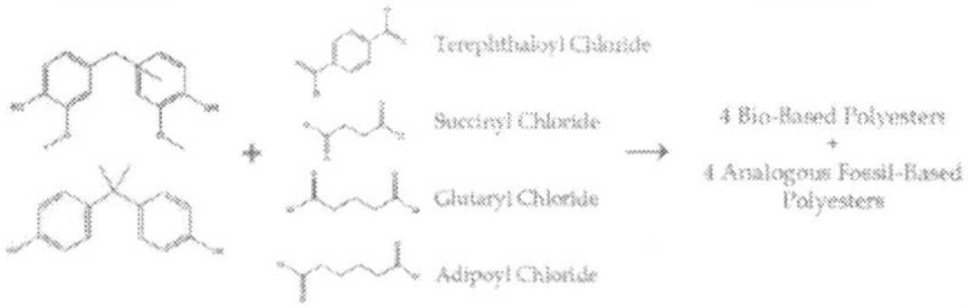

Characterization

- The polyesters were characterized with FTIR, ^{1}H-NMR, DSC and TGA techniques

Objectives

- Confirm polymer structures through NMR and FTIR
- Observe effects of DAC chain length and structure on thermomechanical properties
- Determine structure-property relationships of bio-based polyesters to fossil-based analogues
- Assess the validity of bio-based polyesters as alternatives to current industry standards

Results

FTIR Spectra of Polyesters and BG Monomer

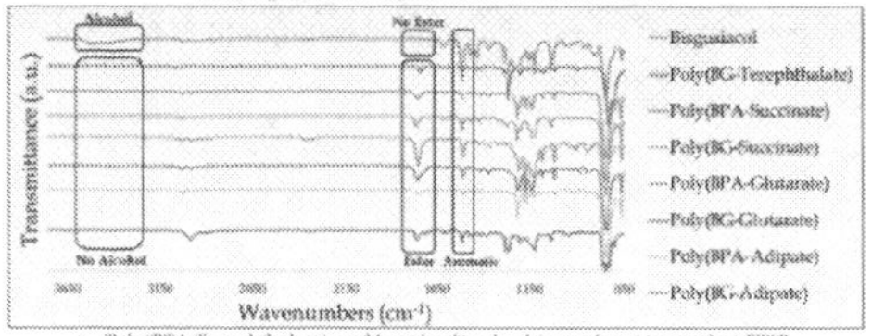

- Poly(BPA-Terephthalate) could not be dissolved in a solvent to conduct FTIR
- Conducted on Thermo Scientific Nicolet 6700

^{1}H-NMR of Poly(BPA-Glutarate)

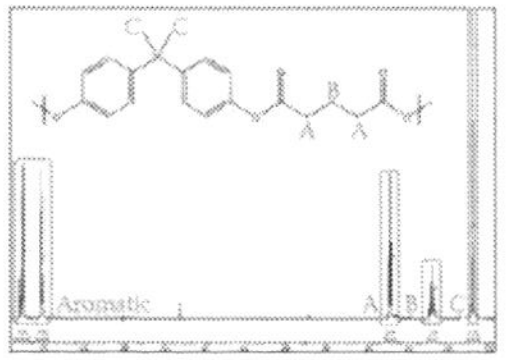

^{1}H-NMR of Poly(BG-Glutarate)

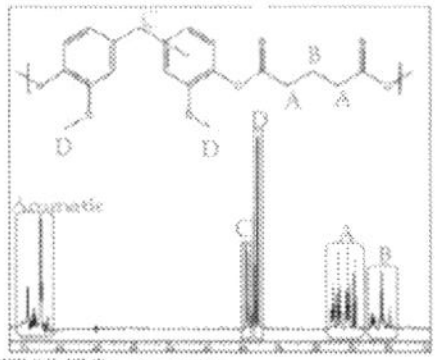

- Conducted on Varian 400 Mhz FT-NMR Spectrometer

Differential Scanning Calorimetry (DSC)

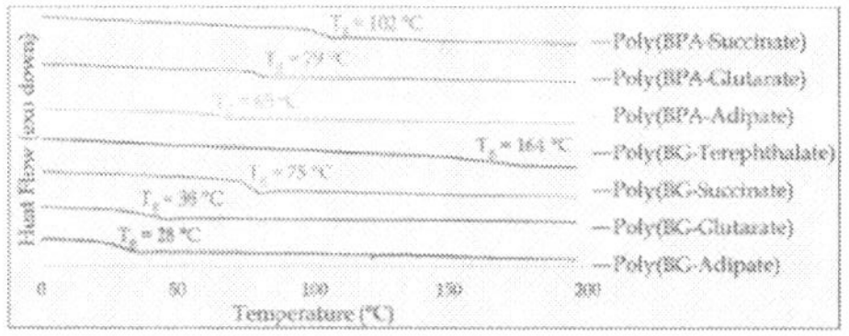

- T_g is glass transition temperature
- Poly(BPA-Terephthalate) showed no visible T_g up to 300 °C
- Conducted on Texas Instruments DSC2500

Thermogravimetric Analysis (TGA)

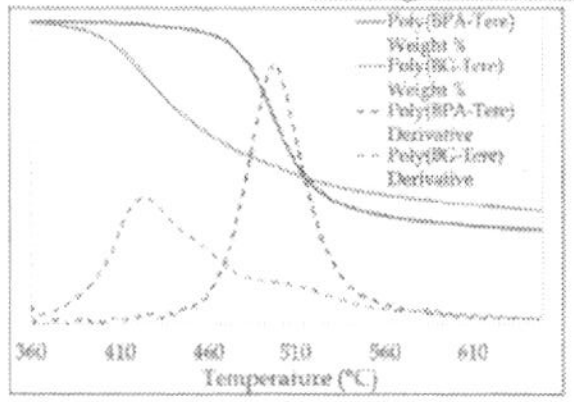

Polymer	T_{max} (°C)	T_{half} (°C)
Poly(BPA-Terephthalate)	498	513
Poly(BG-Terephthalate)	424	503
Poly(BPA-Succinate)	412	413
Poly(BG-Succinate)	396	397
Poly(BPA-Glutarate)	422	422
Poly(BG-Glutarate)	411	394
Poly(BPA-Adipate)	286	375
Poly(BG-Adipate)	367	361

- T_{max} is the temperature at which degradation is occurring at the fastest rate
- T_{half} is the temperature at which half of the sample has been degraded
- Conduced on Texas Instruments Q5500

Results

Polymer	M_n (g/mol)	Polydispersity	Soluble in THF
Poly(BPA-Terephthalate)	-	-	x
Poly(BG-Terephthalate)	38,000	2.16	✓
Poly(BPA-Succinate)	21,700	1.96	✓
Poly(BG-Succinate)	3,700	1.81	✓
Poly(BPA-Glutarate)	18,900	2.55	✓
Poly(BG-Glutarate)	1,800	2.16	✓
Poly(BPA-Adipate)	73,100	2.23	✓
Poly(BG-Adipate)	1,200	2.24	✓

- Based on polystyrene standards ranging from 742 g/mol to 2x10^6 g/mol
- Conducted on Waters 2695 Separations Module

Conclusions

- BG was successfully used as a replacement for BPA in the synthesis of polyesters
- BG polyesters had slightly lower glass transition temperatures and thermal degradation temperatures than BPA polyesters
- Glass transition temperature and thermal stability increases as aliphatic diacid chain length decreases
- BG polyesters display increased solubility in common solvents over BPA polyesters

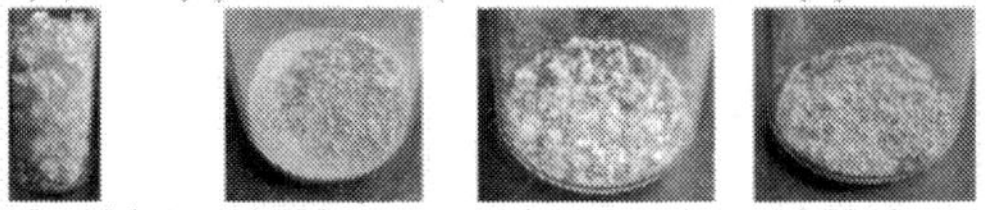

Poly(BG-Terephthalate) Poly(BG-Succinate) Poly(BG-Glutarate) Poly(BG-Adipate)

Future Work

- Conduct repeat analysis of GPC
- Investigate the affects of reaction stoichiometry on molecular weight
- Polymerize BG with 2,5-furandicarboxylic acid to form an aromatic, fully bio-based polyester
- Test degradability of polyesters

References

1. Mauck J.R., *Synthesis Of Highly Bio-based Epoxy Resins, Polycarbonates, Polyesters, And Novel Monomeric Materials Derived From Lignocellulosic Biomass.* (2016).
2. Hernandez, E.D., et al., *Synthesis and Characterization of Bio-based Epoxy Resins Derived from Vanillyl Alcohol.* ACS Sustainable Chemistry & Engineering, 2016.
3. Hong, H., et al., *Experimental Data Extraction and in Silico Prediction of the Estrogenic Activity of Renewable Replacements for Bisphenol A.* Int J EnvironRes Public Health, 2016. 13(7).

Acknowledgements

- Sustainable Materials Research Group
- Prof. Joseph F. Stanzione, III
- Dr. John J. La Scala
- Silvio Curia, Ph.D.

ARL · Rowan University · SERDP

510076 Adhesive-Free, Bonded Assembly of Epoxy Composite Structures Using Stoichiometric Offset Resins

Abstract:

Abstract

Airframe manufacturers increasingly rely on advanced composite materials to improve efficiency and safety while reducing manufacturing costs. The challenge is that the Federal Aviation Administration (FAA) requires mechanical fasteners or comparable arrest features in adhesively bonded joints to comply with Federal Aviation Regulations (FARs). Commercial aircraft commonly contain several miles of adhesively bonded joints and therefore thousands of redundant fasteners. Large-scale composite structures are assembled in one of three ways: secondary bonding, co-bonding, and co-curing. Co-bonded and co-cured products require the use of molds to support the preform, and often an autoclave which limits the size and complexity of the composite structure that can be fabricated in a single cure process. An adhesive-free technique, Co-cure-ply, is proposed to assemble large-scale composite structures without material discontinuities and uncertainties in the structural performance of bonded joints. The Co-cure-ply technique enables thermoset epoxy resins to be cured in a two-step process. The method involves two complimentary epoxy resin formulations: one designated as hardener rich (HR) with an excess of the hardener component and one designated as epoxy rich (ER) with an excess of the epoxide monomer. The method encompasses a stoichiometric offset approach, delaying the epoxy/diamine crosslinking reaction and eliminates material discontinuities by allowing resin interdiffusion across the joining interface during the assembly of a composite structure (analogous to what occurs between plies of conventional prepreg). Previous work investigated the degree of stoichiometric offset and various cure cycles to maximize mixing at the co-cured interface. In previous experiments, the fracture toughness was measured by double cantilever beam (DCB) test with values ranging from 250 to 650 J/m^2 which is commensurate with aerospace structural adhesive systems. This report focuses on preparing and testing resins with greater stoichiometric offset and developing additional mechanical testing methods to measure the shear and tensile strength in addition to the mode I fracture toughness of the Co-Cure-Ply fabricated composites. Characterization methods such as infrared spectroscopy, rheology and electron microscopy, will be performed to access the interface of co-cured specimens.

Copyright © American Institute of Chemical Engineers. All rights reserved.

510929 Using Monomer Chemistry to Predict If Dose Rate Effects Will Occur during EB-Polymerization

Abstract:

Through radiation polymerization the formation of polymer is initiated by exposing a batch of monomer to an initiation energy, like gamma radiation, light or charged particles, that drives the initiation and propagation of large chains composed of repeating monomer units. Electron-beam-irradiation (EB) has been shown as a promising method for radiation initiation of monomers that can be controlled by altering the samples operation voltage and belt speed, changing the processes dose rate. Dose rate effects (DREs) are variations in polymer properties (e.g., conversion, glass transition temperature, molecular weight, etc.) due to changes in the samples dose rate which can hinder EB polymerization growth in industrial settings. Because the dosage of a sample often changes during scaling this has led to unpredictable or varying properties DREs do not occur in all formulation chemistries or under all processing conditions, which makes predicting their occurrence challenging. This lack of understanding is especially problematic when predicting final polymer properties during industrial scale-up from a pilot line. Building off from previous work from affiliated graduates there are many factors still unexplored in predicting factors that cause dose rate effects and being able to predict those effects for materials at different operating conditions. This project explored and tested effects caused by monomer chemistry and compared DREs experienced by comparable monomers. Dynamic mechanical analysis was used to determine and compare the glass transition temperature of several acrylate monomers polymerized at different dose rates. Monomer chemistry was carefully chosen to determine the impact of size and number of abstractable hydrogens on the magnitude of the DREs. Results show that increasing the number of abstractable hydrogen plays a key role in reducing DREs. Results from this study will allow scientists to more easily identify formulation chemistries less prone to DREs, making industrial scale-up for electron beam polymerization more predictable and reliable.

Copyright © American Institute of Chemical Engineers. All rights reserved.

510711 Surface Control: High-x Block Copolymers for Microelectronics

Abstract:

Microelectronic device features continue to decrease in size, pushing the limits of photolithography and therefore requiring increasingly expensive multi-patterning techniques to achieve production of next generation devices. An alternative bottom-up approach to higher resolution patterning uses self-assembly of block co-polymers to increase the frequency of a pre-pattern image created with traditional photolithography. We have developed silicon containing block co-polymers as high-chi (χ) materials that allow production of smaller features. This work focuses on a block co-polymer of poly(4-pentamethyldisilyl styrene)-b-poly(2-vinylpyridine) (PDSS-P2VP) that was produced by anionic polymerization. This polymer has excellent etching selectivity under both oxidative and reductive reaction ion-etching conditions.[1,2] The χ value for PDSS-P2VP was estimated by measuring the order-disorder transition temperature of the polymer and found to be significantly higher than a block co-polymer that can create 10 nm patterns. Neutral top and bottom surfaces were identified by the island and hole technique previously reported, and the block co-polymer was thermally annealed between these layers.[3] Different processing conditions, such as O_2 plasma etch time, were adjusted to demonstrate production of oriented 10nm lamellae.

[1] Bates C. M., et al. *Science* (2012), 338(6108), 775.

[2] Azarnouchea, L., et al. *J. Vac. Sci. Technol. B* (2016) 34 (6), 061602/1-061602/10.

[3] Maher, M. J., et al. *Chemistry of Materials* (2014), **26(3)**, 1471-1479.

Copyright © American Institute of Chemical Engineers. All rights reserved.

509971 Recycling of Benzoxazine-Based Carbon Fiber Reinforced Polymers Via Catalytic Oxidation

Abstract:

Carbon fiber reinforced polymers (CFRP) are structural materials used in the automotive and aerospace industry due to their high specific properties; often outperforming their traditional metallic counterparts. However, there are no standard approaches for recycling end-of-life CFRP composite products or reusing scrap production waste. Current techniques involve pyrolysis or mechanical grinding of the CFRP, which lower the final thermomechanical properties of the recycled product. In this study, we propose a solvolytic approach to degrading a commercial blended benzoxazine/epoxy resin under mild conditions. The thermosetting resin is polymerized, characterized, and then catalytically de-polymerized via hydride abstraction with a ruthenium catalyst. These results pave the way for sustainable, low-cost recycling of CFRP composites.

Copyright © American Institute of Chemical Engineers. All rights reserved.

510976 Use of Bioadvantaged Materials As Replacements of Petroleum Based Materials

Abstract:

Until now, the use of renewable or "green" materials around the world has been limited due to their higher production costs when compared to petroleum derived materials. However, due to the price volatility of petrochemical derivatives and to an ever-increasing interest of consumers to become ecologically responsible, there has been a growing demand for the use of environmentally friendly materials produced from biological sources, other than petroleum. When these materials are polymerized or modified, the products can possess similar or better characteristics to their petroleum-based counterparts. Of these, those that ultimately reach the marketplace must offer an "advantage", some feature that presents a value opportunity that would be otherwise missed using the petrochemical alternative. Our group focuses, primarily, on using low value feedstocks, e.g. soybean oil, and transforming them into higher value bioadvantaged materials that are to be used in different applications such as pressure sensitive adhesives, paints and coatings, asphalt modification, etc. Our group developed a method to synthetize thermoplastic elastomers using acrylated epoxidized soybean oil (AESO), a bioadvantaged replacement of butadiene, both alone and with styrene with the use of controlled radical polymerization techniques.

Copyright © American Institute of Chemical Engineers. All rights reserved.

510354 Synthesis of Fe/C Composite Materials from Spent Coffee Grinds and Ferric Ammonium Citrate

Abstract:

Coffee is the worldâ€™s second largest traded commodity, and its consumption inevitably results in waste, including spent coffee grinds (SCG). In this work, SGC is considered as a precursor for activated carbon (AC) production, with simultaneous Fe impregnation via ferric ammonium citrate (FAC). Three Fe:C mole ratios were considered (1:10, 3:10 and 5:10). SCG carbonization occurred in 0.5 liters per minute (LPM) N_2 at up to 600 °C for 1 h. Physical activation of the recovered char occurred in 0.5 LPM CO_2 at 800 °C for 30 min with a ramp rate of 10 °C/min. Products were characterized using scanning electron microscopy with energy dispersive x-ray spectroscopy, x-ray diffraction, pH of the point of zero charge, thermogravimetric analysis, bulk elemental analysis, and N_2 adsorption. Prepared carbon support were porous, with BET surface areas as high as 180 m^2/g. Bulk iron loadings reached 2.4 wt%, with partially reduced Fe_3O_4 as the primary impregnated iron species. High nitrogen content in SCG and FAC jointly contribute to nitrogen-doping of Fe/C products. Performance testing for environmental applications was completed using volatile organic compounds and hexavalent chromium. Achieved contaminant removals are attributed to adsorptive and/or reductive properties of Fe/C materials. In sum, this work utilizes a plentiful industrial waste to produce tailored Fe impregnated ACs that exhibit promising performance for gaseous and aqueous pollutant removal.

Copyright © American Institute of Chemical Engineers. All rights reserved.

510761 Novel Solid-State Processing of Polymer/Carbon Nanotube Composites for Effective Removal of Contaminants from Water

Abstract:

Many areas of the world today lack access to clean water and in the upcoming years we face growing water stresses due to an increase of micropollutants in water sources. Recent studies demonstrate the ability of carbon nanotubes (CNTs) to more effectively remove a wide-range of contaminants from water compared to the activated carbon. However, â€œbundlingâ€, or clumping, of individual tubes observed in commercial CNTs hinders their contaminant removal efficiency due to limited adsorption-ready sites. By processing the carbon nanotubes with a polymer matrix using solid-state shear pulverization, effective debundling of CNTs can occur, resulting in composite materials that have an enhanced ability to remove target contaminants. While maintaining the same ratio of mass of carbon nanotubes to volume of contaminated solution, the removal efficiency of the polymer-carbon nanotube and the unprocessed carbon nanotubes were compared. The debundling effect was characterized by BET surface area measurement, while the contaminant removal was quantitatively monitored using UV-Vis spectroscopy.

Copyright © American Institute of Chemical Engineers. All rights reserved.

510118 Mechanical Properties of Dual-Electrospun Polymeric Composites

Abstract:

Student: Kaitlyn Whipple

Supervisors: Matthew Ali, Prof. Patrick T. Mather

Supported by STEM Scholars Program and National Science Foundation (NSF)

Polymer composites combine two or more polymers in a single material, that can achieve physical properties which are intermediate to those of the two individual components. Very recently, a new approach has been introduced on the preparation of polymer blended composites using dual-jet electrospinning, followed by hot compaction in a press.* Dual-electrospun blends of poly(e-caprolactone) (PCL) and polylactic acid (PLA) composites demonstrated exceptionally high tensile strength and strain-to-failure (yield strength approaching 100 MPa and 40% strain-to-failure). To investigate potential toughening of PLA, the present work examined dual-electrospinning of PLA with the thermoplastic elastomer, Pellethane 5863-80A®. Furthermore, scanning electron microscopy (SEM) and differential scanning calorimetry (DSC) were used to analyze the morphology and crystallinity of both electrospun composites, respectively. We observed a strong positive relationship between increased pressure, crystallinity, and tensile strength in the PCL-PLA composites. The dependence of thermoplastic elastomer incorporation on mechanical properties â€“ particularly toughness â€“ will be reported for comparison.

*Tumbic, J., Romo-Uribe, A., Boden, M., & Mather, P. T. (2016). Hot-compacted interwoven webs of biodegradable polymers. *Polymer, 101*, 127-138. doi:10.1016/j.polymer.2016.08.057

Copyright © American Institute of Chemical Engineers. All rights reserved.

510995 Challenging the Autoinjector: Investigating Mechanical Degradation of High-Molecular-Weight Polymers in a Contraction Flow

Abstract:

The addition of high-molecular-weight polymer to solvent, even in dilute amounts, can significantly change the response of a fluid to an extensional flow. This behavior is important when formulating models and assessing the ability to print, spray, jet, and spin complex fluids. Generally, macromolecular solutions exhibit large resistance to stretching deformations known as extensional viscosity.

Molecular degradation can make additives useless in liquid solutions and change the characteristics of the fluid itself. Degradation can occur when non-Newtonian fluids flow through regions of high extension, such as a contraction. As a fluid flows through a radially decreasing geometry, the molecules in the solution can be mechanically broken.

A common contraction is drug delivery via an injector through a needle. The recipient will request that the injection would be done as quick as possible and with the smallest needle possible. This request translates to higher shear rates and greater changes in radii that are experienced by the solution, all of which favors molecular degradation. To better understand how pharmaceuticals should be delivered, specifically using an autoinjector, characterizing the viscoelastic behavior provides detailed information on how these fluids should be handled.

We perform a variety of molecular and rheological characterization on model polymer solutions before and after flowing them through a contraction flow. The pressure drop across the contraction is also measured for comparison to theoretical and empirical models.

Copyright © American Institute of Chemical Engineers. All rights reserved.

509989 Development of Model Elastomeric Photomechanical Systems and Their Characterization

Abstract:

Photomechanical liquid crystalline (LC) polymers have blossomed into an exciting field of innovation in recent years due to their potential for large, wireless actuation. It has been demonstrated that photochromic monomers can be integrated into a variety of these systems in a programmed way that allows for control over the material response. Many research groups have demonstrated significant photomechanical deformation of azobenzene functionalized LC polymers. However, these materials often require complicated syntheses that are difficult to scale and normally only produce movement in thin samples, restricting their usefulness for mechanical actuation. Another evident drawback to current systems is the almost exclusive use of azobenzene derivatives that absorb light primarily in the ultraviolet to blue range. At these wavelengths azobenzene experiences competitive absorption with the surrounding organic polymer network. In addition UV light has adverse effects for living tissue, and causes photo decay in polymers over time. Recently, a thiol acrylate Michael addition reaction was introduced as a simple and scalable procedure for producing liquid crystal elastomers (LCEs). The goal of the present work is to functionalize these LCEs with red to NIR absorbing dyes or dopants. To build up capability with this material, commercial azobenzene monomers and gold nanoparticles were incorporated into the LCE. Samples were characterized through photomechanical cantilever bending tests and via dynamic mechanical analysis. Future work will involve characterization of LCEs that incorporate a red-NIR azobenzene derivative that is currently under development. If successful, these light responsive systems hold great potential for utilization in biomedical systems and for the development of morphing surfaces, artificial muscle actuators, and microvalves.

Copyright © American Institute of Chemical Engineers. All rights reserved.

510017 Characterization of UV-Cured Polymers Containing Modified Soybean Oil for Adhesion Applications

Abstract:

The adhesive properties and applications of a cured co-monomer mixture containing acrylated epoxidized soybean oil (AESO) were studied. In previous research, the film applications of the co-monomer mixtures was investigated. From this research it was gathered that they had a strong adhesion to glass. The co-monomers used in this study included styrene, polyester acrylate oligomer (CN2303), and tetrahydrofurfuryl acrylate (THFA). These co-monomers were mixed with AESO in different ratios. A photoinitiator, Irgacure 2022, was also added to the AESO/co-monomer mixture. The mixture was injected onto a glass slide, which was then pressed to another glass slide. The glass slide samples were cured under UV light for four, six, eight, or ten minutes. The samples underwent tensile testing to determine the breaking point of the adhesive. For ease of testing, each glass slide was fitted with a plastic piece to increase the friction of the hold of the tensile testing machine. The bio-based adhesive showed positive results that gave high levels of force required to break the overlapping point. Additionally, in some cases the glass slide broke before the adhesive failed.

Copyright © American Institute of Chemical Engineers. All rights reserved.

510140 Curing Epoxy with a Variety of Hardeners: Reaction Kinetics and Mechanical Characteristics

Abstract:

828 EPON Resin is mixed with varying ratios of the amine-based hardeners diethannolamine (DEA) and dibutylamine (DBA). Changing the ratio of hardeners changes the amount of hydroxyl groups available during the curing reaction. Cure kinetics at a variety of cure temperatures are measured using a TAM Air isothermal calorimeter. After curing, various material properties are measured. Yield stress and young's modulus are found through compression tests in an Instron, while the glass transition temperature, Tg, is found using a DSC Q2000.

Sandia National Laboratories is a multimission laboratory managed and operated by National Technology and Engineering Solutions of Sandia LLC, a wholly owned subsidiary of Honeywell International Inc. for the U.S. Department of Energyâ€™s National Nuclear Security Administration under contract DE-NA0003525.

Copyright © American Institute of Chemical Engineers. All rights reserved.

511036 Characteristics of Adhesion in 3D Printed Composites

Abstract:

Development of fused filament fabrication (FFF) printers has introduced the possibility of printing multiple plastics during a single print. In addition, new thermoplastic polymers are being introduced to filament printing. At this point, there has been very little research on the adhesion between these new materials and materials that have been used extensively for 3D prints.

Thermoplastic polyurethane (TPU) is a novel material that has recently been profiled to print through 3D printers. This plastic is classified as an elastomer, giving it characteristics to flex and bend. Using this plastic in combination with a hard plastic such as Acrylonitrile butadiene styrene (ABS) or Acrylonitrile Styrene Acrylate (ASA), composite parts can now be printed with soft-to-hard gradients. To understand the strength of adhesion between materials, ASTM D1002 lap shear tests were conducted, with ABS or ASA as the substrate (bar) and TPU acting as the adhesive (joint). These lap shear bars were printed using 3ntrâ€™s A4v3 3D printer, a FFF printer introduced to the market as a mid-ranger price point printer with print qualities of higher priced industrial grade printers. The 3ntr printer has three extruder nozzles, giving the ability to print with multiple materials on the same print.

During the evaluation of lap shear experiments, it was discovered that the adhesion between plastics differs on the configuration of the print. It was found that TPU had better adhesion to the bottom substrate during print and delamination occurred strictly at the joint at the top substrate. The ultimate goals of the project are to determine the adhesion between TPU and the hard plastics and how the order the plastics are printed (TPU onto cool hard plastic or hot hard plastic onto cool TPU) effect the adhesion and mechanical properties.

Data on Bond Strength vs. plastic combination will be presented for plastics TPU, ABS, and ASA at print profile temperatures T_{TPU} = 218°C, T_{ASA} = 240°C, and T_{ABS} = 240Â°C. This bond strength is considered the minimum bond strength between plastics printed, as the ability to delaminate the TPU from the ABS or ASA is restricted to only one side. Thermal characterization of plastics will also be presented (Glass Transition Temperatures (T_g), Melt Temperatures (T_M) and Crystallization Temperatures (T_{xtal})), as well as the viscosity (Î¼) of these plastics at temperatures within Â±10Â°C of the print temperatures.

Copyright © American Institute of Chemical Engineers. All rights reserved.

509764 Impact of Concentration and Temperature on the Crosslinking of Butyl Rubber Using Triethylenediamine

Abstract:

A viscoelasticrubber must be crosslinked into a network. This removes the translational mobility of the chains but not the segmental mobility and result in an elastic solid. Crosslinking of butyl rubber, which is a copolymer composed mostly of isobutylene and a small fraction of isoprene, is difficult due to the low number of unsaturated bonds. Traditionally, the rubber is crosslinked via vulcanization, a process that requires high temperatures and a curative package composed of several chemical reagents. In this work, an ester-functionalized butyl rubber was examined formed by grafting pentafluoro propiolate onto the butyl rubber chain. This group is electron-withdrawing and made the unsaturated bonds more susceptible to crosslinking. In addition, two geminal C=C bonds were introduced to further enhance the reactivity. A base, triethylenediamine (TED), was used to crosslink the butyl rubber. The present crosslinking reaction can be carried out at temperature much lower than conventional vulcanization temperatures. Variation of the concentration of triethlyenediamine was found to affect the molecular weight between crosslinks. The crosslinking density initially increases with increasing base concentration and then decreases beyond a threshold value. This is believed to be a result of a side reaction becoming increasingly important as TED concentration is increased. As expected, crosslinking occurred faster at higher temperatures. The tensile strength of the rubber increased with increasing TED concentration to the same threshold value previously discussed then began to decrease.

Copyright © American Institute of Chemical Engineers. All rights reserved.

511041 Glass Transition Temperature Broadening of Syrene/n-Butyl Acrylate Gradient Copolymer and Polyphenylene Oxide Blends

Abstract:

Over the past several decades, there has been great interest in polymeric materials that exhibit damping properties to minimize sound and vibrations. The acoustic properties of polymers are closely coupled with its glass transition, which is the transition from a glassy state to a rubbery state that occurs over a temperature regime specific to that polymer. A broad glass transition is a highly desirable feature, because the temperature range over which the material exhibits good damping characteristics is also broadened. My research focuses on broadening the glass transition behavior of a styrene-based polymer blend. I hypothesized that unusually broad glass transitions would be observed when synthesizing styrene/*n*-butyl acrylate gradient copolymers to lower the onset of the glass transition regime, and then blending the copolymers with polyphenylene oxide (PPO) to raise the endset of the glass transition regime. I expect the glass transition to broaden with increasing amounts of *n*-butyl acrylate and PPO in the polymer blend, until a limit of miscibility after which the components of the polymer blend will phase separate. Experiments have shown good miscibility and glass transition breadths already exceeding breadths previously reported for blends of polystyrene homopolymer and PPO, indicating the great potential of these novel materials for damping applications.

Copyright © American Institute of Chemical Engineers. All rights reserved.

510511 Chemically Recyclable Polyurethane Foams from Sustainable Polymer

Abstract:

Polyurethanes (PUs), a versatile class of polymers, commonly appear as rigid or flexible foams for widespread applications in transportation, automotive, and consumer industries. The attributes that make PU foams desirable, namely their low density and durability, pose inherent issues in their end-of-life processing. In order to mitigate the widespread practice of incinerating or landfilling these thermoset foams, we present the synthesis of a sustainable PU foam that can be chemically recycled. Poly(-methyl--caprolactone)-based foams can undergo a straightforward depolymerization process to afford the starting monomer in high purity and yield. We also demonstrate that these foams can be synthesized to exhibit a range of mechanical properties. By adjusting the isocyanate index and surfactants in the formulations, we create an array of foams with different compressive strengths, cell morphologies, and densities. PU foams from poly(-methyl--caprolactone) thus present a robust and sustainable alternative to traditional petroleum-based PUs.

Copyright © American Institute of Chemical Engineers. All rights reserved.

Rowan University

Synthesis & Characterization of Molecularly Hybrid Bisphenols Derived from Lignin & Cashew Nutshell Liquid: Application in Thermosetting Resins

Amy E. Honnig[1], Kayla R. Sweet[1], Claire M. Breyta[1], Julia H. Reilly[1], Alexander W. Bassett[1], John J. La Scala[2], Joseph F. Stanzione, III[1]
[1]Department of Chemical Engineering, Rowan University, Glassboro, NJ 08028
[2]U.S. Army Research Laboratory, Aberdeen Proving Ground, MD 21005, United States

ARL

Introduction

- Commercial vinyl ester (VE) resins contain bisphenol A (BPA) derivatives, a known human endocrine disruptor and a petrochemical feedstock derivative[2]
- Lignin: a major byproduct of the pulp and paper industry with a high aromatic content[5]
- Vanillyl Alcohol: reduced form of vanillin, a derivative of lignin[3]
  - A viable platform chemical for the production of bisphenols[3]
- Cardanol: phenolic lipid from cashew nutshell liquid (CNSL) with a long C15 alkyl chain that naturally can be saturated or unsaturated[4]
  - May impart hydrophobicity, thermal resistance and mechanical toughness to overall polymer networks[4]
- Coupling vanillyl alcohol and cardanol has created a promising bisphenol for use in VE resin formulation

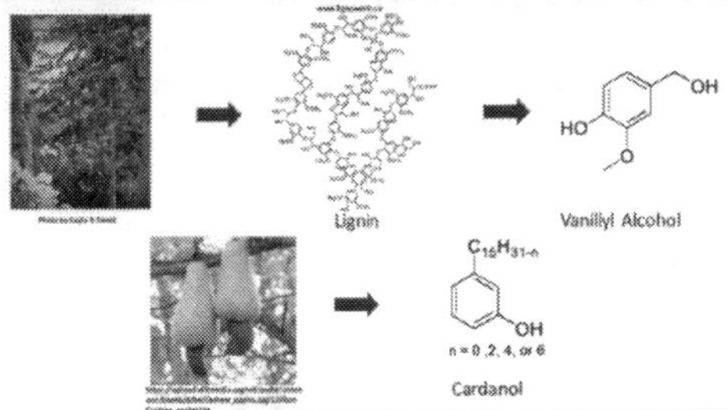

Synthesis of Bisphenols & VE Monomers

Synthesis of Vanillyl Alcohol Cardanol (VAC) and Vanillyl Alcohol Hydrogenated Cardanol (VAHC)

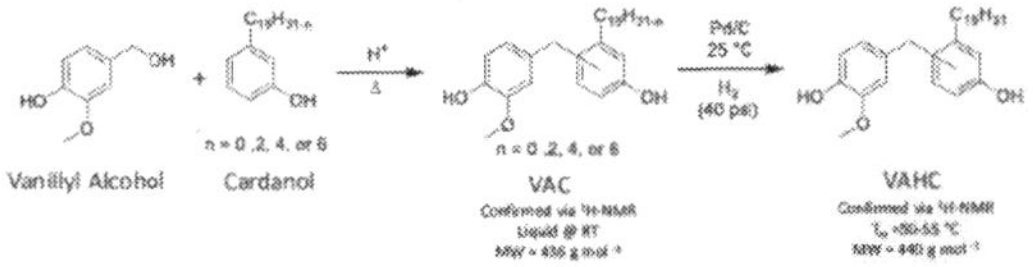

Synthesis of VAC Dimethacrylate (VACDM) & VAHC Dimethacrylate (VAHCDM)

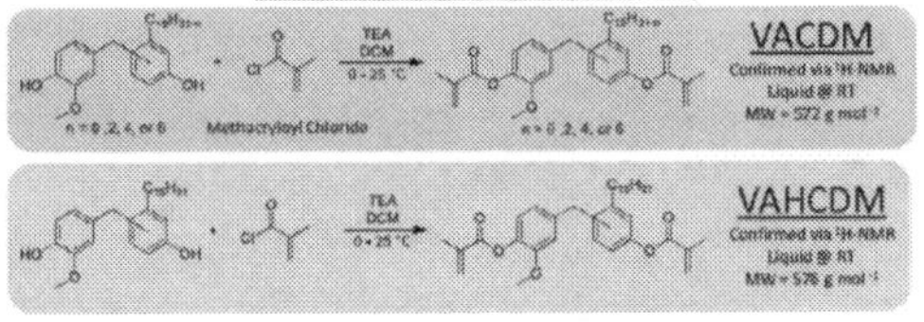

Synthesis of Vinyl Ester 828 (VE828)[2]

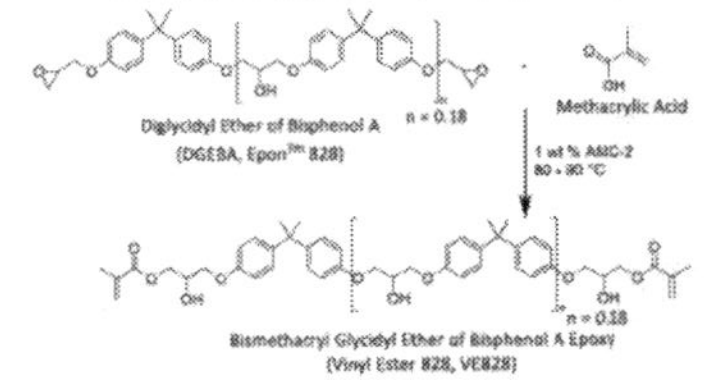

Polymerization and Cure

Cure Procedure

- Synthesized VE resins blended with 35 wt% Styrene
- 1.5 wt % Trigonox® 239 added to blends
- Cure Cycle: in an argon atmosphere, 90 °C for four hours, 160 °C for two hours
- Post cure samples cut into 1.5 mm and 2.5 mm thick samples to run DMA

$$\text{Extent of Cure } (\%) = \frac{\left(\frac{ABS_{reactive}}{ABS_{ref}}\right)_{uncured} - \left(\frac{ABS_{reactive}}{ABS_{ref}}\right)_{cured}}{\left(\frac{ABS_{reactive}}{ABS_{ref}}\right)_{uncured}} \times 100$$

System	Extent of Cure %
VE828 35wt% Styrene	99.4 ± 0.3
VACDM 35 wt% Styrene	97.8 ± 0.2
VAHCDM 35 wt% Styrene	98.7 ± 0.2

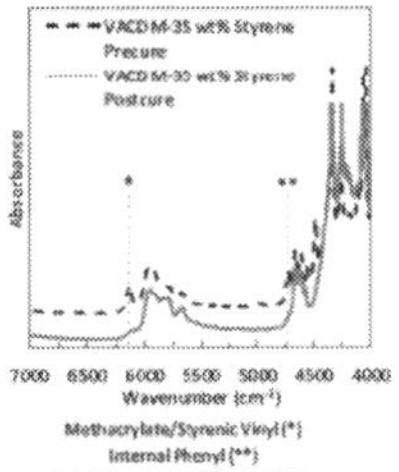

Methacrylate/Styrenic Vinyl (*)
Internal Phenyl (**)
Spectra Vertically Offset for Clarity

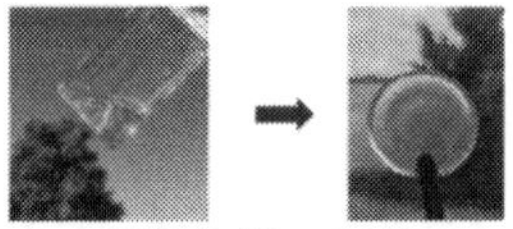

VAHCDM 35 wt% Styrene Resin Cure

References

1. Ha, S., et al, ChemSusChem. 2010, 3, 1078-1084.
2. La Scala, J. J.; Orlicki, J. A.; Winston, C.; Robinette, E. J.; Sands, J. M.; Palmese, G. R., The use of bimodal blends of vinyl ester monomers to improve resin processing and toughen polymer properties. Polymer 2005, 46 (9), 2908-2921.
3. Hernandez, E. D.; Bassett, A. W.; Sadler, J. M.; La Scala, J. J.; Stanzione, J. F., Synthesis and Characterization of Bio-based Epoxy Resins Derived from Vanillyl Alcohol. ACS Sustainable Chemistry & Engineering 2016, 4 (8), 4328-4339.
4. Voirin, C.; Caillol, S.; Sadavarte, N. V.; Tawade, B. V.; Boutevin, B.; Wadgaonkar, P. P., Functionalization of cardanol: towards biobased polymers and additives. Polym Chem-Uk 2014, 5 (9), 3142-3162.
5. Calvo-Flores, F. G.; Dobado, J. A., Lignin as Renewable Raw Material. ChemSusChem 2010, 3 (11), 1227-1235.
6. McAninch, I. M.; Palmese, G. R.; Lenhart, J. L.; La Scala, J. J., DMA Testing of Epoxy Resins: The Importance of Dimensions. Polymer Engineering and Science 2015, 55 (12), 2761-2774.

Acknowledgements

- Stanzione Sustainable Materials Research Group
- Rowan University Department of Chemical Engineering
- Financial Support of Army Research Laboratory through W911NF-14-2-0086 & W911NF 16-2-0225
- Drexel University Department of Chemical and Biological Engineering
- PPG

Thermal Analysis & Viscoelastic Properties

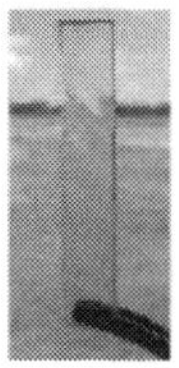

Cured VACDM 35 wt% Styrene

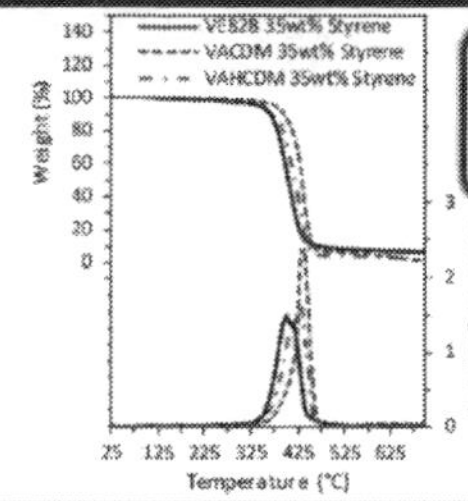

TA Instruments Discovery TGA 550
- 10 mg sample
- Platinum pan
- N_2 atmosphere
  - 40 mL min⁻¹ balance gas flow rate
  - 25 mL min⁻¹ sample gas flow rate
- Heating ramp rate = 10 °C min⁻¹ to 700 °C

	IDT (°C)	$T_{50\%}$ (°C)	T_{max} (°C)	Char Content (%)
VE828 35wt% Styrene	344 ± 6	403 ± 3	401 ± 3	1.0 ± 0.4
VACDM 35 wt% Styrene	374 ± 2	430 ± 3	432 ± 5	2.3 ± 0.8
VAHCDM 35 wt% Styrene	347 ± 3	419 ± 1	429 ± 3	1.3 ± 0.5

TA Instruments Q800 DMA
- Single Cantilever
- Frequency = 1.0 Hz
- Deflection amplitude of oscillation = 7.5 µm
- Poisson's ratio = 0.35
- Heating ramp rate = 2 °C min⁻¹ from -50 °C to 250 °C

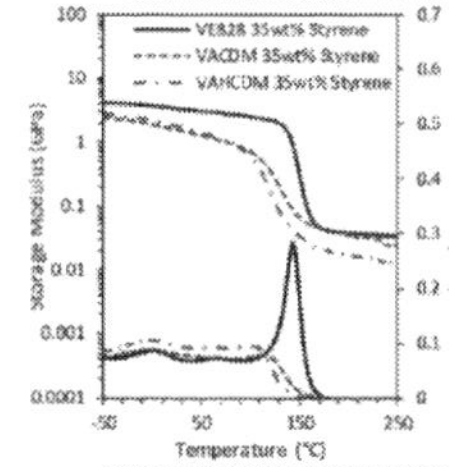

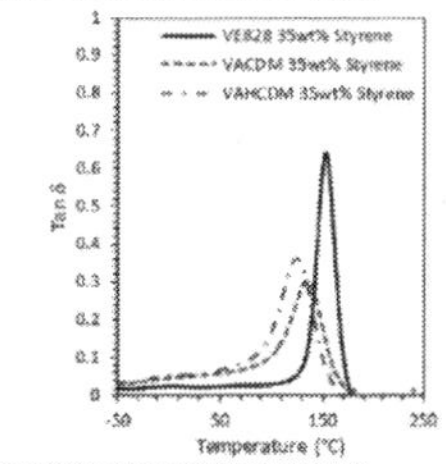

System	E' @ 25 °C Gpa	E' (T) Inflection (°C)	Peak of tan δ (°C)	ρ @ 25 °C (g cm⁻³)
VE828 35wt% Styrene	2.8 ± 0.3	140.9 ± 2.4	152.9 ± 2.0	1.157 ± 0.005
VACDM 35 wt% Styrene	1.7 ± 0.2	118.4 ± 1.3	136.9 ± 1.9	1.083 ± 0.003
VAHCDM 35 wt% Styrene	1.7 ± 0.1	107.2 ± 0.8	127.7 ± 2.0	1.070 ± 0.001

Conclusions and Future Work

Conclusions
- Coupled vanillyl alcohol and cardanol show potential to mitigate the use of BPA based VE resins with some compromise in properties
- Thermal stability and T_g are greater with saturated chain than unsaturated

Future Work
- Evaluate rheology, fracture toughness, cure kinetics, contact angle, and toxicity
  - VACDM and VAHCDM are qualitatively lower viscosity liquids than VE828

509874 Modeling the Effect of Core-Shell Structure on Upconverting Nanoparticle Emission Intensity

Abstract:

The phenomenon of upconversion is useful in many applications, such as biological imaging, due to its ability to produce light with greater energy than that of the incident photons. In order to design nanoparticles with maximum emission intensity, it would be useful to create accurate models of the upconversion phenomenon in a nanoparticle and run simulations while varying important reaction parameters. Prior models have been used to describe upconversion, but these models have focused on quenching of excited states at the surface. These models do not consider the migration of energy from the center of the nanoparticle to the surface, which significantly reduces the emission intensity from the nanoparticle. In this project, a reaction-diffusion mathematical model was used to model and simulate the photophysical processes that give rise to upconversion. The process of upconversion was represented by a "reaction", while the process of energy migration in the nanoparticle was modeled as "diffusion". MATLAB was used to simulate the model under varying parameters, using finite-element analysis to solve the partial differential equations. First, the accuracy of the model was tested using a simpler two-state ytterbium model with instantaneous surface reactions. The results obtained from the model were verified analytically. Next, these simple cases were simulated to determine the effect of surface reactions (which represent surface quenching of upconversion emission) on the concentration of excited states in the nanoparticle. Finally, a more complicated scenario (involving series reactions, which is more closely related to upconversion) was simulated with this model, using both a core model and active core-inactive shell model to determine whether the latter provided an enhancement in upconversion emission intensity. The simulation results demonstrate that the core-shell model provides a greater average concentration of the doubly excited state of Er^{3+} within the nanoparticle, which would result in a significant enhancement in the upconversion emission intensity from the nanoparticle.

Copyright © American Institute of Chemical Engineers. All rights reserved.

509905 Detection of Lead Contamination in Water Using Fluorescence of Functionalized Gold Nanoparticles

Abstract:

The goal of this research was to determine whether functionalized gold nanoparticles can be useful as a sensitive residential fluorescent sensor for lead contamination in drinking water. In the future, such nanoparticles could be dispersed into a membrane visible through a window inside a household tap filter. A large body of work exists in the literature concerning use of gold nanoparticles to detect lead. Almost all of these depend on a visible color shift, which is less sensitive compared to fluorescence and can confuse the consumer. The 5 nm diameter gold nanoparticles (GNPs) were functionalized with 11-mercaptoundecanoic (MUA) acid. Controls and MUA-GNPs were exposed to lead ion/water concentrations as low as the EPA drinking water limit, 15 ppb. The fluorescent response was tested using excitation/emission matrix spectroscopy (EEMS). Corrections were performed on the data for inner filtering and Raman and Rayleigh scattering of water. The GNPs were shown to detect lead at varying concentrations, enhancing the fluorescent intensities of fluorophores at some concentrations and quenching at others. This result is possible if the lead ion was changing the state of aggregation of the colloidal solution in a concentration-dependent manner for the ranges tested and concentration therefore affected the distance at which the ion was adsorbed/spaced from the GNP surface. Dynamic light scattering results will also be presented

Copyright © American Institute of Chemical Engineers. All rights reserved.

510700 Colloidal Crystal Growth on Chirped Grating Surfaces

Abstract:

In this work, we report the lattice structures of monodisperse colloidal crystals grown on a chirped grating surface via confined convective self-assembly. The grating was fabricated by using a laser interference lithography technique. The periodicity of the substrate ranges from 660 nm to 11,000 nm. Employing a chirped grating allows for analysis of a large periodicity range on a single substrate. When the grating periodicity is much larger than the diameter of the colloidal particle, the grating simply served as a compartment for the particles which resulted in formation of chain-like structures. Conversely, when the periodicity of the grating is commensurate to the diameter of the particles, the gratings influenced the crystal lattice structure formed. The crystal lattice structure transitioned from a square lattice to a triangular lattice through a rhombic intermediate structure. The crystal structures were optically characterized by laser diffraction with a corresponding SEM image.We have also shown that the monolayer crystal lattices can serve as templates for growth of unique structures such as rectangular lattice, chains of particle pairs or triplets, graphite-structured lattice, etc.

Copyright © American Institute of Chemical Engineers. All rights reserved.

510035 Battering Bacteria Against a Barrier

Abstract:

The introduction of active matter into a colloidal system leads to ill-defined parameters in the behavior of formed colloidal solids. In order to better understand the influence of active matter in industrial and medical applications we form flow-stabilized colloidal solids (FSS) with both active and passive particles. We hypothesize that these FSS will illuminate the influence of particle shape and motility of active matter on the dynamics. Three strains of bacteria, motile and non-motile *Escherichia coli* and *Staphylococcus epidermidis*, were chosen to isolate the influence of activity from that of particle shape. In order to form FSS, a bacterial suspension flows through a quasi-2D microchannel. The bacteria encounters a barrier perpendicular to the direction of fluid flow where shear and compressive forces cause pile formation. These bacteria are suspended in a nutrient broth to sustain the bacteria with an added polymer, polyvinylpyrrolidone, which reduces adhesion to the barrier, the walls of the channel, and other bacteria. It has been previously shown in colloidal experiments that above a critical flow rate triangular, solid-like piles form on the barrier. When using non-motile *E. coli* we observe formation of piles. We have observed that the motile bacteria avoid the barrier thus preventing pile formation; illustrating the influence of motility on pile formation.

Copyright © American Institute of Chemical Engineers. All rights reserved.

510701 Thermodynamic Analysis of Hexagonal Boron Nitride Crystal Growth from Metal Solutions

Abstract:

Hexagonal boron nitride (hBN) has many unique properties such as a strong interaction with thermal neutrons, a wide energy bandgap (>6.0 eV), and a graphite-like structure. These properties make it a good candidate for compact, efficient, and low-cost neutron detectors, deep UV light emitting diodes, and in graphene-based electronics as a substrate and dielectric. Thus, producing high quality hBN single crystals would have major implications for national defense (detection of nuclear weapons by neutron detectors), health (sterilization of water by radiation from deep UV LEDs), and computing (high-speed graphene electronics).

Currently, we are producing hBN single crystals by precipitation from a molten nickel-chromium solvent at temperatures between 1450 °C and 1550 °C under 1 bar of flowing nitrogen. To date, we have grown single crystals up to 3mm in diameter and hundreds of microns thick. However, the residual impurity concentrations in these crystals is still too high for many electronic and optoelectronic device applications. Specifically, the carbon and oxygen concentrations are typically on the order of 10^{19} cm^{-3} and 10^{20} cm^{-3} respectively. To solve this, the chemical reactions involving carbon and oxygen (the main impurities) and simultaneous phase equilibrium were analyzed using thermodynamics to determine how process conditions may be adjusted to reduce the impurity concentrations. Aside from all of the elements required for hBN crystal growth (B, N, Ni and Cr), hydrogen was also included in the study to test its potential as a reducing agent. Ideally, hydrogen will react with metal oxides and carbides to form volatile oxygen-containing or carbon-containing species, which would be removed in the system exhaust. For this analysis, chemical equilibrium constants were calculated from Gibbâ€™s energy of formation data available from the JANAF Thermochemical Tables. All relevant reactions were coupled, and the equilibrium condition was solved for through an iterative method based on extents of reactions. The vapor phase was treated as an ideal gas and activity coefficients for components in the liquid phase were calculated from excess Gibbâ€™s Energy expressions which were determined via the Calphad method.

We predict that, if the temperature is kept above 1400Â°C and the ambient partial pressure of water vapor in the system is kept below ~10^{-6} bar, all of the oxides may be reduced by hydrogen. The maximum partial pressure of water vapor increases with temperature to a value of ~10^{-5} bar at 1700Â°C. This range of conditions is practically feasible, thus adding hydrogen is a viable option for the removal of oxygen. Boron oxide is the most stable oxide (as compared to the oxides of nickel and chromium), so will be the most difficult to reduce with hydrogen. However, under normal process conditions, B_2O_3 will readily evaporate, providing an alternative mechanism for the removal of oxygen. The reaction of hydrogen with carbon to form volatile methane would provide a mechanism for the removal of carbon, but is thermodynamically unfavorable, and is thus unlikely to succeed. Other potential reactions for the removal of carbon are being considered, but have not been thoroughly studied.

Copyright © American Institute of Chemical Engineers. All rights reserved.

VIRGINIA COMMONWEALTH UNIVERSITY

Fabrication of of Gold-Loaded Nanoreactors by Polymer Directed Self Assembly

Tien Vuong, Andrew Harrison, and Christina Tang
Department of Chemical and Life Sciences Engineering
Virginia Commonwealth University, Richmond, VA 23284-3028

VCU School of Engineering

Motivation

Liquid phase chemical processing is commonly used for production of fine chemicals and pharmaceuticals. Large amounts of organic solvents are required generating large amounts of hazardous waste. Multifunctional polymer nanoreactors containing gold nanoparticle catalysts that combine reaction and separation could improve efficiency and minimize hazardous waste.

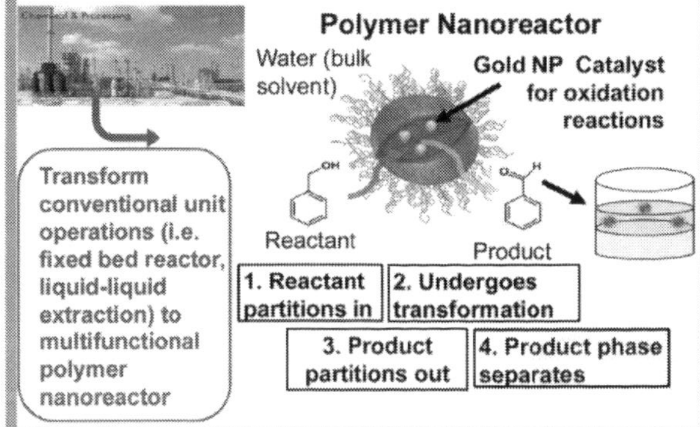

Project Goals

This focus of this project is fabrication of gold-loaded nanoreactors. Specifically, we aim to:

- Tune nanoreactor size between 50 and 300 nm
- Tune size of the nanoreactor and the catalyst loading (# of gold nanoparticles per nanoreactor) independently
- Vary core properties: i.e. solid core (e.g. polystyrene) or liquid core (e.g. trioctylamine)

Approach

Nanoreactor Self-Assembly

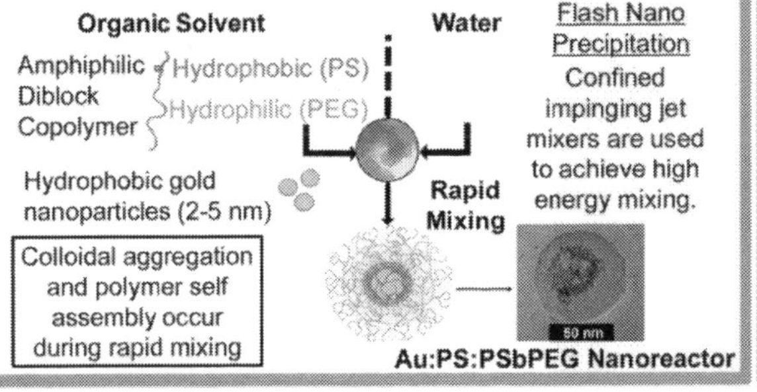

Results

Tuning Nanoreactor Properties

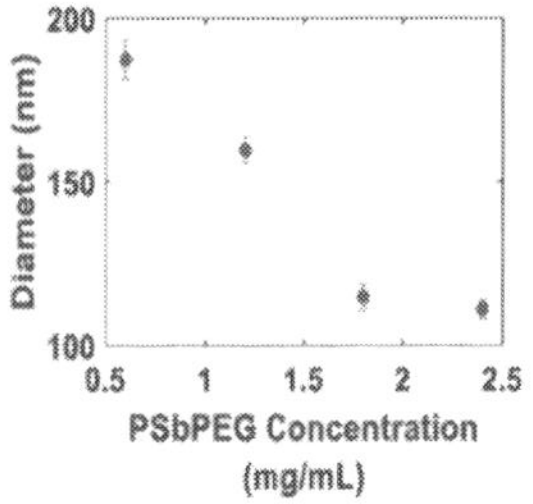

Effect of block copolymer concentration

Increasing the copolymer concentration (PSbPEG):

- Decreases the size of the nanorectors
- Decreases the catalyst loading

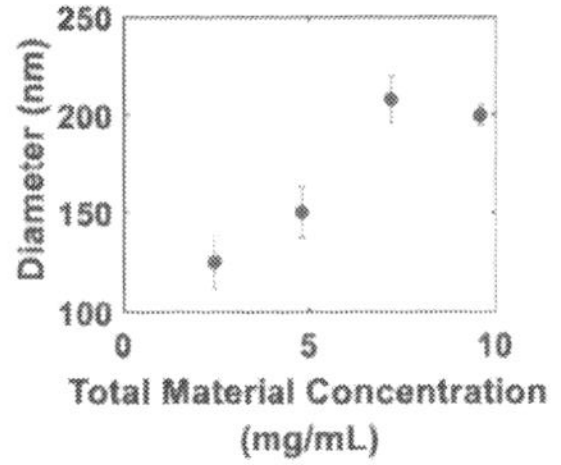

Effect of the total solids concentration

Increasing the total material concentration :

- Increases the nanoreactor size without affecting catalyst loading

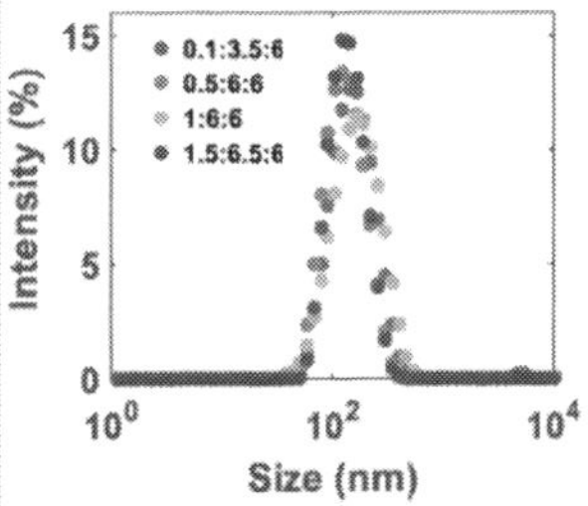

Effect of core material concentration

Increasing the ratio of gold nanoparticles to co-precipitant:

- Increases the catalyst loading (nominal) without significantly affecting the nanoreactor size

- Block copolymer concentration affects both the size of AuNRs and the catalyst loading.
- The size of AuNRs can be tuned independent of catalyst loading by changing the total material concentration.
- Gold loading (nominal) can be tuned independent of nanoreactor size by changing the ratio of gold nanoparticles to co-precipitant.

Results

Nanoreactor Stability

(Precipitants):Stabilizer	Ratio (mg:mg:mg)	Stability	Peak 1 (nm)
Au:PS:PSbPEG	0.5:5.5:6	months	136 ± 1
Au: Vit E:PSbPEG	0.5:5.5:6	months	165 ± 6
Au:ToA: PSbPEG	0.5:5.5:6	3 days	160 ± 3
Au:Dode:PSbPEG	0.5:5.5:6	1 days	164 ± 2
Au:PS:ToA:PSbPEG	0.5:4:1:6	at least 13 days	112 ± 1
Au:Ps:Dode:PSbPEG	0.5:4:1:6	at least 13 days	111 ± 3

- The PDI for all samples was <0.2
- Gold nanoreactors are consider to be stable if: ΔD ≤ 10nm
  (D: Diameter of AuNRs)

- Gold nanoreactors with Polystyrene (PS) or Vitamine E (Vit E) are stable in aqueous solution for months.
- Gold nanoreactors with Trioctylamine (ToA) or Dodecane (Dode) are not stable in the solution.
  - Co-precipitating ToA with PS increases stability of gold nanoreactors to at lest 13 days.

Conclusions and Future Work

- Polystyrene nanoreactors loaded with gold nanoreactors have been prepared by encapsulating pre-made gold nanoparticles with various hydrophobic co-precipitants.
- Gold-loaded nanoreactors between 50 nm and 300nm were achieved by varying the formulation parameters.
- The nanoreactor size and gold nanoparticle loading were varied independently.
- Currently, we are investigating the effect of the co-precipitant on nanoreactor stability and quantifying gold loading.

Acknowledgements

- VCU School of Engineering DURI program, CLSE Department.
- Dr. Turner, VCU chemistry.
- This material is based upon work supported by the National Science Foundation under grant no. CMMI-1651957.

511054 Ligand-Mediated TiO2 Functionalization of Silica Nanoparticles for Efficient and Selective Natural Product Separation

Abstract:

Introduction: Post synthesis titania (TiO_2) functionalization of silica nanoparticles (SNPTs) is a facile, yet robust process to obtain a stable system capable of binding natural products of therapeutic interest, such as flavonoids. However, these particles are not efficient in utilizing all of the TiO_2-sites due to clustering or layer formation, and may not be selective in what they bind. Here, we present an approach to produce isolated TiO_2-sites for maximum efficiency by directing single TiO_2 groups towards silica surface using the polar head-group of a TiO_2-surfactant complex. Separately, specificity can be obtained using TiO_2-flavonoid complex during "one-pot" functionalization with TiO_2 sites created by a soft molecular imprint of the desired compound. The use of complexes instead of TiO_2 precursor alone for functionalization should promote dispersion of TiO_2 throughout the system and allow for engineering selectivity towards target metabolites. The ability to improve the accessibility and selectivity of TiO_2 on the particle surface is necessary in designing an advanced and versatile nano-carrier for natural products.

Methods: Silica nanoparticles were synthesized and functionalized simultaneously in a "one-pot" approach following traditional Stöber synthesis of silica nanoparticles, followed by addition of TiO_2 precursor complexed with either a surfactant or flavonoid to the newly formed silica particles. The surfactant/flavonoids were removed from the particle surface by calcination, and the resulting particles were characterized by scanning/transmission electron microscopy, UV-Vis absorbance, N_2 adsorption, and chemical analysis. Particle-flavonoid interactions, capacity, efficiency and selectivity were quantified by equilibrium adsorption isotherms of the flavonoids to differently functionalized nanoparticles.

Results and Discussion: Monodisperse SNPTs were successfully synthesized, consisting of 110 nm particles with different TiO_2 amounts and complexing ligands. Flavonoid adsorption studies demonstrated that binding efficiency is higher for these particle compared to the literature results with post-synthesis functionalization. While selective binding due to imprinting was not observed, binding isotherms of different flavonoids reveal in selective binding based on the structure of the flavonoid, which can be further utilized in a multistep extraction process. This work provides the framework for the design of TiO_2 functionalized silica nanoparticles for the efficient and selective binding of target molecules.

Conclusion: We have developed an approach to obtain well-isolated TiO_2 binding sites on silica nanoparticle surfaces using a one-pot synthesis with simultaneous ligand-guided functionalization. These nanoparticles were demonstrated to efficiently utilize titania for adsorption of flavonoid molecules due to the superior accessibility of surface TiO_2. Moving forward, these concepts will be transferred to mesoporous silica nanoparticle systems to improve the capacity while exploring additional routes of surface functionalization to achieve targeted specificity.

Copyright © American Institute of Chemical Engineers. All rights reserved.

510973 Characterization of Molecularly Imprinted Particles Modified with Various Hydrophobic Monomers

Abstract:

Two potential platform technologies for the oral delivery of protein therapeutics were synthesized and tested. Particles were characterized for protein binding capacity to develop an optimal hydrogel composition for the oral delivery of a model protein therapeutic in response to binding free trypsin in the small intestine.

A displacement mechanism for protein delivery was verified equilibrating the hydrogels in solution with cytochrome c, the model therapeutic, before incubating the loaded hydrogels with trypsin, the target protein. The displacement mechanism showed more success with imprinted polymers than non-imprinted polymers. A poly(methacrylic acid-*co*-N,N-methylene bisacrylamide-*co*-acrylamide) (P(MAA-*co*-MBA-*co*-Aam)) composition was modified with hydrophobic monomers or peptides to increase affinity for the trypsin molecule and improve on the oral delivery mechanism.

Multiple formulations of hydrogels with various hydrophobic monomers in different proportions were synthesized using free radical polymerization. The successful incorporation of hydrophobic monomers into the microparticles was verified through Fourier Transform Infrared Spectroscopy (FTIR). Initially, in non-competitive solutions a composition substituted with 5 mol% benzyl methacrylate for acrylamide showed the greatest binding capacity for trypsin. However, in competitive solutions with other proteins present, adding hydrophobic monomer to a hydrogel decreased the specificity of the microparticles for trypsin. Thus, adding hydrophobic monomers increases the particlesâ€™ affinity for a protein in solution, but does not increase the specificity of the particles for a specific protein, trypsin.

Multiresponsive P(MAA-*co*-MBA-*co*-Aam) particles were synthesized and surface modified with peptide ligands. The peptide ligands were incorporated using EDC-NHS zero-length links between the carboxylic acid groups of the polymer and free primary amines on the peptide. The reaction of the peptide was confirmed by primary amine assays. Modifying the hydrogels with peptides specific for interactions with trypsin increased the competitiveness of the particles for trypsin when compared to unmodified hydrogels.

Ultimately, modifying the original P(MAA-*co*-MBA-*co*-Aam) structure with peptides increased the selectivity these particles had for trypsin, the target protein, while incorporating hydrophobic monomer into the polymer network decreased the selectivity the hydrogels had for trypsin.

Copyright © American Institute of Chemical Engineers. All rights reserved.

510141 Effect of Polymer Excipient on Metastable Succinic Acid Particle Formation Via Monodisperse Droplet Evaporation

Abstract:

Effect of Polymer Excipient on Metastable Succinic Acid Formation via Monodisperse Droplet Evaporation

2017 AIChE Annual Meeting

Ted Egnaczyk and Dr. Ryan C. Snyder

The crystallization tendencies of Active Pharmaceutical Ingredients (API) have a great impact on their properties, uses, and drug performance. There are several characteristics of these molecules of interest that directly affect their bioavailability in the body, and therefore their effectiveness. One of which is water solubility. As new pharmaceutical compounds are continually created, the active ingredients have increased in complexity, tending towards decreased water solubility, and in turn decreasing bioavailability in the body. One method of positively influencing the water solubility of model API ingredients is through forming amorphous dispersions. One way to form amorphous dispersions, and to create more energetic metastable polymorphs, is by including a polymer additive in solution. Including a polymer additive, or excipient, in a given formulation is known to influence the polymorph that is formed. Additionally, relatively small amounts of polymer excipient in solution may create a large variation in the crystallization tendencies of a model API compound.

Succinic acid is used as the model organic compound in this study. This dicarboxylic acid has a very high propensity to crystallize. Succinic acid is very well studied and there exists significant data to compare results. This molecule is known to have two metastable polymorphs, alpha-succinic acid and beta-succinic acid. The addition of polyvinylpyrrolidone (PVP) as a polymer excipient was shown in the past to cause the formation of the alpha form of succinic acid. A polymer excipientâ€™s purpose is to disrupt the regular crystallization patterns of a molecule and cause metastable polymorphs to form or the particle to become amorphous. Monodisperse droplet evaporation is implemented using a Vibrating Orifice Aerosol Generator (VOAG). In this context, the formation of succinic acidâ €™s alpha metastable polymorph was studied as a function of the amount of PVP polymer excipient added. Scanning Electron Microscopy (SEM) and X-Ray Diffractometer (XRD) are used to investigate the morphology and identity of metastable succinic acid polymorphs present in each processed sample. Isopropyl alcohol is used as the solvent for all particle solutions. In this work, I look to quantify the impact of PVP by considering the amount of it required to form the alpha metastable polymorph.

Copyright © American Institute of Chemical Engineers. All rights reserved.

510261 Efficient Tuning of siRNA Dose Response By Combining Mixed Polymer Nanocarriers with Simple Kinetic Modeling

Abstract:

Small interfering RNA (siRNA) therapeutics have emerged as a means to post-transcriptionally silence genes and treat a wide range of genetic diseases. For clinical applications, siRNA must be encapsulated within nanoscale delivery vehicles to survive *in vivo*. These nanocarriers must be stable enough to deliver siRNA into targeted cells while simultaneously promoting siRNA release within these cells. Cationic polymers are a promising nanocarrier method for siRNA delivery in which the polymers electrostatically self-assemble with siRNA to form polyplex structures. However, many cationic polymer nanocarriers lack efficient siRNA delivery capabilities and controlled release mechanisms. To regulate siRNA release, our laboratories previously demonstrated the use of photo-responsive diblock copolymers to provide a method for modulating siRNA release *via* light-mediated charge reversal of the cationic polymer block, which breaks apart the polyplex structure and releases siRNA only upon the photo-stimulus. Herein, we explore the ability to modulate polyplex stability and gene silencing activity using mixtures of these photo-responsive diblock copolymers with two different cationic block lengths. The polymer with the shorter cationic block is shown to promote more siRNA release and form a polyplex smaller in diameter, while the polymer with the longer cationic block is shown to have increased polyplex stability and increased overall positive surface charge. By mixing the two polymers, the contradictory demand for stability and release in polyplex systems is addressed, and the polyplex diameter and surface charge are balanced to enhance uptake into cells. Through formulation of polyplexes from a 50/50 mixture of the two different diblock copolymer lengths, the level of gene knockdown easily was optimized to achieve the maximum level of 70% GAPDH protein silencing for a single dose in NIH/3T3 cells. In addition to maximizing gene silencing by tuning polymer composition, simple kinetic modeling was employed to create dosing schedules based on siRNA, mRNA, and protein concentrations in the cells over the dosing process. This modeling allowed an increase to 84% GAPDH protein silencing upon applying a second dose of polyplexes. Thus, this work demonstrates that pairing advances in biomaterial design with simple kinetic modeling provides new insight into gene silencing dynamics and a strategy for efficient tuning of polymer composition to maximally control gene silencing through polyplex siRNA delivery.

Copyright © American Institute of Chemical Engineers. All rights reserved.

510623 Potential of Mesoporous Silica Nanoparticles As Small Molecule Delivery Platform Against Pathogenic Bacteria

Abstract:

Nanoparticles are increasingly being used in an array of applications such as pharmaceuticals, environmental protection, and electronics due to their unique characteristics such as the ability to exhibit technologically important size-dependent properties such as quantum confinement effect in semiconductor particles, surface plasmon resonance in some metal particles and superparamagnetism in magnetic materials. Mesoporous silica nanoparticles (MSNs) in particular are being researched extensively as potential cargo delivery systems due to their large surface area to volume ratio, tunable pore size, large pore volume, and the range of possible surface functionalization options they have. They are emerging as promising candidates for delivery of therapeutic small molecules against antibiotic resistant pathogenic bacteria of Escherichia coli, Salmonella typhimurium and Staphylococcus aureus. In this study, we present findings on the use of engineered MSNs loaded with gentamicin small molecule. The MSNs were synthesized using surfactant templating approach and surface functionalized with polyethyleneimine (PEI) to permeabilize the outer membrane (OM) of bacteria and facilitate antibiotic uptake. The nanoparticles were characterized using dynamic light scattering (DLS), zeta potential measurements and transmission electron microscopy (TEM). The amount of gentamicin loaded into the nanoparticle and release rates were quantified using Nanodrop spectrophotometer based on UV-Vis absorption. The findings show that MSNs can slowly release gentamicin and kill E. coli K12 T7 cells in vitro according to both colony formation and kinetic of killing assays. Together, the findings show that, MSN is a viable antibiotic small molecule delivery platform which can be explored for complex in vivo applications.

Copyright © American Institute of Chemical Engineers. All rights reserved.

510934 Fabricating Non-Close Packed Colloidal Monolayers for Ion Irradiation Templates

Abstract:

Due to their unique electrical, mechanical and thermal properties, anisotropic nanoparticles are desirable components for next generation applications yet there are few processes capable of fabricating nanoparticle impregnated coatings suitable for the manufacturing environment. Our work seeks to use templated substrates to induce ordering nanoparticle thin films during robust and scalable coating processes. Specifically, we are developing new masking techniques for the production of the templated substrate. We are using lift-up soft lithography for fabricating non-close packed (ncp) colloidal monolayers onto silicon substrates, which then serve as the template for ion irradiation. The first steps to creating this monolayer are obtaining a spin-coated poly(vinyl alcohol) (PVA) thin film of ~200 nm and a close-packed colloidal monolayer using a peltier heater. We achieved a thin film of PVA with the desired thickness by examining the effects of the RPM and viscosity, measured by a rheometer, on the thickness of the film which was evaluated by atomic force microscopy (AFM). Results showed that as the RPM increases and the viscosity decreases, the thickness decreases. To form the colloidal monolayer, a drop of silica spheres suspended in ethanol was placed onto a silicon wafer and mounted onto a peltier element. The entire system was placed inside a closed box, controlled at room with fixed tilt, temperature, and humidity. We preferred using piranha etch to clean the substrate, allowing for minimum defects.

Copyright © American Institute of Chemical Engineers. All rights reserved.

510737 Localized Surface Plasmon Spectroscopy on Individual Self-Assembled Pd-Au@Silica Heterodimers

Abstract:

Single particle measurements characterize the heterogeneities present within batches of nanocrystals, revealing properties obscured by ensemble measurements. By understanding the properties of individual nanocrystals, specific subpopulations can be targeted as materials for catalysis, separations, and energy storage. This project uses localized surface plasmon spectroscopy to detect chemical changes in single palladium nanocubes (PdNCs); single particle dark-field spectroscopy is used to detect shifts in the localized surface plasmon resonance (LSPR) of plasmonic nanocrystals located near the PdNCs, revealing information about transformations in the PdNC.

LSPRs arise from the resonant optical properties of plasmonic nanoparticles under illumination. Incident light generates a collective oscillation of electrons along the nanoparticle surface, which is resonant at a specific wavelength (the LSPR wavelength). The LSPR is highly sensitive to the composition, shape, and size of the plasmonic nanoparticle, as well as its local electromagnetic environment. As a result, spectral shifts and intensity changes in the LSPR can detect real-time chemical changes in a variety of sample environments, such as in-vitro measurement of biomolecule binding kinetics. In this project, the LSPR shifts of a plasmonic nanocrystal are applied to probe chemical changes in a nearby catalytic nanoparticle (PdNC).

The plasmonic nanocrystals compared are gold nanorods (length: 74 ± 6 nm, width: 21 ± 1 nm) and gold nanospheres (50 – 100 nm). Both were synthesized in aqueous solutions of cationic surfactant (CTAB) with a seeded growth method. Similar reaction chemistry was used to synthesize PdNCs with edge lengths of 13 ± 4 nm, 41 ± 5 nm, and 66 ± 7 nm. As synthesized, both the plasmonic nanocrystals and PdNCs were cationically surface charged. Electrostatic self-assembly was enabled by applying a 6 ± 1 nm silica layer to the plasmonic nanocrystal surface. This inverted the surface charge and prevented alloying between the plasmonic nanocrystals and PdNCs.

For each combination of PdNC size and plasmonic nanocrystal tested, PdNC:plasmonic nanocrystal ratios were varied between 2:5 and 3:1, and CTAB concentrations were varied between 10 uM and 1,000 uM. The synthetic yield of these heterodimers were characterized using ensemble solution-phase UV/Vis measurements, and SEM and TEM images of drop cast samples. Dark field scattering spectra were taken on single heterodimers and correlated to SEM images, revealing different sensitivity to the chemical changes in the PdNCs for the Au nanorods and Au nanospheres.

Copyright © American Institute of Chemical Engineers. All rights reserved.

509249 Nitrogen Doped Graphenated Carbon Nanotubes for Electrochemical Reduction of Carbon Dioxide

Abstract:

Electrochemical reduction of atmospheric carbon dioxide with renewable energy has the potential to provide clean fuel to meet the worldâ€™s growing energy demands while reducing the global anthropogenic carbon footprint. This study investigates a novel nitrogen-doped carbon nanomaterial as a promising metal-free electrocatalyst for CO_2 reduction. Graphenated carbon nanotubes (gCNTs) grown with a single step microwave plasma-enhanced chemical vapor deposition (MPECVD) process were doped with nitrogen via post-growth thermal annealing in ammonia. Scanning electron microscopy (SEM) was used to characterize the morphology of nitrogen-doped gCNTs (N-gCNTs) and x-ray photoelectron spectroscopy (XPS) characterization was used to quantify N-doping levels at various treatment temperatures from 650Â°C to 950Â°C. Graphitic and pyridinic forms of nitrogen present in these N-gCNTs were quantified by deconvolution of the XPS N 1s region.

Copyright © American Institute of Chemical Engineers. All rights reserved.

509469 Various Effects on the Growth of Tetrathiafulenve Salt Nanowires

Abstract:

Tetrathiafulvalene (TTF) salts such as $(TTF)Br_{0.76}$ are a suitable material for the synthesis of one dimensional (1D) nanostructures. It exhibits rod formation when electrocrystalized at certain parameters on gold substrates. Unlike bulk TTF, salt derivatives such as $(TTF)Br_{0.76}$ exhibit high electrical conductivity which can be utilized with its nanocharacteristics for many uses including gas sensing. Due to the large surface area to volume ratio of nanorods, surface effects become predominant. Gas interactions at the surface become measurable and allow for potentially cheap, sensitive, organic gas sensors to be produced. By maximizing the length, quantity, and conductivity of $(TTF)Br_{0.76}$ nanorods, the sensitivity, selectivity, and response time can be greatly improved, further increasing its potential in industry and environmental safety as well as public safety and military defense.

Different sensor patterns and different growth parameters will be studied in a design of experiment (DOE) to discover optimal growth conditions. These conditions will then be used in further studies evaluating the technologies potential in a commercial product.

Copyright © American Institute of Chemical Engineers. All rights reserved.

509704 Fabrication of Janus Particles

Abstract:

Fabrication of Janus Particles

Baseemah Rucker,[1,2] Zohreh Jalilvanh,[1] and Ilona Kretzschmar[1]

[1] *Dept. of Chemical Engineering, City College of New York*

[2] *Dept. of Chemical Engineering, Hampton University*

Abstract

Janus particles are spherical particles that have two faces with equal surface areas and different properties. Janus particles were fabricated by semi coating monolayers of silica particles with gold. Janus particles can be coated with a variety of different materials. When a Janus particle coated with platinum is submerged in a hydrogen peroxide solution one face of the Janus particle reacts with the hydrogen peroxide solution, which propels the particle forward. The side of the particle that is coated with platinum reacts with the hydrogen peroxide and give off water and oxygen which acts as the fuel for the propulsions. The purpose of this research is to use different laboratory techniques to fabricate 4 micron semi coated Janus Particles.

Introduction

Janus particles are spherical particles with asymmetrical surface properties due to differences in the chemical compositions of their surface.[1] Janus particles are named after the ancient Roman god Janus, who had two faces with one looking in the past and one looking in the future. Recently, Janus particles have gained a lot of attentions for their potential industrial applications such as use in display technology, magnetic storage, and drug delivery. Janus Particles are also used in academic research settings as molecular surfactants, building blocks of complex superstructures, chemical and biological sensors and self-motile colloidal particles.[3] Recently, Janus particles have been considered as potential drug carriers because of their unique propulsion behavior.

In some solutions, Janus particles have been observed to show self-propulsion behavior owing to a region of their surface giving off a fuel resulting in a forward propulsion of the particle. This property could be very useful in many different settings, if the direction of the particle motion could be controlled. However, Brownian rotation, i.e., the tendency of a microscopic particle to rotate randomly, causes Janus particles to be propelled in random directions hampering their applicability. However, recent work has shown that interactions with walls and cracks have a guiding effect on the particles' motion because of hydrodynamic interactions.

The purpose of this research project is to use the convective assembly technique to Fabricate Janus Particles of 4 micron silica semi coated in gold.

Results and Discussion

Close-packed monolayers of silica particles have been made using the monolayer technique describe above. To make the solution of silica particles, dry silica powder was measured out on a scale and water was added resulting in a weight fraction of 50%. The 50 wt% solution was sonicated for 30 minutes and centrifuged at 4.4 rpm for 8 minutes. Afterwards, the water above of the particles was removed and 3 drops of deionized water were added to the solution. The solution was added into the wedge and the infusion rate of the syringe pump was set to 10 mL/min. At this rate, multilayers formed. Reducing the infusion rate to 6 mL/min resulted in a mix of monolayer and multilayers. Next, the infusion rate was kept the same, but two more drops of water were added to the solution reducing the weight fraction to ~30%. At 30wt%, the submonolayers were formed with particles broadly disperse on the substrate. At this point, the infusion rate was increased to 7.5 mL/min and monolayers were successfully produced (see **Figure 1**).

Figure 1. Close-Packed Monolayer of SiO_2 particles.

Some areas of the particle layers showed non-close packing and were wiped of the silicon wafer with a cotton swab. Once cleaned, the wafers with close-packed regions were transferred to the PVD system (**Figure 2**).

Figure 2. Close-packed Monolayer of SiO_2 in PVD machine before coating

A 5 nm layer of titanium was first evaporated followed by the evaporation of 15 nm of gold yielding particles with a 50% gold coating. Both layers are rather thin and thus the gold layer appears transparent to the naked eye (see **Figure 3**). However, inspection of the monolayer sample under the optical microscope reveals the gold layer (see **Figure 4**).

Figure 3. Close-packed Monolayer of SiO_2 in PVD machine after coating process.

Figure 4. Close-Packed Monolayer of SiO_2 particles coated with 5 nm Ti and 15 nm Au.

Methods and Materials

Janus Particle Fabrication

Janus particle fabrication occurs in two steps; (i) monolayer formation and (ii) physical vapor deposition to create the cap. Closed packed monolayers are made using a syringe pump. Two glass slides are placed at an angle and one glass slide is moved using the syringe pump mechanism while the other is helps steady. A solution of particles with a specific weight percentage of particles in water is then added in the wedge between the two glass slides and the top glass slide is moved across the bottom glass slide spreading the particle solution across the silicon oxide wafer. A microscope is used to identify the regions of the silicon wafer that contain close-packed monolayers and the regions with multilayers and dispersed particles are wiped off. Close-packed monolayers are needed to prepare patchy particles, i.e., particles with less than 50% coating.

The wafers containing the close-packed monolayers are then transferred to a physical vapor deposition (PVD) system and one side of the particles will be coated with platinum at a reduced pressure of 10^{-4} mbar. Once coated, the Janus particles will be sonicated off the substrate and stored in solution in a glass vial.

Conclusions

Close-packed monolayers of silica particles have been made with convective assembly technique. Those particles were then sequentially coated with titanium and gold in a PVD machine making Janus particles. Next steps include the evaporation of platinum, which requires higher temperatures and the testing of the Janus particles in a 3 wt% solution of aqueous hydrogen peroxide to test if they will show propulsion behavior as a result of hydrogen peroxide decomposition. Once particle activity has been established, the substrate will be modified with polymer brushes that that arte intended to create a molecular step to direct the movements of the Janus particles. Next, the particle solution will be put on the modified silicon oxide surface in order to determine if the Janus particles in the solution are able to sense the molecular properties of the polymer brushes on the modified surface and be guided along the surface of the substrate.

Developing techniques to guide Janus particles can lead to development in drug delivery and chemical sensors.

Reference and Notes

1. Bradly, Melanie; Rowe, Joanne. *The Royal Society of Chemistry* **2009**, *5*

2. Axel H. E. Müller*;* Andreas Walther *Royal Society of Chemistry* **2008**, *4*

3. Okubo, Masayoshi ;Tanaka, Takuya; Okayama, Masaru; Kitayama, Yukiya; Kagawa, Yasuyuki *Langmuir* **2010**, *26(11)*

4. Vyas, Mukesh; Schneider, Konrad; Nandan, Bhau; Stamn, Manfred *Royal Society of Chemistry* **2008**, 4, 1024-1032

Copyright © American Institute of Chemical Engineers. All rights reserved.

511024 BSA-Protected Gold Nanoclusters for Detection of Pb(II) in Water

Abstract:

Lead (Pb) is among the top most common contaminants in drinking water, generally in its divalent oxidation state (Pb^{2+}), arising from several industrial processes including lead-based pipes or solders in water supply systems, gasoline and paint. It's toxicological manifestations on humans include severe damage to central and peripheral nervous system, anemia and other renal, cardiovascular and reproductive effects. A recent report revealed that approximately 18 million people in the U.S. live in communities where the water systems are in violation of the EPA's action level of 0.015 mg/L. Up to date, however, there is no easy-to-use and cost-effective technology for detection and quantification of Pb^{2+} in water for household usage. Gold nanoclusters (Au NCs) with sizes < 2 nm display intense photoluminescence due to strong quantum confinement and thus hold incredible promise for optical sensor applications. In this work, bovine serum albumin (BSA) was used to stabilize Au NCs in a simple one-pot and "green" chemical synthesis. First, Au^{3+} ions were reduced in the presence of BSA to form BSA-Au ion complexes. Second, upon addition of NaOH, the complexes were further reduced and aggregated into clusters consisting of 25 gold atoms (Au_{25}). The resulting material is highly luminescent and red-emitting when excited at a wavelength of 470 nm. The BSA-Au NCs were further demonstrated as a "turn-off" sensor in the presence of Pb^{2+} with the potential to enable rapid at-home detection of lead in drinking water.

Copyright © American Institute of Chemical Engineers. All rights reserved.

511135 Sol-Gel Synthesis of TiO2 Nanosheets

Abstract:

Philip Whong

Rutgers University

Department of Chemical and Biochemical Engineering

Research Abstract: Sol-Gel Synthesis of TiO_2 Nanosheets

In the last decades, the research progress on the design and development of TiO_2 based materials has been instrumental, especially due to the r~~r~~einforced interest in semiconductor materials for photovoltaics, photocatalysis, and electrochemistry. More specifically, crystalline rutile or anatase are very promising supports for both metal and metal oxide catalysts. Contrary to rutile, anatase polymorph of TiO_2's crystalline structure, exhibits higher specific surface area. However, the main challenge in the field of synthesis of high surface area TiO_2 is its tendency to aggregate even at relatively mild temperatures.

To improve the functionality of TiO_2 nanoparticles, a variety of emerging synthesis methods attempt to enhance TiO_2 nanoparticle surface area and structure. For example, TiO_2 has been synthesized through, colloidal synthesis, microemulsion method, and pyrolysis. In our approach, TiO_2 nanoparticles have been synthesized from Titanium Isopropoxide precursor via a modified sol- gel route. Sol-gel synthesis is an attractive method of preparing nanomaterial TiO_2 as it uses low temperatures for reaction synthesis. Non-toxic biomass templates were incorporated in our sol- gel method with varying synthesis temperatures for preparation of TiO_2 nanoparticles. Results showing that TiO_2 nanoparticles with improved surface area were produced. Samples showed improve surface area with BET values greater than 150 m^2/g. SEM results revealed unique morphology of TiO_2 nanosheets incorporated into a porous network. In addition, characterization techniques such as Raman and XRD were performed on the TiO_2 samples, revealing the formation of anatase structure. New Raman features were found that could probably ascribed either to TiO_2 particles with crystallite size less than 20nm or small oxyhydroxide domains. Further studies will explore varying concentrations of precursors, calcination temperatures, and reaction times for optimal synthesis parameters for higher surface area.

Copyright © American Institute of Chemical Engineers. All rights reserved.

509798 Insertion of Ionic Liquid into Silica Nanopores for Increased Catalytic Performance

Abstract:

The confinement of ionic liquids in nanoporous materials has been of recent interest for use as tunable microenvironments for catalysis. Ionic liquids 1-(3-trimethoxysilylpropyl)-3-methylimidazolium chloride) and (1-(3-trimethoxysilylpropyl)-3-methylimidazolium chloride) were loaded into orthogonally oriented hexagonally close packed cylindrical silica pores via soaking and tethering methods respectively. In the first method, porous silica films were soaked in a 10 wt.% ionic liquid solution then vacuum dried. Tethering was performed by submerging the silica film in an ionic liquid and chloroform solution and refluxing at 60 â—¦C for 26 hours. FTIR was performed on soaked films that underwent a second step of soaking in deionized water. Through FTIR we showed that the ionic liquid is not easily washed from the silica surface in an aqueous environment. Electrical impedance spectroscopy showed a large increase in the resistance of silica films loaded with ionic liquid via soaking in comparison to silica films without ionic liquid. Films loaded with ionic liquid through tethering showed a much lower increase in resistance. This lower electrical resistance achieved through tethering may indicate better transport through the silica pores due to special structuring of the ionic liquid. This may be useful in applications using these films as micro-environments for catalysis.

Copyright © American Institute of Chemical Engineers. All rights reserved.

509934 Observations of Three Ringed Networks of BZ Droplets

Abstract:

Synchronization is a huge part of the cardiovascular system. When a person has a cardiac arrest or a heart murmur, the rhythm of their heart is changed. Because the heart is a complex network to study, we use a simpler network with similar qualities to study these patterns. The Belousov- Zhabotinsky reaction or BZ reaction is a nonlinear chemical oscillator. Just like the heart, BZ has a specific synchronization pattern. Over a period of time, this coupled oscillator continually changes from a red reduced state to a blue oxidized state. For this research, we wanted to observe a ring network of BZ. As a wave traveled in one direction in this network, we wanted to find a way to switch the direction of the wave by inhibiting a single drop. Using a microfluidic drop maker, droplets of a consistent size were then observed in one of two environments: capillaries or silicon chips. For capillaries, precut capillaries filled with these droplets were attached to a glass slide using an epoxy mixture. For silicon chips, the BZ droplets were carefully squeezed onto the desired etching, sealed with a glass slide and placed in a special clamping device. For both methods, observations were recorded using the Programmable Illumination Microscope, or PIM. In capillaries, we observed two isolated rings spontaneously switch direction after two droplets in the ring simultaneously oscillated. In silicon chips, a new sloshing state was observed. Instead of completing a full cycle through all three droplets, the ring continually switches direction. As a future direction, we want to continue engineer a method to control the direction of these rings. We also want to find and examine the parameters that cause the sloshing state to occur.

Copyright © American Institute of Chemical Engineers. All rights reserved.

510618 A Metal-Organic Framework from Adenine

Abstract:

A cheap and easily manufactured porous material capable of adsorbing gases such as H_2 and CH_4 could be put to use towards a â€œgreenerâ€ future for energy storage and pollution control. Metal-organic frameworks (MOFs) are comprised of metal ions coordinated with organic ligands that self-assemble under various synthetic conditions, producing porous structures that can be fine-tuned to yield desired properties. The common nucleic acid, adenine, has been found to serve as an adequate ligand when coordinated with copper ions to form crystalline metal-organic frameworks. It is shown that multiple modes of coordination occur from the pyridinic-N, pyrrolic-N, and the amine moiety. The structure of this material was solved by single-crystal X-ray diffraction, and its purity confirmed by powder X-ray diffraction experiments. The composition of the material is $\{[CuClAd_2DMSO_2]$Â·$(DMSO)(MeOH)\}_X$ and was found to have a surface area of 85 m^2/g from a nitrogen isotherm experiment.

Copyright © American Institute of Chemical Engineers. All rights reserved.

511128 Structural Effects of Proton Conducting Ionomers on Water Uptake

Abstract:

The proton exchange membrane fuel cell (PEMFC) is a type of fuel cell that is widely used in transport, stationary, and portable applications. It consists of an anode, a cathode, and an electrolyte composed of an ionomer. Ionomers are used to carry charges for protons, separate reactant gases, and prevent electrons from passing through. Additionally, they also serve as the electrolyte film in catalyst layers that bind to the catalyst sites, where electrochemical activity takes place. However, the ionomer is confined to a nanoscale size in the catalyst layer, where particles are not well understood. The mobility of water and polymer chains are constrained due to severe confinement and deviates from the bulk-like behavior seen in polyelectrolyte membranes. Using sodium lignin sulfonate, perfluoroimide aicd (PFIA), and 825 EW ionomer, we determined how the chemical structure, film thickness, and humidity affects their water uptake. We used the quartz crystal microbalance (QCM) to measure the frequency of the crystal containing the polymer film at various relative humidity. This was used to calculate the mass change from the Sauerbrey equation, which represents the amount of water absorbed by the film at set conditions. The percent water uptake from the results of these polymers show that they are associated with a series of factors such as acidity, water-ionomer interactions, and swelling. These factors are all affiliated with proton mobility and impact the proton conductivity of the cell.

Copyright © American Institute of Chemical Engineers. All rights reserved.

510336 Flex to Stretch Electronics

Abstract:

Effective Mechanical and Electrical Bonding of Stretch-to-Flex Electronics

Kunal Mondal[1], Steven Erlenbach[1], Siyuan Ma[2], Andrew Fassler[2], Jim Holbery[2], Michael D. Dickey[1]

1Department of Chemical and Biomolecular Engineering, North Carolina State University

Engineering Bldg 1

911 Partners Way

Raleigh, NC 27606

2 Applied Science Group, Microsoft Corporation

Redmond, WA

Abstract

Stretchable electronic devices maintain electrical connections when subjected to stress or strain and are therefore useful for enabling electronics with new mechanical properties. However, these devices often require power or electrical signals from non-stretchable components. Connecting flexible and stretchable substrates is a possible solution that maintains the electrical connection within a medium that can withstand deformation and prolonged use. The ability to create a robust electrical connection between these mechanically disparate components may enable new types of hybrid devices. In this study we present a basic method to adhere and thereby fabricate such new hybrid devices. The adhesion at the interface arises from surface chemistry that forms strong, covalent bonds. The utilization of liquid metals as the conductor provides stretchable interconnects because liquid metals are both conductive and fluidic. We characterized the mechanical and electrical properties of these hybrid devices. While we focus on silicone elastomers and liquid metals, there are numerous other methods of fabricating stretchable devices that share many properties with the devices in this work.

Copyright © American Institute of Chemical Engineers. All rights reserved.

510941 Iron Oxide Impregnated Membrane for Sulfate Radical-Based Oxidation of Trichloroethene

Abstract:

This study investigates an iron-activated oxidative reaction with sulfate radicals to degrade trichloroethene (TCE) using a composite membrane. The optimization of this degradation pathway was conducted by comparing three different oxidation states of iron (Fe^{3+}, Fe^{2+}, and Fe^{0}), which are possible to incorporate in a variety of Fe nanoparticles. Performance of Fe-NPs was determined by measuring TCE degradation in a solution phase. The membranes consisted of polyacrylic acid (PAA) and graphene oxide (GO), deposited onto a polyvinyl difluoride (PVDF) backing. Diluted acrylic acid was polymerized using a UV initiator. Two types of membranes were made: a "mix" membrane where PAA and GO are deposited simultaneously, and a "sandwich" membrane where PAA is deposited between two GO layers, creating a substrate layer capable of iron deposition. Membrane characterization included permeability, contact angle, salt rejection ($NaSO_4$ and $CaSO_4$), pH responsiveness, and persulfate oxidation. Fe(III) systems showed 0.3 mol Cl^- generated per 1mol $S_2O_8^{2-}$, which is a measure of TCE degradation efficiency. Efficiencies for systems with iron oxide and zero-valent iron were calculated, but require further analysis. When persulfate was passed through the "mix" membrane and "sandwich" membrane, persulfate concentration was reduced by 22.73% and 26.5%, respectively. In a control experiment, a GO only membrane reduced the concentration by 14.63%. This study demonstrates the successful use of impregnated membranes as a medium for the oxidative sulfate radical-based degradation pathway. This research was supported by NSF KY EPSCoR program and by NIH-NIEHS-SRP.

Copyright © American Institute of Chemical Engineers. All rights reserved.

511105 Effect of Processing on Electrochemistry and Surface Chemistry of PGM-Free Catalyst for Oxygen Reduction Reaction

Abstract:

Rising oil prices, limited reserves of oil, and evidence of global warming due to human contribution and carbon emissions motivate the need for alternative energies. For automobiles, there is interest in creating a completely electric vehicle that is not dependent on fossil fuels. Fuel cells are one type of technology that could address this. A fuel cell converts chemical energy directly to electrical energy, using hydrogen or another liquid (e.g. methanol, ethanol, etc from renewable sources such as biomass) as a fuel and produces only water as the final product. Fuel cells still face challenges for large scale implementation because of platinum, therefore another electrocatalyst that is active, durable, and cost-efficient must be found. Platinum-group-metals-free (PGM-free) catalysts based on a low-cost transition metal (e.g. Fe, Mn, Co, and Ni) on a carbon-nitrogen rich support were found to be promising alternative. The catalysts are synthesized using the sacrificial support method (SSM). After pyrolysis of the organic precursor with the metallic salt, the template is etched to create open-ended pores and well-defined catalyst morphology. Improvement of catalyst synthesis relies on an understanding of how the catalyst changes by each processing step during preparation, and the relationship between catalytic activity, surface chemistry, and morphology. Rotating ring disk electrode (RRDE) experiments and x-ray photoelectron spectroscopy (XPS) were conducted on samples taken from each step of the preparation process. After these measurements are obtained, correlations between surface chemistry, catalytic activity, and processing were investigated. The results discuss the effect of pyrolysis, etching, and grinding on the catalyst performances for future improvements.

Copyright © American Institute of Chemical Engineers. All rights reserved.

511048 Building an Electrical/Thermal Actuation Device Using Homogenously Dispersed Multi-Walled Carbon Nanotubes

Abstract:

Carbon Nanotubes (CNTs) act as great thermal conductors which have the potential to be utilized in many engineering systems. The issue with incorporating CNTs into systems is their lack of growth reproducibility. The purpose of this study is to investigate the effect of an applied transverse electric field on the alignment of CNTs dispersed in Toluene. The experimental variables isolated in this study is the concentration of CNTs in the solution, the voltage of the applied transverse electric field, and the voltage of the applied internal electric field. The applied transverse electric field was varied between 0V and 20V in increments of 5V, the internal voltage of the system was varied between 0V and 20V in increments of 1V, and then the resulting resistance was calculated using LabVIEW. In the second set of testing, it was determined that increasing the concentration of CNTs in a system, with an applied transverse electric field, contributed to higher levels of solution resistance in the direction perpendicular to the transverse current.

Copyright © American Institute of Chemical Engineers. All rights reserved.

511155 3D-Printing a Transparent Part for Use with Nanoparticle Sensors

Abstract:

The rapidly-evolving technology of 3-D printing offers many exciting possibilities and applications. This research expands the possibilities to include 3-D printed nanoparticles that may be used in sensors. Nanoparticle sensors are currently in use today, but the ability to 3-D print these sensors would improve the manufacturing process. Nanoparticle sensors could be created in any shape or size desired, without the need for a molded part. In this case, iron-core/gold-shell nanoparticles were compounded with Amphora transparent polymer filament and extruded into a spool for use in a Makerbot Replicator 2X. The resultant printed part maintains the sensing properties of the nanoparticles, namely, their ability to rotate light in the presence of magnetic field. Joined with a polarized lens and a magnet, the printed part makes an effective qualitative demonstrator. Future work is to quantify the detection abilities, and develop reproducible and uniform nanoparticle concentrations within the filament.

Copyright © American Institute of Chemical Engineers. All rights reserved.

510974 Investigation of Fractal Formation Phenomena in TIPS/Pentacene System Solution-Sheared on Kapton Substrate

Abstract:

Kapton is a semi-crystalline polyimide tape developed by Dupont that is commonly used for electronic applications. TIPS-Pentacene (6,13-Bis(triisopropylsilylethynyl)pentacene) is an organic molecule commonly used for charge transport when crystallized in solution via confinement through the method of solution-shearing in the evaporative regime of flow onto a smooth substrate. In this investigation, a continuation of previous studies of solution-shearing thin-film organic semiconductors composed of solutions of TIPS-Pentacene in Toluene on various substrates was extended to using Kapton as a primary substrate. Trials performed with Kapton as a substrate resulted in fractal formation approximating Sierpinski Triangles in the films, a phenomenon that was unknown to the authorâ€™s knowledge. The investigation so far has involved parameter searches to evaluate the sensitivity of the system, ImageJ analysis of fractal dimension and whether it correlates significantly with any shearing parameters including shearing temperature, shearing rate, and solution concentration. Further, recent experiments have focused on controlling fractal formation and analyzing the role of crystallization molecule and fluid dynamic effects in inducing this phenomenon. In situ high-speed cameras have been employed as a means of attempting to capture and observe the motion that drives the fractal formation. Varying the shearing molecule among TIPS-Pentacene, TCNQ (Tetracyanoquinodimethane), and Caffeine was additionally employed to determine the dependence of fractal formation on the shearing molecule, though no consistent fractals have been observed for films that do not include TIPS-Pentacene. FTIR was conducted to measure the thickness of the TIPS-Pentacene films and appeared to display variations in thickness corresponding to the fractal locations, suggesting the fractals were essentially void of film. The methods used in this study derive from previous strategies of Giri et al. in regards to using a solution-shearing method to yield aligned crystalline films of organic semiconductors with TIPS-Pentacene as the molecule and Silicon Oxide as the substrate.[1] Control samples of TIPS-Pentacene sheared on Silicon Oxide wafers as well as glass and aluminum foil resulted in no fractal formation. Results and current understanding of the system suggest surface energetic effects as well as fluid dynamics phenomena as the primary factors for the fractal formation, though further investigation is required for comprehensive understanding.

1. Giri, G.; Park, S.; Vosgueritchian, M.; Shulaker, M. M.; Bao, Z. *Advanced Materials* 2013, *26* (3), 487â€“493.

Copyright © American Institute of Chemical Engineers. All rights reserved.

511017 The Effect of Fluorination on the Interfacial Properties of Alcohols and Carboxylic Acids

Abstract:

The Effect of Fluorination on the Interfacial Properties of Alcohols and Carboxylic Acids

Molecular dynamic simulations are used to investigate the effect of the fluorination on surfactant-like moleculesÕ physical properties and interfacial behavior at air-water and octanol-water interfaces. The project includes comparisons between the properties of molecules such as methanol, pentanol, and octanol to their fluorinated counterparts and various C8-fluorotelomers. Interfacial tension and octanol-water partition coefficients are calculated in an effort to predict the biological and environmental impact of these molecules. Understanding the partitioning and interfacial behavior of these molecules is expected to inform the development of alternative surfactants that could, potentially, serve the same purpose with lower potential for bioaccumulation.

To date, we have collected preliminary results for the octanol-water partitioning coefficients [1] and surface tension behavior of 8-carbon fluorotelomers. The Adaptive Biasing Force (ABF) method was used to calculate the change in free energy for hydration and solvation [2-4] of the C8-fluorotelomers and the partitioning coefficient was then modeled by the equation

Ultimately, the change in free energies are calculated as the difference between the two plateaus. Values for the change in free energies and the octanol-water partition coefficients are recorded in Table 1.

Table 1: Calculated octanol-water partition coefficients

Molecule Name			
Perfluoromethanol (CF_3OH)	2.375	1.719	0.999
Perfluorooctanol ($C_8F_{17}OH$)	-7.761	-2.195	1.002
7-1FTOH ($C_8F_{15}H_2OH$)	6.919	-3.230	0.996
6-2FTOH ($C_8F_{13}H_4OH$)	7.040	0.462	0.997

Simulations were performed in NAMD, version 2.9[5]. For the fluorotelomer surface tension calculations, an interface 50 slab of TIP3P water molecules, with a density of 1 g/cm^3, in the center of a box was used[6]. Calculations were only made for TIP3P water model at 300K. Figure 3 represents the surface tension as a function of concentration of the fluorotelomer at the TIP3P water interface.

References:

1. Bhatnagar, N., G. Kamath, and J.J. Potoff, *Prediction of 1-octanolÐwater and airÐwater partition coefficients for nitro-aromatic compounds from molecular dynamics simulations.* Physical Chemistry Chemical Physics, 2013. **15**(17): p. 6467-6474.

2. Darve, E. and A. Pohorille, *Calculating free energies using average force.* The Journal of Chemical Physics, 2001. **115**(20): p. 9169-9183.

3. Darve, E., M.A. Wilson, and A. Pohorille, *Calculating free energies using a scaled-force molecular dynamics algorithm.* Molecular Simulation, 2002. **28**(1-2): p. 113-144.

4. Rodriguez-Gomez, D., E. Darve, and A. Pohorille, *Assessing the efficiency of free energy calculation methods.* The Journal of chemical physics, 2004. **120**(8): p. 3563-3578.

5. Phillips, J.C., et al., *Scalable molecular dynamics with NAMD.* Journal of computational chemistry, 2005. **26**(16): p. 1781-1802.

6. Vega, C. and E. De Miguel, *Surface tension of the most popular models of water by using the test-area simulation method.* The Journal of chemical physics, 2007. **126**(15): p. 154707.

Copyright © American Institute of Chemical Engineers. All rights reserved.

509840 Bone (collagen/hydroxyapatite) Tissue Engineering Optimization: By "Green" Electrospinning

Abstract:

In the human body, tissues are mainly composed of a nano-fibrous structure of collagen which makes up part of the extra cellular matrix (ECM). There are many structural proteins that compose the organic matrix, but collagen corresponds to 90% of the total protein content in bone ECM. Of the different types of collagen, the most abundant in bone tissue is type I collagen. Its mineral phase, which represents 70%–90% of the bone, is primarily composed of hydroxyapatite. Electrospinning is an attractive method for the generation of polymeric nano-fibrous mats to be used for applications in biomedical engineering and tissue regeneration. It offers the capacity of emulating the morphological characteristics of tissue matrix in geometry, chemistry, mechanical properties, orientation and scale. Since the ECM of bone is composed of an organic phase, composed of elastic proteins that provide fracture resistance and a mineral phase composed of inorganic mineral which contributes to the rigidity in bones; it is ideal to prepare a collagen/hydroxyapatite (COL/HA) electrospun scaffold for bone tissue engineering applications. The fabrication of electrospun bone tissue scaffolds that mimics the extracellular matrix of bones have been confined to the use of toxic solvents such as trifluoroacetic acid, and 1,1,1,3,3,3,-hexafluoro-2-propanol in its preparation, thus have been limited. These toxic solvents limit or reduce the potential use of the scaffold in biomedical applications. In our work, a nano-fibrous collagen/hydroxyapatite scaffold was prepared by electrospinning using a mild solvent. To accomplish this, hydroxyapatite was dispersed into a collagen/acetic acid/water solution, which was later electrospun to generate composite nanofibers. Furthermore, electron microscopy revealed that it is possible to modulate the fiber diameter, porosity and fiber alignment by varying the supplied voltage, volumetric flow rate, the RPMs of the motor and Hydroxyapatite concentration. The versatility of this technique allows for the fine-tuning of the final properties of the scaffold by controlling the different operational variables. Moreover, both X-ray and infrared spectroscopy confirm the presence of the hydroxyapatite embedded in the collagen fibers or adhered to the surface of the fiber. This phenomenon was observed to be dependent on the diameter of the particle. Additionally, the inclusion of hydroxyapatite does not alter the native collagen structure. Lastly, in vitro studies showed these composite nanofibers support pre-osteoblast adhesion, which indicates that this non-toxic or "green" electrospun ECM mimetic structure of bone tissue could be used to generate nanocomposite scaffolds with potential biomedical applications such as tissue regeneration

Keywords: electrospinning, collagen, Hydroxyapatite

Copyright © American Institute of Chemical Engineers. All rights reserved.

509915 Preparation and Characterization of Shape Memory Assisted Self Healing Coatings

Abstract:

Shape memory assisted self-healing (SMASH) coatings have the ability heal nearly completely after damage with the application of heat, making them useful for corrosion resistance in steel structures. This project sought to expand upon existing research to develop an efficiently healing and transparent SMASH coating that could be used in technological applications, such as on cell phone or computer screens. The materials tested were comprised of electrospun poly(Îµ-caprolactone) (PCL) and poly(methyl methacrylate) (PMMA) fiber mats coated in epoxy with glass transition temperatures (T_g) of 50Â°C and above. The healing efficiency of the samples was tested by damaging each coating with a mechanized scratch tester (Qualtech Mar Tester Model QPI-AST880) and healing them over varying times and temperatures in an oven. The transparency of each sample was observed visually before and after heating, a technique that was later refined by electrospinning onto glass plates. Once the most ideal combination of material, time, and temperature is found, further experimentation will be conducted to determine whether the ideal epoxy/PCL, or epoxy/PMMA, combination is also one that can withstand multiple damage-healing cycles to the same spot without becoming opaque.

Copyright © American Institute of Chemical Engineers. All rights reserved.

510917 Development of a Coarse-Grained Model to Study DNA Interactions with Nanomaterials

Abstract:

Hybrid nanostructures composed of single-walled carbon nanotubes (SWCNT) coated with single stranded DNAs (ssDNA) are of great interest due to their various applications in the nanotechnology and biomedical field. In particular, short oligonucleotides in biofluids can be promising disease biomarkers, while carbon nanotubes prove to be an ideal material for implantable biosensors. Although ssDNA-coated-SWCNT particles have been a focus of recent studies, the mechanism and energetics of DNA-nanotube binding process are not yet fully understood. To address this, we developed a coarse-grained model to study DNA-SWCNT interactions. Adopting an existing 2nd generation of 3-Site-Per-Nucleotide (3SPN.2) coarse-grained DNA model, in which base, sugar and phosphate groups are each represented by one site, our new model incorporates coarse-grained SWCNTs with 2:1 mapping of carbon atoms. The interaction potentials between DNA sites and CNT sites are characterized by standard 12-6 Lennard-Jones potential, for which parameters are based on all-atom MD simulation results, and parameters are differentiated for different bases (A,T,G,C). The simplicity of this model improves computational affordability, while capturing the essential features of DNA-CNT interactions observed from all-atom MD simulations. Employing umbrella sampling, we focus on analyzing the relationship between adsorption free energy and two important parameters, namely, length of the oligonucleotide and SWCNT surface coverage. Results indicate that the absolute value of adsorption free energy increases linearly with the length of oligonucleotides (of the same base type, e.g., T3, T6, T12), and that the free energy for oligonucleotides of the same length are sequence-specific (e.g., T12 versus A12). Analysis of nanotube surface coverage by various oligonucleotides shows that the absolute value of adsorption free energy decreases quadratically with increasing the DNA coverage (the number of strands of pre-adsorbed ssDNA on nanotube), and that the decrease in free energy is dependent on the length and sequence of both the analyte oligonucleotide and pre-adsorbed ssDNAs.

Copyright © American Institute of Chemical Engineers. All rights reserved.

509800 Electrospinning and Degradation of Poly(lactic-co-glycolic acid) (PLGA) As a Potential Wound Healing Dressings

Abstract:

Chronic wounds affect more than 6.5 million people in United States with expenditures of more than 25 billion dollars for treatments and other interventions. In spite of this, a significant amount of cases have bad prognoses, of which a majority lead to amputations. The objective of this research is to fabricate poly(lactic-co-glycolic acid) (PLGA) electrospun fibrous mats, study the degradation kinetics in phosphate-buffered saline, and analyze the time-dependent release of therapeutics with potential applications for wound healing. The fibrous mats can be used as wound dressings, which provide excellent mechanical support for cell migration, tissue growth, and as an agent of local drug delivery.

Copyright © American Institute of Chemical Engineers. All rights reserved.

510838 Development of a Low Cost, Low-Force Mechanical Testing Device and Its Application in Measuring Mechanical Properties of Polyelectrolyte Capsules for Engineered Tissue Assembly

Abstract:

Tissue engineering muscle is a promising approach to restoring function in cases of volumetric loss due to cancer or trauma. We previously developed polylelectrolyte GAG-chitosan microcapsules as a platform for modular tissue engineering. We subsequently modified the capsule technology to generate directly extruded polyelectrolyte hollow fibers encapsulating mesenchymal stromal cells, and proposed to evaluate their utility in engineering skeletal muscle tissue. However, the hollow fibers require significant enhancement to their mechanical properties before deployment into the muscle problem. In this study, we explored how the addition of photocrosslinkable acrylate groups to chitosan affected the mechanical properties of the GAG-chitosan polyelectrolyte membranes found in the microcapsules and hollow fibers. A convenient method of quantifying the membrane mechanical properties and the effects of photocrosslinking involves measuring the burst strength of spherical polyelectrolyte capsules. To generate capsules, 0.5 mL of 1.5 wt% carboxymethylcellulose/4 wt% chondroitin 4-sulfate solution was dispensed dropwise into a rapidly stirred solution of glycidyl-methacrylate modified chitosan (GMA-chitosan) to form microcapsules 3-4 mm in diameter. Microcapsule membranes were subsequently exposed to long-wave UV light to induce photocrosslinking of the GMA moieties. Since the costs of low-load mechanical testing systems are prohibitive, a suitable system was designed and assembled in house from commercially available components. Specifically, a 100 g load cell and a 10 cm stroke linear actuator were purchased, together with manufacturer-supplied USB interface boards. Microcapsule rupture strength was measured following a modified standard. Microcapsules 4 mm in diameter were prepared, with the microcapsule placed exactly in the middle of the load cell plate. Each capsule specimen was compressed by the linear actuator. Failure occurred under uniaxial compression at a controlled strain rate. The rupture strength of the capsule was then calculated from the maximum load recorded by the load cell. Results to date indicate that this system is able to measure capsule rupture forces with a high degree of precision. Derivatization of chitosan with GMA was found to significantly increase capsule rupture strength even without any photocrosslinking of the GMA groups. Additionally, rupture force pre-crosslinking was found to increase with the degree of GMA substitution. These observations suggest that hydrophobic interactions between non-crosslinked methacrylate groups may be playing a significant role in strengthening capsule membranes. Microcapsules formed with GMA-chitosan at a 5% degree of substitution ruptured at 1.72 g before crosslinking compared to 7.98 g after crosslinking. Interestingly, GMA-chitosan capsules with a 15% degree of substitution ruptured at 23.77 g pre-crosslinking, but showed reduced strength (12.06 g) after photocrosslinking. The testing system is currently being used to evaluate the detailed effects of degree of GMA substitution on the burst strength in order to optimize capsule membrane properties.

Copyright © American Institute of Chemical Engineers. All rights reserved.

510379 Transdifferentiation of Mesenchymal Stem Cells into Schwann Cell-like Phenotypes Using Electrical Stimulation

Abstract:

Transdifferentiation of Mesenchymal Stem Cells into Schwann Cell-like Phenotypes using Electrical Stimulation

Maxsam S. Donta[1], Merjem Mededovic[1], Emily Kozik[2], Metin Uz[1], Donald S. Sakaguchi[2, 3], Jonathan Claussen[4] and Surya K. Mallapragada[1, 2, 4]

[1] Department of Chemical and Biological Engineering, Iowa State University, Ames, Iowa 50011

[2] Neuroscience Program, Iowa State University, Ames, IA 50011

[3] Department of Genetics, Development and Cell Biology, Iowa State University, Ames, IA 50011

[4] Division of Materials Science and Engineering, Ames Laboratory (DOE), Ames, Iowa 50011

Mesenchymal stem cells (MSCs) are stem cells with multipotent properties and as such, can be differentiated into a multitude of different phenotypes including neuroglial cells. The phenotype can be manipulated during the differentiation process by such factors as chemicals, electrical stimulation, or environmental cues; one or a combination of the aforementioned factors may be used to influence the final phenotype.[1,2] In order to examine the effects of differing electrical stimuli on MSC differentiation, electrical stimuli were applied to genetically engineered brain-derived neurotrophic factor (BDNF)-hypersecreting MSCs through different devices – two-dimensional flexible inkjet-printed graphene interdigitated electrode circuits as well as three-dimensional gelatin-based conduits with conductive graphene rods – in order to differentiate them into Schwann cell-like and neuronal cell-like phenotypes. To explore these effects, electrical stimulation of either 25, 50, and 100 mV, at 50 Hz was applied for 10 min/day and for 10 days to MSCs seeded on each respective device. Successful differentiation was evaluated through Immunocytochemistry (ICC) analysis using Schwann cell markers s100, s100β, and p75 as well as the neuronal markers TuJ1, Map2, and GFAP; paracrine activity using enzyme-linked immunosorbent assay (ELISA) was evaluated to detect secreted amounts of BDNF, glial cell line-derived neurotrophic factor (GDNF), and nerve growth factor (NGF). ICC results indicate highly efficient differentiation of MSCs into Schwann cell-like phenotypes and a low degree of differentiation into neuronal phenotypes. As voltage was decreased from 100 mV to 25 mV, the percentage of MSCs that differentiated into neuronal phenotypes increased. Voltage was differed for only the 2D circuits; the 3D gelatin conduits stimulation studies were conducted at 100 mV and exhibited s100, s100β, and p75 staining. ELISA results revealed that GDNF release was very low (100-1000 pg/mL) for all voltages, as was expected due to its neuronal-supporting nature. BDNF releases were higher (20,000-25,000 pg/mL) because the cells were genetically engineered to secrete BDNF. NGF had the highest secretion levels (65,000-75,000 pg/mL), which is indicative of Schwann cell-like phenotypes due to NGF's role in repairing nerve tissue. Furthermore, the NGF levels measured were much higher in the electrically-stimulated cells than the control cases with releases measured at three to four times higher. Moreover the NGF levels secreted from electrically-stimulated cells showed higher or at least similar paracrine activity compared to chemically differentiated cells. The trends were consistent in the 2D devices and the 3D gelatin conduits; however, the 3D conduits with electrical stimuli exhibited lower neurotrophic factor release, which can be attributed to a difference of electrical field between the circuits and conduits. The NGF release in the gelatin conduits was also not as relatively high compared to the chemically differentiated control, but release was still almost twice that as the untreated control cells. Methods performed with the aforementioned devices were contrasted with each other as well as with a control, which utilized established chemical differentiation procedures as well as the same ICC and ELISA tests. BDNF-secreting MSCs that are treated with only electro-stimulus are expected to exhibit equal, if not superior rates of differentiation for both 2D and 3D models than traditional chemical differentiation shown in our previous works.[1,2] Results from these experiments have the potential to revolutionize the way that peripheral nervous system damage is treated.

References

1. Das, S. R.*; Uz, M.*; Ding, S.; Lentner, M. T.; Hondred, J. A.; Cargill, A. A.; Sakaguchi, D. S.; Mallapragada, S.; Claussen, J. C., Electrical Differentiation of Mesenchymal Stem Cells into Schwann-Cell-Like Phenotypes Using Inkjet-Printed Graphene Circuits. Advanced Healthcare Materials 2017, 6 (7). (* indicates equal contribution)
2. Uz, M.; Buyukoz, M.; Sharma, A. D.; Sakaguchi, D. S.; Altinkaya, S. A.; Mallapragada, S. K., Gelatin-based 3D conduits for transdifferentiation of mesenchymal stem cells into Schwann cell-like phenotypes. Acta Biomaterialia 2017, 53, 293-306.

Copyright © American Institute of Chemical Engineers. All rights reserved.

510937 Optimization of Therapeutic Lipid-Coated Mesoporous Silica Nanoparticle Vaccines

Abstract:

Monophosphoryl lipid A (MPL) is the biologically active portion of lipopolysaccharide (LPS), a major constituent of the outer membrane of gram negative bacteria. MPL, which is used clinically as an adjuvant in vaccines, stimulates antigen-presenting cells (APC) by targeting Toll-like receptor 4 (TLR-4). While MPL has similar adjuvant activity to LPS, it is 1000 times less toxic. Activated APC secrete cytokines and have elevated surface expression of co-stimulatory molecules. Our research seeks to create a nanoparticle-based vaccine that utilizes the well-characterized adjuvant properties of MPL in combination with the model antigen ovalbumin to stimulate APC and to deliver antigen that can be processed and presented by APC in association with major histocompatibility complex to ovalbumin transformed cancer cells. We propose to use lipid-coated mesoporous silica nanoparticles (LC-MSN), a platform that we have successfully used for delivery of diverse cargo, including proteins and nucleic acids.

Our LC-MSN technology builds on the success of liposomal delivery systems by increasing nanoparticle stability and loading capacity through addition of a mesostructured silica core. MPL integrates into the lipid bilayer and is presented on the surface of the LC-MSN for easy binding to surface TLR-4 on APC. Here we present studies aimed at optimizing multivalent surface presentation of MPL to maximize the immunogenic response to LC-MSN. We have created LC-MSN using MSN with different surface chemistries, diverse liposomal preparation procedures, and variable loading ratios of ovalbumin to MSN cores.

Copyright © American Institute of Chemical Engineers. All rights reserved.

511046 Assessment of Drug Diffusion of Sulfathiazole and Metronidazole from Polyurethane Films

Abstract:

Urinary and cardiac catheters, made of polyurethane, are used in patients for extended periods of time. As a result of a catheterâ€™s location and prolonged use, the catheter is vulnerable to contaminants that lead to infections by microorganisms such as Escherichia coli and Candida spp. A polymer that can prevent the development of these microbes, without compromising the functionality of the catheter, would be an ideal solution to this problem. The purpose of this experiment is to develop a polymer that contains an antimicrobial commonly used to treat the various causes of catheter-induced infections.

Polyurethane is the preferred material for catheter production due to its flexibility and biocompatibility with humans, and thus became the material of choice for trials. Sulfathiazole and metronidazole have a history of being prescribed for various infections, including urinary tract infections. While both pharmaceuticals are soluble, metronidazole has a greater ability to dissolve in water. Solubility is an important factor for analyzing the concentration of drug releasing from the polyurethane.

Sample films were made by solvent casting. Polyurethane and the drug (sulfathiazole and metronidazole, separately) were dissolved in a solution of dimethylformamide and tetrahydrofuran. The resultant solution was constantly stirred for 24 hours, after which it was poured into molds, and the molds were placed in an oven at 37°C for 24 hours. The dried films were released from the molds and stored in a desiccator. In order to confirm that drug can diffuse out of the polyurethane, a precisely weighed portion of the drug-loaded polymer film was immersed in distilled water constantly stirred at 200 RPM and 37°C for 8 hours. Every hour, 5 mL aliquot was taken out and tested with a Shimazdu UV-2450 spectrophotometer to measure the absorbance at 270 nm for sulfathiazole and 319.5 nm for metronidazole. The concentrations of aliquots taken out periodically were calculated from the absorbance values and the calibration curve made with standard aqueous solutions of each drug with different concentrations, ranging from 1.25 µg/mL to 100 µg/mL.

Drug-loaded polymer films were made with two different sulfathiazole concentrations, 0.3 wt/wt% and 1.5 wt/wt%. Metronidazole was only made with a 1.5 wt/wt% concentration. Both sulfathiazole concentrations show similar release mechanisms which follow the Korsemeyer-Peppas kinetic model of diffusion; metronidazole follows the same kinetic model. From the similarity in release rates and kinetic modeling, it can be concluded that the diffusion behavior of the medications are independent of its concentration in polyurethane.

Copyright © American Institute of Chemical Engineers. All rights reserved.

510506 Determining How the Molecular Structure and Coassembly Interactions of Peptide Amphiphiles Influence the Energy Landscapes of Their Supramolecular Assembly

Abstract:

Peptide amphiphile (PA) molecules that incorporate a beta-sheet forming sequence supramolecularly assemble into micelles or one-dimensional fibers in solution. It has previously been discovered that changing the salt concentration of PAs alters their energy landscapes and results in supramolecular assemblies with changed structure and properties. Additionally, it is known that the molecular structure of the PA molecule affects the structure and properties of the supramolecular assemblies. This work investigates PAs with minor alterations to the beta-sheet forming sequence; first we looked into how these minor alterations affect the energy landscapes and then we looked into how coassembling these similar PAs at different proportions affects the energy landscapes.

Four PAs with minor differences in their beta-sheet forming sequence were characterized. The first PA consisted of a palmitoyl tail followed by three valines, three alanines, and three lysines in that order (V3). One edit had a missing valine (V2), one had an extra valine (V4), and the final edit had glycines instead of valines (G3). After characterizing the four PAs individually, V3 and V2 were coassembled at various proportions and characterized. The methods to probe the energetics of the various assembly states and the landscapes through which the states are connected have only been developed recently. The Nile Red assay was used to determine the transitions from monomers to micelles and from micelles to fibers. Circular Dichroism was used to observe transitions from beta sheets to random coil and vice versa. Dynamic Light Scattering was used to derive the energetic barrier between assemblies by fitting the rates of change of scattering counts at varying temperatures to an Arrhenius plot. Atomic Force Microscopy and cryo-Transmission Electron Microscopy were used to confirm morphology. Finally, computation simulations were compared with experimental results.

It was found that a missing valine resulted in a more dynamic assembly without sharp transitions in assembly. The beta-sheet strengths can be tuned by doping in small proportions of a stronger or weaker beta-sheet PA. We can now measure the transition beta-sheet strengths at which the supramolecular assembly changes. Interestingly, the coassembly of V3 and V2 resulted in thermodynamic and kinetically-trapped products that differed from the pure PAs. Finally, the results also suggested that small amounts of V3 seemed to stabilize the V2 fibers and that the V2 is likely nucleating off the V3 micelles or fibers. These findings will allow for more control in the structure and properties of self-assembling amphiphilic materials and could potentially lead to advancements in artificial extracellular matrix design.

Copyright © American Institute of Chemical Engineers. All rights reserved.

510993 PEGDA Hydrogel Network Model Evaluation: Correlating Mechanical Properties to Molecular Structure

Abstract:

Poly(ethylene glycol) diacrylate (PEGDA) hydrogels are important biomedical materials with particular use in drug delivery and tissue engineering. Modelling the underlying polymer network of these gels is of key importance in understanding the mechanical, diffusive, and swelling properties that dictate hydrogel behavior and performance in application. The gels are often assumed to have a network structure that lead them to follow neo-Hookean elastic models, from which network parameters such as crosslink density and network mesh size are calculated. However, several groups have used scattering techniques to demonstrate significant heterogeneity of PEGDA hydrogel networks. Because such specialized techniques are not feasible to be applied to biomedical hydrogel networks generally, it is still common to calculate network parameters assuming the validity of the neo-Hookean model despite the underlying heterogeneity. Thus a systematic study of PEGDA hydrogel mechanical properties was performed to evaluate the validity of the neo-Hookean model on PEGDA hydrogels and the ability to predict network structure from macroscopic mechanical tests using both the neo-Hookean and Mooney-Rivlin models. Mechanical properties of 700 Da and 5000 Da PEGDA hydrogels (PEGDA700, PEGDA5000) were determined. Hydrogels of PEGDA concentrations from 7 - 30 wt. % were prepared and synthesized via photoinitiation at a wavelength of 312 nm with the photoinitiator lithium phenyl-2,4,6-trimethylbenzoylphosphinate (LAP). Hydrogels were molded into disks, swollen to equilibrium in deionized water, and evaluated using uniaxial compression tests to determine stress versus strain relationships as well as the elastic modulus (E) and shear modulus (G) for each gel under neo-Hookean assumptions. The swelling degree (Q) was also determined from the ratio of wet to dry weights of each gel. Gel properties were shown to vary widely across polymer concentration with neo-Hookean elastic moduli increasing with concentration and ranging from 81.0 kPa to 2620 kPa and 51.6 kPa to 1030 kPa for PEGDA700 and PEGDA5000, respectively. Swelling degree decreased with composition and ranged from 9.20 to 3.00 and 14.7 to 5.46 for PEGDA700 and PEGDA5000, respectively. Further evaluation using the Mooney-Rivlin network model highlighted the heterogeneity of these network structures, and suggests that these gels are deviating from neo-Hookean behavior. Gels with low polymer content approached neo-Hookean behavior and behaved as apparent Mooney-Rivlin solids, while high polymer content gels behaved similarly to networks that follow constrained polymer chain models with notable transition behavior for hydrogel networks with compositions near the polymer overlap volume fraction. These results suggest that low molecular weight PEGDA hydrogels deviate significantly from the neo-Hookean model and correlation of moduli to PEGDA network parameters is limited.

Copyright © American Institute of Chemical Engineers. All rights reserved.

510957 Controlled Hydrogen Peroxide Release Induces Angiogenesis

Abstract:

Regenerative engineering is an exciting field that combines expertise from materials science, developmental biology, and stem cell science to achieve complex tissue repair. While many advances have been made in this field over the past quarter-century, engineered solutions capable of achieving higher-order tissue regeneration have yet to be translated to the clinic. Nowhere is this more apparent than in bone regeneration for which commercially available solutions improve small fracture healing but are sub-optimal for the large volume repair necessary for non-union bone defects due to traumatic injury or tumor removal. One reason for this bottleneck is that most technologies focus on the use of polymeric scaffolds loaded with proteinaceous growth factors which have significant drawbacks including their cost, fragility, poor processability, potential immunogenicity, and high dosage requirements. Additionally, most current models solely focus on the regeneration of osteoblasts and fail to recognize the importance of facilitating vascular network development which is imperative for new bone tissue to thrive. An alternative to protein growth factors are simple signaling molecules which are ions, redox reagents, and gases that have been found to possess potent bioactivity. Specifically, calcium and phosphate ions have been found to induce osteogenesis in mesenchymal stem cells (MSCs) and hydrogen peroxide has been found to induce angiogenesis in endothelial progenitor cells (EPCs). It was hypothesized that controllably delivering hydrogen peroxide can allow for the regeneration of high-quality vasculature which can be leveraged in the future to facilitate complex bone tissue repair.

To controllably deliver hydrogen peroxide, we evaluated the release profiles of several different peroxide generating molecules over the course of 7 days. EPCs were then exposed to a gradient of these molecules to determine their angioinductive potential. At selected time points, Von Willebrand Factor (vWF) fluorescent staining was conducted and assessed via confocal microscopy and flow cytometry to determine if the EPCs were differentiating down the endothelial lineage. Cell viability and proliferation were measured using commercial colorimetric and fluorometric assays. This study uncovered the therapeutic window where hydrogen peroxide can induce angiogenesis in EPCs.

Copyright © American Institute of Chemical Engineers. All rights reserved.

509251 Effect of Silk-Based Hydrogel Topography on Intestinal Epithelial Cell Morphology and Wound Healing In Vitro

Abstract:

Recent advances in the field of biomaterials have suggested that cells cultured on substrates resembling the native tissue mechanical properties, matrix and growth factor composition, and topography can adopt phenotypes that more closely resemble the *in vivo* tissue compared to cells cultured on non-mimetic constructs. Understanding the effect of culture substrate on *in vitro* tissue formation is important for bioengineering applications that include mechanistic studies of healthy tissue function and development of disease models. In this work, Caco-2 adenocarcinoma cells were seeded on flat and crypt-like topographies of 3D-printed cytocompatible hydrogels derived from silk fibroin protein. Silk hydrogels were selected for their tunable mechanical properties, their ability to support intestinal cell culture, and their relatively slow degradation profile, which was hypothesized to enable the maintenance of substrate topography during culture. 3D-printed acellular silk hydrogels were found to maintain their shape for over eight weeks in culture medium. Cells were seeded on the silk hydrogel substrates, where it was found that cells both attach and proliferate on substrates with flat and intestinal crypt-like topographies. Future work includes investigating the effect of hydrogel topography on cell morphology using microscopy with visualization of junction molecules by immunostaining. In addition, the effect of topography on the closure of disrupted epithelium will be assessed using a wound scratch assay. It is anticipated that the topographical studies presented will assist in developing more physiologically relevant tissue models that may be used to investigate wound healing in the colon.

Copyright © American Institute of Chemical Engineers. All rights reserved.

509511 3D Printing of Biodegradable Elastomers for Osteoarthritis Repair

Abstract:

3D Printing of Nor-PGS as Biodegradable Elastomer

Claude King III, Jonathan Galarraga, Dr. Yi-Cheun Yeh, Dr. Chris Highley, Dr. Jason Burdick

Abstract

The repair of cartilage defects is currently under an extensive amount of research; cartilage is avascular and aneural, making it difficult for self-repair to occur *in vivo*. There are a number of therapies that are currently employed to repair cartilage defects sites such as microfracture, autologous chondrocyte implantation, and mosaicplasty.[1] These options utilize stem cells or autologous cells/tissue from the host to repair defects. However, each of these approaches have only demonstrated limited success in clinical application. Engineered scaffolds are a more promising solution towards repairing cartilage defects because they can be designed with properties that mimic native cartilage. When designing scaffolds for applications in tissue engineering, properties such as material mechanics, biocompatibility, biodegradability, porosity, and surface chemistry should be engineered to support cell growth and regeneration of native tissue.[2] For example, the elastic moduli of implants should be sufficiently high to provide mechanical integrity to defects; in parallel, interconnected porosity is required to allow adequate space for tissue growth.[2] With a wide range of properties, scaffolds can be created through different methods; however, 3D printing will be utilized in the approach herein described because it allows full control over the scaffold fabrication process. 3D-printing is of growing interest in the cartilage repair field because it can be exploited to print patient-specific defect geometries. However, 3D-prinitng of biomaterials into free standing objects is a complicated barrier for the translation of this technology. In conventional extrusion-based printing, printed constructs require a base to print on and are restricted to layer-by-layer fabrication. Free-form embedded printing is necessary to create patient specific geometries that can be extruded in any direction without undergoing stress relaxation due to body forces such as gravity. To achieve this goal, a sacrificial support material that acts as a Bingham plastic may be employed to permits deposition of material with spatial control. Specifically, the use of support material allows for constructs to be printed anywhere in three-dimensional space such that a larger range of scaffold geometries are accessible. Towards translation, printing patient-specific geometries will consist of scanning a specific defect geometry to obtain an accurate 3D-model of the site. The 3D shape can then be processed via computer-aided design (CAD) and saved into an STL file so that it can be readily printed. The work herein described focuses on the development of a biodegradable elastomer (Nor-PGS) that can be printed into a Carbopol (Lubrizol) support to form clinically relevant scaffold geometries.

References

1. Brittberg, M., Peterson, L., Sjouml;gren-Jansson, E., Tallheden, T., & Lindahl, A. (2003). Articular Cartilage Engineering with Autologous Chondrocyte Transplantation: A Review of Recent Developments . *The Journal of Bone & Joint Surgery , 85(suppl_3)*, 109-115.

2. Hutmacher, D. (2006). Scaffolds in tissue engineering bone and cartilage. *The Biomaterials*, 175-189. doi:10.1016/b978-008045154-1/50021-6

Copyright © American Institute of Chemical Engineers. All rights reserved.

511120 Tissue Simulant Materials for Studying Blast-Induced Traumatic Brain Injury Mechanims

Abstract:

Blast-induced traumatic brain injury (bTBI) has become a common problem for military personnel in recent years. The mechanisms of the tissue damage present in bTBI remain poorly understood, but may be involved in the development of post-traumatic stress disorder (PTSD). More knowledge of these mechanisms could allow researchers to develop better ways to treat and prevent bTBI. To do this, accurate representation of the brains response to bTBI need to be performed. Thus, brain tissue properties need to be biomimetically represented. This study characterized the physical properties of several materials intended for use as layered cranial phantoms in shock tube and open field blasts for gross observations and post-blast analysis. The data verified the tissue phantoms possessed qualities close to actual brain tissue and post-blast analysis of the phantoms showed possible damage at the interfaces between material layers. These findings provide a promising direction for research in the mechanism of bTBI.

Copyright © American Institute of Chemical Engineers. All rights reserved.

511138 The Effect of Cell Geometry on the Phenotypic Behavior of Valvular Interstitial Cells

Abstract:

The Effect of Cell Geometry on the Phenotypic Behavior of Valvular Interstitial Cells

Derek M. Montoya[1,2], Olivia A. Bell[1,2], Matthew N. Rush[1,3], and Elizabeth Hedberg-Dirk[1,2]

[1]Center for Biomedical Engineering, [2]Department of Chemical and Biological Engineering,

[3]Nanoscience and Microsystems Engineering,

University of New Mexico, Albuquerque, NM

Valvular Interstitial Cells (VICs) are the primary cells of the aortic heart valves responsible for valve formation, maintenance, and disease progression. Using well-defined surface chemistries, our lab has established two distinct VIC phenotypes for study of heathy/activated and diseased behavior. When activated, elongated VICs may spontaneously undergo differentiation into an osteoblast-like phenotype, with a rhomboidal geometry, which is responsible for calcification of the heart valve.[1] It has previously been shown that changes in cell geometry, such as those exhibited during healthy and diseased states in VICs, can affect cell fate through changes in cytoskeletal signaling.[2] To better understand which factors, surface chemistry or cell geometry, provide greater influence to activated or osteoblastic VIC phenotypes, we will examine the effect of pattern geometry on the phenotypic behavior of VICs grown on functionalized surfaces. Using micro contact printing (μCP), we can pattern surfaces which control the shapes, sizes, and spacing between cells to test for changes in the VIC phenotype. Additionally, we wish to further explore the relationship between the focal adhesions and integrin expression present in healthy vs diseased VICs and the influence on cytoskeletal changes; specifically, we wish to examine how these cytoskeletal changes influence cell geometry.

References: 1) Rush, M.N., *Acta Biomaterialia*, 28 (2015), 76-85. 2) Chen, C.S., *Science*, 276 (1997), 1425-1438.

Copyright © American Institute of Chemical Engineers. All rights reserved.

510982 Exploring the Phase Behavior of Peg-Grafted Polyelectrolytes for Intracellular Drug Delivery

Abstract:

The cellular and endosomal membranes represent a significant delivery barrier in the administration of therapeutics whose targets lie in the cytosol. To address this challenge, several investigators have developed cationic and pH responsive self-assembling polymers to facilitate the uptake and endosomal escape of cargo into the cytosol. To date, these approaches have largely leveraged cationic and pH responsive homopolymers or block copolymers, and as such design rules for other polymer architectures for maximizing the delivery efficiency of therapeutics to the cytosol have been poorly explored. Here we synthesize a library of poly[(poly(ethylene glycol) methyl ether methacrylate)-*co*-(butyl methacrylate)-*co*-(2-(diethylanimo)ethyl methacrylate)] (PEGMA-c-DB) graft copolymers for evaluation as mediators of cytosolic delivery. Polymers are synthesized *via* one pot RAFT polymerization, and PEG graft length, total chain molecular weight, and PEGMA weight percent are systematically and combinatorically varied to determine the effects of these changes on the self-assembled polymer architecture in aqueous media. By evaluating self-assembled particle yields, hydrodynamic radii, and with TEM analysis, we demonstrate that similar to in block copolymer architectures, large weight percents of PEGMA typically yield micellar self-assembled nanoparticulates, while smaller PEGMA weight percents typically yield wormlike and vesicular structures. However, unlike linear architectures, our PEGMA-c-DB system was capable of supporting vesicular self-assembly at molecular weights as large as M_w = 25,000, which we also demonstrate to be desirable for facilitating efficient endosomal escape. Collectively, these data demonstrate that PEGMA-c-DB polymers are useful formulation of endosomolytic vesicles, which are promising candidates for versatile cytosolic delivery of cargos.

Copyright © American Institute of Chemical Engineers. All rights reserved.

510268 Femtosecond Direct Writing of PDMS Channels and Its Application in Tissue Engineering

Abstract:

Femtosecond Direct Writing of PDMS Channels and its

Application in Tissue Engineering

Nannette Hernandez[1], Sabrina McCarthy[1], Alessandro Enrico[2], Anna Herland[3],

Frank Niklaus[2] and Göran Stemme[2]

[1] Department of Chemical Engineering, The City College of New York, CUNY, New York, NY 10031, USA

[2] Department of Micro and Nanosystems, School of Electrical Engineering, Royal Institute of Technology, Stockholm, Sweden

[3] Department of Physiology and Pharmacology, Karolinska Institute, Stockholm, Sweden

Fabrication of channels in elastomers such as PDMS is crucial for microfluidics and tissue engineering. Conventional approaches for patterning PDMS exploit moulding and subsequent crosslinking. Two conventional approaches are 1) 3D moulding which require special or multistep demoulding procedures and 2) multi layered design moulding where PDMS is patterned and then bonded to another flat or patterned layer of material such as PDMS or glass. However, the former approach results in limited 2D geometric patterns while the latter requires a mould to cast the PDMS as well as demoulding which may be time and resource expensive. Here we propose femtosecond laser direct writing of channels in PDMS. It is possible to ablate anywhere in the depth of a PDMS block by adjusting the location of the focal point in the material.

A 1040nm femtosecond laser was used for direct writing of channel-like microstructures in a block of 1:10 PDMS. Parameters such as repetition rate and power were optimized to achieve these channel-like microstructures. For this purpose, it was observed that multiple laser passes in the same position facilitates the removal of the ablation debris inside the channel. Moreover, it was observed that the diameter of channels decreased with increasing processing depth in the material. In light of these results, femtosecond laser direct writing results in a simple, precise fabrication process in PDMS with minimal post processing and without need for moulding and demoulding.

Copyright © American Institute of Chemical Engineers. All rights reserved.

510247 Surface Engineering of Nanostructured Electrodes and Electrolytes for Solid-State Battery Applications

Abstract:

Lithium ion secondary batteries are widely used to power portable electronics such as laptops, mobile phones, etc. and have become an integral part of our daily lives. As technology advances, batteries with a high capacity and long lifetime are demanded for increasingly diverse applications. Solid-state lithium ion batteries using solid electrolytes offer a promising avenue to meet these demands for the following reasons. First, solid-state batteries are easy to miniaturize and can be produced in a thin-film form. Second, solid electrolytes are more stable than any other electrolyte system and circumvent the well know safety issues with lithium ion batteries. However, current solid-state battery capacity is limited by significant interfacial impedance due to the nature of the solid electrolyte/electrode interface. Realization of a commercially viable solid-state battery will require significant progress in minimizing interfacial impedances.

Incorporating nanotechnology processes into the lithium ion battery system has helped enhance the performance of several lithium ion battery chemistries. Nanostructuring a lithium ion batteryâ€™s anode and cathode allows for vastly increased electrode surface area available for the battery system, thereby reducing the impact of interfacial impedance. High surface area nanostructured electrodes can be produced by using an RF magnetron sputter coater to deposit the electrode material onto a nanoporous Anodized Aluminum Oxide (AAO) membrane with nanopores of 200 nm in diameter. This research uses a nanostructured SnO_2 anode against a bulk $LiCoO_2$ cathode and a polyethylene-oxide (PEO) based solid polymer electrolyte. The electrolyte is similarly nanostructured by confining it in the nanopores of an AAO membrane. By melting a film of the electrolyte onto an AAO membrane under vacuum, the polymer is forced into the AAO nanopores, forming a nanocomposite electrolyte membrane. The confinement of the electrolyte increases its ionic conductivity to levels comparable to liquid electrolytes.

A functioning room-temperature solid-state lithium ion cell has been developed using the nanostructured SnO_2 anode, $LiCoO_2$ cathode, and PEO nanocomposite electrode. The cell was investigated using galvanostatic cell cycling and scanning electron microscopy (SEM). The cell achieved a specific discharge capacity of 125 mAh/g (89% of theoretical capacity) and a capacity retention of 20% after 20 cycles.

Copyright © American Institute of Chemical Engineers. All rights reserved.

510246 Correlating Structural Changes with Electrochemical Activity in the Defect Perovskite Type ReO3

Abstract:

Rhenium Oxide (ReO_3) was investigated for use as a novel cathode material for rechargeable battery applications. ReO_3 was synthesized by dissolving Re_2O_7 in methanol in a round bottom flask, followed by heating to 250 °C to form ReO_3 nanoparticles. Upon evaporation of methanol, the ReO_3 particles were deposited on the bottom of the flask. The particles were characterized using Transmission Electron Microscopy (TEM) and X-Ray Diffraction (XRD) pattern analysis. The material was shown to adopt the defect perovskite structure with a vacant A-site. Guest ion insertion was then studied electrochemically via Cyclic Voltammetry (CV) and Galvanostatic Cycling with Potential Limitation (GCPL). Both lithium and sodium was inserted during electrochemical studies. CV scans were used to study phase changes in the material upon ion insertion and GCPL was used to characterize cell performance in Swagelok and three electrode cells. Thick film ReO_3 electrodes were used with Swagelok cells and microelectrodes were used for characterizing redox processes in three electrode cells. Either lithium or sodium hexafluorophosphate was used as the electrolyte, depending on the ion being inserted. The electrolyte solvent was ethylene carbonate (EC) and dimethyl carbonate (DMC). Structural changes during ion insertion were studied using a Swagelok in-situ cell with a beryllium window during XRD. The diffraction patterns show the distortion of the crystal structure as guest ions are introduced. Upon the intercalation of a single lithium ion per unit cell, the crystal distorts and contracts. The Re octahedra were shown to twist inwards upon lithiation by a single Li ion. Upon insertion of a second ion the unit ReO_3 cell returns to the original volume. Despite returning to the original cell volume, the structure maintains distortion and octahedral twist. Sodium intercalation was also studied using CV and GCPL to characterize phase changes and cell performance. Sodium insertion during battery cycling was studied using in-situ XRD. Compared to lithium intercalation, far less structural distortion was observed possibly owing to sodium ion size matching the interstitial space of the ReO_3 cell.

Copyright © American Institute of Chemical Engineers. All rights reserved.

511042 Crystalline Transition Metal Cathode Materials for Rechargeable Aluminum Batteries

Abstract:

Brian Chen, Ankur Jadhav, Robert J. Messinger,

Department of Chemical Engineering, The City College of New York;

160 Convent Avenue, New York, NY 10031

<u>Title</u>: Crystalline Transition Metal Cathode Materials for Rechargeable Aluminum Batteries

The expanding use of electric vehicles and grid-scale storage of renewable energy sources requires novel electrical energy storage systems. Lithium-ion batteries have dominated the portable electronics market; however, rechargeable aluminum batteries have several advantages over lithium-based batteries. Trivalent aluminum ions can transfer three electrons each upon Faradaic reaction, yielding almost three times the volumetric capacity compared to Lithium ions. Aluminum is also the most abundant metal in the Earthâ€™s crust and consequently has a low cost. However, one of the outstanding challenges lies in the limited number of cathode materials that are known to reversibly intercalate trivalent aluminum ions.

This work highlights two crystalline transition metal oxide and sulfide structures as intercalation cathode materials for rechargeable aluminum-ion batteries: the spinel-phase Mn_2O_4 and the chevrel-phase Mo_6S_8. Best practices for synthesizing pure spinel and chevrel phases will be discussed, as well as how to incorporate them into composite cathodes. Electrochemical analyses of the cathode materials, using cyclic voltammetry (CV) and galvanostatic cycling, will be presented and analyzed in detail. X-ray diffraction (XRD), scanning electron microscopy (SEM), and ^{27}Al magic-angle-spinning (MAS) solid-state NMR measurements yield insight into the intercalation mechanisms and the molecular and structural changes that the materials undergo upon electrochemical cycling. The results yield insights into the challenges and opportunities of designing crystalline intercalation cathode for rechargeable aluminum batteries.

Copyright © American Institute of Chemical Engineers. All rights reserved.

510848 Rational Synthesis of Lead Bromide Perovskites with Various Crystallite Sizes

Abstract:

The goal of this project was to synthesize methylammonium (MA), formamidinium (FA), and cesium (Cs) lead bromide single crystal and microplate perovskites with high phase purity. Perovskites are a family of semiconductors with the specific crystal structure ABX_3 (A=MA, FA, or Cs; B=Pb; X=Br). They have desirable properties including long diffusion lengths, easy fabrication, and low trap-state density. In order to accurately measure their fundamental properties, the perovskites must be synthesized with minimal defects and high phase purity. Single crystal and microplate perovskites offer fewer grain boundaries, which can improve these properties relative to bulk crystals.

Microplates were synthesized by a dissolution recrystallization method. A lead precursor was spin coated or drop cast onto a glass slide and then reacted for several hours in an ABr solution to form a thin of perovskite microplates. Scanning electron microscopy was used to determine the sizes of the microplates and assess the microscopic coverage of the films, which changes depending on the reaction conditions. All samples consisted of microplates ranging from 1 to 5 microns in size, and the coverage varied depending on the precursor deposition. The single crystals were grown over the course of several days using an antisolvent vapor diffusion method. The resulting crystals were millimeters in size.

The XRD patterns for the microplates and the single crystals match well with the accepted patterns found in literature. The peak width is inversely proportional to the crystallite size, so the full width at half maximum is greater for the microplates than for the single crystals. Initial Cs microplate samples showed some phase impurities, but these were avoided by reducing thermal processing in subsequent iterations.

We have demonstrated the synthesis of single crystal and microplate MA, FA, and Cs lead bromide perovskites, confirmed by XRD and SEM. Single crystal growth by the anti-solvent diffusion method requires a high precursor concentration (0.2 to 0.6 M) in the initial solution. The microplate shape, size, coverage, and purity depends on the reaction times and temperature, reactant concentrations, and precursor deposition method.

Copyright © American Institute of Chemical Engineers. All rights reserved.

510549 Characterizing Upconversion Efficiency and Photophysics in CdSe(Te)/CdS/CdSe Nanostructures for High-Efficiency Photovoltaics

Abstract:

Semiconductor nanostructures that upconvert low-energy light to high-energy light have demonstrated properties advantageous for various applications, such as bioimaging and high-efficiency photovoltaics. By modifying the spatial dependence of the band structure within these nanostructures through heterostructure engineering, we can control carrier recombination pathways for preferential emission of high-energy photons. We aim to optimize this design for photovoltaic use by engineering CdSe(Te)/CdS/CdSe nano-rods, where CdSe(Te) serves as an absorber quantum dot and CdSe as an emitter quantum dot in a CdS rod. In the absorber, quantum confinement allows for two-photon excitation of holes, which can then migrate to the emitter for recombination. Increasing the Tellurium (Te) doping in the absorber increases the structuresâ€™ absorption bandwidth, a salient feature for high-efficiency photovoltaics. However, its effects on device efficiency, or upconversion quantum yield (UCQY), is unknown. To study this effect, we implement a procedure for determining the quantum yield of luminescent colloidal solutions to study rod-like CdSe(Te)/CdS/CdSe double quantum dot nanostructures with varying Te doping. Photoluminescence (PL) from the nanostructures is collected in an integrating sphere and coupled to a spectrometer via fiber bundles. The system is calibrated using rhodamine 101, a known 100% quantum yield standard, and by measuring the incident power of the exciting laser. The internal and external quantum yield of each sample are then calculated by comparing the PL of the sample to the standard. Photophysics responsible for trends in quantum yield is probed with (a) time-resolved photoluminescence (TRPL) using a pulsed laser excitation source and time-correlated avalanche photodiode detector and (b) dependence of upconversion photoluminescence intensity on excitation laser power. We find that nanostructures with the lowest Te doping in the core quantum dot, at 4% Te, produce the highest upconversion quantum yield, approximately 0.001%, while the highest Te doping of 40% produces the least efficient structures. However, neither demonstrate the quadratic power dependence expected of two-photon upconverters, indicating the presence of thermally-assisted excitation processes. A sample with 10% Te doping shows the greatest quadratic quality in power dependence studies, and TRPL indicates that it also produces the highest ratio of core/absorber dot exciton lifetime to emitter dot exciton lifetime, possibly due to greater quantum confinement in the cores of this sample. Although increasing Te doping diminishes upconversion quantum yield, relationships between two-photon quality and quantum yield are not evident.

Copyright © American Institute of Chemical Engineers. All rights reserved.

509342 (Plasma-Enhanced) Atomic Layer Deposition Infilling of Nanocrystal Networks

Abstract:

In this study, we present infilling of nanocrystal networks using of atomic layer deposition (ALD) and plasma-enhanced atomic layer deposition (PEALD) as effective methods to produce thin-film materials with desirable electrical and optical properties. Using a scanning transmission electron microscope (STEM), we acquire spatially-resolved energy dispersive x-ray (EDX) and electron energy loss spectra (EELS) to assess the uniformity of infilling nanocrystal networks using both PEALD and ALD. By infilling ZnO nanocrystal networks with TiN using PEALD, we fabricate transparent conducting thin-films with possible applications in photodetection devices, photocatalysis, and photovoltaic energy conversion. Further, we provide justification for previous investigations of the insulator-to-metal transition by demonstrating uniformity of various element concentrations throughout both insulating and metallic ZnO nanocrystal networks infilled with Al_2O_3.

Copyright © American Institute of Chemical Engineers. All rights reserved.

511056 High Throughput Screening Platform for Cesium-Lead Perovskite Nanocrystal Synthesis

Abstract:

Colloidal organic/inorganic perovskites have great potential as photoactive compounds in solar cells and light emitting diodes. Scale-up of perovskite nanomanufacturing will require methods to precisely tune perovskite properties. The recent advent of colloidal synthesis in flow offers a promising method for controlling synthesis conditions (residence time, mixing, temperature, pressure, etc). In turn, synthesis in flow enables the creation of uniform nanoparticles with tunable band gap, photoluminescence, and crystal structure. This poster presents a novel modular microfluidic reactor platform for the rapid and automated synthesis of highly photoluminescent cesium lead perovskite nanocrystals at room-temperature. A fiber-optic UV spectrometer and orthogonal light sources (for fluorescence and absorption) are connected to a translational flow cell to capture spectra at varying residence times (while maintaining flow rates and mixing constant). Furthermore, the reactor platform is connected to a LabView script, which enables testing of a large parameter space in a short amount of time (up to 30,000 spectra per day). The script utilizes an automatic alignment algorithm to ensure the translational stage is accurately aligned with the 20 sampling ports along the reactor.

Using the platform, the kinetics of the $CsPbBr_3$ synthesis reaction are elucidated at timescales as early as 200 ms in single- and multi-phase regimes. In multi-phase experiments at low flow rates, a local variance algorithm is applied to solely capture spectra of individual liquid plugs. It is shown that increasing flow rate at a constant residence time results in greater blue-shifts of emission wavelength for multi-phase systems than single-phase systems, indicating that the nucleation and growth kinetics of cesium-lead perovskite reactions are highly dependent on mass transfer. Furthermore, the demonstrated microfluidic approach offers greater control over mass transfer across different scales than batch methods, which results in perovskite nanocrystals with uniform properties. This reactor platform represents an important step to developing scalable methods for continuous nanomanufacturing of perovskites for photoactive applications.

Copyright © American Institute of Chemical Engineers. All rights reserved.

511101 Optimization of Performance of Fe-N-C Catalysts on Tantalum-Doped Titanium Dioxide Supports for ORR

Abstract:

The combination of hydrogen and oxygen into water by way of fuel cells suggests a promising avenue for the conversion of stored chemical energy into usable electrical energy. Due to membrane stabilities in electrolyte solutions, proton-exchange membrane fuel cells (PEMFCs) have exhibited the most potential for viability in current applications. The performance of these fuel cells is dictated mainly by the sluggish oxygen reduction reaction (ORR) at the cathode. Traditional ORR catalysts are composed of platinum group metals (PGMs), which are highly active but expensive noble metals. However, reaching comparable activity with less expensive materials serves as an alternative to PGM catalysts. An example of these novel, non-PGM catalysts is the Fe-N-C class of catalysts.

Fe-N-C catalysts are frequently supported by conductive carbon. Due to the highly oxidizing environment of the cathode, carbon corrosion, in which the support is used as a carbon source in the formation of carbon dioxide, can occur. This corrosion decreases activity by inducing agglomeration and detachment of the active sites. For this reason, conductive non-carbon supports investigated in conjunction with Fe-N-C catalysts could potentially serve to increase the stability and longevity of non-Pt catalysts for cathodic ORR. Over the course of this project, synthesis and mass-loading parameters were varied to find optimal activity on oxide supports, creating a novel Fe-N-C @ $Ta_{0.05}Ti_{0.95}O_2$ electrocatalyst.

To evaluate the potential for this TiO_2-based support, synthesis by pyrolysis and performance testing by cyclic voltammetry were completed. Each material was produced by dry mixing and heat treating known precursors for Fe-N-C catalysts, aminoantipyrine (AAP) and iron(III) nitrate ($Fe(NO_3)_3$), in the presence of the Ta-doped TiO_2 support. Certain synthesis parameters were varied, such as the mass ratio of AAP to $Ta_{0.05}Ti_{0.95}O_2$ and the mass ratio of $Fe(NO_3)_3$ to AAP. Both *in situ* and *ex situ* synthesis techniques were studied, as was a sacrificial support method (SSM). Materials underwent a first heat treatment in a H_2/N_2 mixture and a second heat treatment in ammonia (NH_3). Once heat treatments were completed, the products were ground into fine powders and made into inks using a Nafion solution, DI water, and isopropyl alcohol. The inks were sonicated and then applied to a glassy carbon and platinum rotating ring-disk electrode (RRDE) via drop casting. Each catalyst was tested using dual electrode cyclic voltammetry (DECV) with a Pine WaveNow potentiostat in both acid and alkaline electrolyte solutions using a graphite counter and a reference hydrogen electrode while holding the ring at potentials for peroxide oxidation. After performance testing, characterization techniques such as BET, XRD, and XPS were used to further analyze the structural composition of the synthesized catalysts.

A novel set of Fe-N-C catalysts supported by conductive, non-carbon, Ta-doped TiO_2 was synthesized and tested to show considerable ORR activity. Ample opportunity for future work within this project exists, with respect to durability testing and physical characterization to elucidate a structure-property-performance relationship. Overall, this

research demonstrated the viability of this new class of electrocatalyst supports for ORR.

Copyright © American Institute of Chemical Engineers. All rights reserved.

511097 Enhanced Charge Transport through Secondary Nucleation of P3HT

Abstract:

Printable organic electronics have developed as a cost-effective alternative to conventional inorganic electronics due to the solution processability of semiconducting polymers. The hypothesized ideal structure consists of interconnected, densely packed, aligned fibrillar domains on the mesoscale. Methods to induce an ideal morphology are limited by the self-assembling properties of poly-3-hexylthiophene (P3HT). We aim to use a type of secondary â€œseedâ€ nucleation by mixing amorphous and treated P3HT to achieve the ideal microstructure. Treatment incorporates exposure of P3HT in solution to ultraviolet irradiation followed by solution aging ultimately resulting in the formation of nanofibers. Three solutions were made with 20%, 50%, and 80% treated P3HT with the balance amorphous polymer. A thin film of solution was deposited on a bottom gate, bottom contact transistor through a blade-coating technique. Using a semiconductor parameter analyzer, it was found that mixing a small amount of treated P3HT with a larger amount of amorphous polymer resulted in high charge mobility across the field effect transistor. The maximum charge carrier mobility resulted from a mixture of 20% treated P3HT suggesting that high amounts of nucleated polymer leads to precipitation and inferior performance. UV-vis spectroscopy of solution and film support this data with repeatable trends of increasing aggregate fraction with increasing amount of treated P3HT. Consistent with high charge mobilities, atomic force microscopy (AFM) shows densely packed and aligned fibers for secondary nucleated solutions. Seed nucleation demonstrates a viable method capable of reducing resources required for solution processing by nearly four times compared to the traditional method of treating the entire solution.

Copyright © American Institute of Chemical Engineers. All rights reserved.

510333 Production and Utilization of Biochar from Slow Pyrolysis of UConn Dining Hall Food Waste

Abstract:

As the global human population increases, the amount of waste produced increases as well. There has been a heightened interest in utilizing this waste for a range of purposes. These purposes include the production of useful biochemicals â€“ such as syngas, bio-oil, and biochar â€“ through the slow pyrolysis of organic wastes. Biochar, in particular, has the potential to improve crop yields as a soil additive by adsorbing contaminants. This work specifically focuses on the physical properties of biochar produced under various slow pyrolysis conditions. Food waste from the University of Connecticut dining halls was used as the feedstock. Pyrolysis was carried out at a range of temperatures, from 275 Â°C to 625 Â°C, and residence times ranged from 30 minutes to 2 hours. The resulting biochars were studied and compared based on their surface morphology, surface area, pore volume, elemental makeup, bond structure, and oxidation behavior. Pyrolysis at the lowest temperature produced the greatest yield of biochar, but low pore volume suggests that this biochar would be a poor adsorbent. Biochars produced at higher temperatures and greater residence times demonstrated larger pore volumes, despite lower yields. Selected biochars were tested for adsorption of copper and 1-methyl naphthalene, two common soil contaminants, to determine the potential viability of biochar as a soil amendment.

Copyright © American Institute of Chemical Engineers. All rights reserved.

510638 Visible Spectrum Sensitization of Vaporizable Polyaldehydes for Transient Devices

Abstract:

Materials which disappear or vaporize upon triggering are classified as â€œtransient technology,â€ an emerging field of research with applications in electronic devices and small delivery vehicles. In these applications, a controlled, triggered vaporization of the device is desired when device recovery is not possible. In this project, low ceiling temperature polymers are used to construct transient devices. The ceiling temperature (T_C) of a polymer is the temperature at which the rates of polymerization and depolymerization are in equilibrium; at a temperature below T_C, the polymer form is preferred. The cationic polymerization of poly(ortho-phthalaldehyde) (p(PHA)) forms a cyclic polymer that kinetically stabilizes the macromolecule at ambient conditions, meaning that it will maintain its polymeric form at temperatures well above T_C. Low T_C polyaldehydes, such as p(PHA), can depolymerize and vaporize at ambient conditions by acid-catalyzed depolymerization, making them suitable for transient devices. A strong acid can catalyze depolymerization of the polymer at temperatures above T_C. The irradiation of a p(PHA) film loaded with a photoacid generator (PAG) can create such an acid. The photoacid cleaves the polymer backbone and decomposes the polymer into its monomer form. Most PAG molecules are only sensitive to UV light, making them near-useless in a sunlight exposure. Polycyclic aromatic hydrocarbons (PAHs) absorb visible light and sensitize the PAG to visible light through photoinduced electron-transfer (PET). In this project, the molar ratio of the PAG to PAH was investigated in terms of absorption coefficient, wavelength, energy transfer, and residue remaining after polymer depolymerization. The thermodynamic driving force of the PET reaction with various PAHs is characterized through the Rehm-Weller equation. Results show anthracene to be the most practical PAH due to the fast depolymerization of p(PHA), minimal residue formation, and a colorless polymer film. A 1.2:1 PAH to PAG molar ratio at 1 wt% PAG was shown to be the most efficient mixture for decomposition. The optimized ratio and PAH were used to examine dose linearity (i.e. the tradeoff between intensity and time) and to evaluate the evaporation rates of sensitized copolymer films. Absorbance spectroscopy was used to determine the path length through a polymer film with a given PAH. Utilizing the 1.2:1 ratio at 1 wt% concentration of PAG, the maximum path length of absorbance is 8500 μm in an anthracene-loaded polymer film.

Copyright © American Institute of Chemical Engineers. All rights reserved.

510781 Investigating Micellar Phases in Hydrated Nonionic Surfactants

Abstract:

Some surfactants â€“ amphiphilic molecules that comprise a hydrophilic headgroup and hydrophobic tail â€“ form lyotropic liquid crystal (LLC) phases upon the addition of water. LLCs have applications that range from water desalination membranes and drug delivery to templates for inorganic materials. To effectively design surfactant-water systems for useful applications, we must better understand the detailed molecular features that govern their LLC phase behaviors. While complex micellar phases have been reported in both block polymer and ionic surfactant systems, previous studies of the phase behaviors of aqueous LLCs of non-ionic surfactants lack definitive phase assignments. However, nonionic polyoxyethylene alkyl ether (C_iE_j) surfactants do demonstrate the potential for complex micellar phase formation. Thus, we investigated the ordered aqueous micellar phases of a C_iE_j surfactant using high-resolution small-angle X-ray scattering as functions of varying hydration and temperature. We report the formation of BCC and FCC phases, and most significantly, the observation of Frank-Kasper A15 and Ïƒ phases. From our results, we conclude that nonionic surfactants indeed form complex liquid crystal phases that bridge the gap between polymer and ionic surfactant systems.

Copyright © American Institute of Chemical Engineers. All rights reserved.

510897 Mechanochromic Devices for Strain Sensor Applications

Abstract:

In the field of materials science and engineering, materials that respond to stimuli, such as strain or humidity, and materials that are â€œsoft,â€ or bendable and stretchable, are becoming more commonplace. Mechanochromic devices are a class of technology that change color or transparency properties when under mechanical stress or strain. This phenomenon is largely achieved through changes in the physical surface structure of the device, such as the formation of wrinkles or cracks in the top layer of a multi-layered composite material. For some mechanochromic devices, the composite material is an elastomer substrate topped with two or more layers of rigid films formed from pigments or dyes; the elastomer substrate can also be colored to reduce the number of rigid film layers. By design, mechanochromic devices are soft materials that can bend or stretch. Furthermore, the sensitivity of mechanochromic devices can be controlled via careful design of the device; the method of pre-stretching the substrate before the colored layers are applied can be used to tune the mechanochromic response. Therefore, by controlling the response of mechanochromic devices, mechanochromic devices can be applied as mechanical strain sensors. In a time where many manufacturing fields are moving towards soft technologies, such a strain sensor is appealing. Additionally, due to their relatively facile preparation procedures and conditions, mechanochromic devices can easily be integrated into the design of other soft technologies. The ideal use of the mechanochromic devices in this capacity would be to tune them to change color when a soft technology is being stretched to its point of mechanical failure; this blatant, qualitative indication has the potential to stop users from breaking expensive technology. This has significant implications in fields with materials moving in a soft direction, such as stretchable electronics, which have potential applications in robotics, aerospace, and smart clothing technologies.

Copyright © American Institute of Chemical Engineers. All rights reserved.

510915 Development of Alternative Applications for Artificial Turf through Novel Extrusion Methods

Abstract:

Artificial turf continues to be used as a substitute for grass fields all over the world. Currently there is no recognized recycling method for artificial turf material, which results in a substantial amount of waste when a field is replaced. There is a need for new and innovative ways to recycle the plastic fibers of the artificial turf. Through reprocessing the turf using methods such as Single Screw Extrusion (SSE), Solid-State Melt Extrusion (SSME), Twin Screw Extrusion (TSE), and Solid-State Shear Pulverization (SSSP), several material properties were evaluated to test the strength, flexibility, and hardness of the recycled plastic. By comparing these values to literature values for common polymers, possible applications for the reprocessed turf are explored. We present two different approaches for possible applications for the recycled plastic. The first approach to recycling turf looks to link the production, consumption, and recycling of the turf by utilizing the recycled plastic as the rubber-like infill for new fields. The second approach focuses on small-scale applications in which the recycled plastic is used as 3D printing filament. Ultimately, the most cost effective and impactful use for recycled turf will be determined by evaluating both large and small-scale applications.

Copyright © American Institute of Chemical Engineers. All rights reserved.

510886 Solution Shearing As a Strategy to Control the Morphology of Organic Photovoltaic Active Layers

Abstract:

As global demand for renewable energy sources increases, solution-processable organic semiconductors have tremendous potential to enable large-scale manufacturing of lightweight, flexible and low-cost solar panels for light energy harvesting. The photoactive layers of these devices typically comprise a mixture of an electron donating polymer and an electron accepting small molecule that are co-deposited from solution onto electrodes. One of the main challenges facing the commercialization of organic solar cells is the lack of control over the morphology of the photoactive layer, which in turn leads to poor device efficiencies. Furthermore, at least 90% of the current research field uses batch spin coating methods to deposit the photoactive layer. Our research team has recently developed a novel approach to use shear forces to tune the phase separation between the electron donor, poly(3-hexylthiophene) (P3HT), and acceptor, phenyl-C61-butyric acid methyl ester (PCBM), prior to film deposition. Comprehensive rheological testing has revealed that shearing P3HT/PCBM solutions in the gel state can induce the nucleation and crystallization of both P3HT and PCBM. The strength of the gel phase was found to depend on the solvent type, but was independent of the presence, or absence of PCBM. Interestingly, replacing PCBM with a different small molecule, 6,13-bis(triisopropylsilylethinyl)pentacene (TIPS-Pentacene), significantly interfered with the crystallization of the P3HT. We also examined the incorporation of a popular processing additive, 1,8-octanedithiol (ODT). ODT preferentially dissolves PCBM and has a lower vapor pressure than the host solvent, in this case dichlorobenzene. Due to both of these properties, the presence of ODT was found to enhance the crystallization of P3HT. Rheological testing was performed to observe the effects of ODT on the gel strength of P3HT solution. It was discovered that increasing ODT concentration decreased gel strength in both the sheared system and non-sheared system. Significantly, this shearing process is compatible with continuous processing methods that will decrease the cost of fabricating organic solar cells and propel this technology towards commercialization.

Copyright © American Institute of Chemical Engineers. All rights reserved.

511044 Manufacturing Methods for Photoresist Adhesion in Nano-Antenna Fabrication

Abstract:

Hypothesis

Proper fabrication of prototypes will achieve the desired hardness characteristics to allow best adhesion to the underlying substrate along with the best hardness characteristics.

Summary of Work

Antenna rectification could be a viable source of alternative energy, and can be more efficient than current solar and thermal alternative energy sources. Nanoantenna electromagnetic collectors (NECs) which combines the technologies of antennas, which collects electromagnetic energy to form a charge and alternating current, with rectifiers which converts this alternating current (AC) in to useable direct current (DC). Laboratory scale fabrication is built upon silicon wafer prototypes with SU-8 photoresist being built into the nanoantennas. Contact Photolithography is the mechanism used to pattern the antenna array on the silicon substrate surface using Ultra-violet (UV) light against the substrate while it is in contact with an optical masks. Exposure dosage and post exposure development time also affects the viscosity and density of the photoresist. Alignment of the SU-8 mask and conductive metals is also of great importance for the resistance and linearity of the final prototype so great care must be taken in aligning optical masks onto the substrate. Patterned SU-8 is then wet etched to remove unexposed photoresist to reveal SU-8 pillars embedded in the conductive metal. Rectifying diodes are then added after fabrication. Following fabrication of the prototype and application of diodes, current versus voltage test are performed to obtain variances in charge produced across different temperatures. Further study is required to differentiate legitimate rectification from the thermocouple effect between metals.

Copyright © American Institute of Chemical Engineers. All rights reserved.

511150 Automated Non-Destructive Corrosion Detection for Savy Container Surveillance

Abstract:

SAVY containers are prone to interior corrosion while in use. Non-destructive testing (NDT) for corrosion within SAVY containers will augment current container monitoring procedures. Ultrasonic wall-thickness testing (UT) and eddy current array (ECA) flaw detection were chosen following a feasibility study and consultation with industry experts. UT testing can measure loss in wall thickness as well as signal attenuation due to pitting. ECA uses electromagnetic induction to detect surface and subsurface flaws.

To understand the capabilities and instrumental response of each technology, several test samples were created using 316L stainless steel plates with the same thickness as SAVY containers. In a collaboration with New Mexico Tech, 316L plates were corroded to varying degrees with HCl and FeCl3 in controlled conditions. A custom calibration plate was machined with indentations of varying diameters and depths to simulate different levels of pitting corrosion. Finally, two corroded SAVY containers and a control were made available for testing by the container management team in NPI-2. Each of these different samples were used to determine the accuracy of UT wall thickness measurements, the smallest pit detectable by UT and ECA analysis, and the limitations of each technology.

UT and ECA testing will complement each other in an automated inspection system to generate a color-coded plots of the interior of each container. This data will enable rapid visual inspection and flaw detection, with each container serving as its own control. An automated NDT system will reduce the amount of time workers are in radiation areas, increasing worker safety as well as surveillance cycle efficiency. More frequent surveillance of at-risk containers, creation of a detailed record of each container, and long-term data collection for lifetime certifications will be possible.

Copyright © American Institute of Chemical Engineers. All rights reserved.

510883 Characterization of Melting Point Depression and Phase Change Behavior in [Tmsfn][Tf2N] + CO2 Biphasic Systems

Abstract:

Ionic liquids have become the subject of intense research due to their unique chemical properties including negligible vapor pressure and molecular tunability, with a theoretical quantity of 10^{18} unique cation/anion combinations. However, a significant proportion of ILs have high melting temperatures (>75°C) which prohibit using these novel salt molecules in processes that operate around room temperature (23°C). Previous studies have shown that the melting point of an ionic liquid can be lowered by absorption of a compressed gas. However, further research is needed to quantitatively characterize the depressed melting point. We will present findings for an ongoing study investigating the melting point depression of the model ionic liquid trimethylsulfonium bis(trifluoromethylsulfonyl)imide [TMSfn][Tf_2N] using compressed carbon dioxide. In this study a gravimetric microbalance with a precision of +/- 1 microgram was used to precisely measure the weight of a [TMSfn][Tf_2N] sample under constant carbon dioxide pressure while temperature was slowly increased. Upon melting the sample weight increased significantly due to sudden carbon dioxide uptake caused by the solubility difference between the solid and liquid phases. The depressed melting point of [TMSfn][Tf_2N] was measured under 5, 10, 15, and 20 bar of carbon dioxide. These methods can be readily applied to other systems and the collected data will provide a quantitative characterization of their phase behavior.

Copyright © American Institute of Chemical Engineers. All rights reserved.

510972 Effect of Structural Differences in the Hydrophobic Alkyl Backbone and Functional Groups on the Effectiveness of Solvents for Water Desalination

Abstract:

Previous work from Bajpayee, Luo, Muto and Chen[1], demonstrated a novel water desalination method, directional solvent extraction (DSE), utilizing decanoic acid as the directional solvent (DS). DSE involves the direct extraction of pure water from a saline solution into DS at a moderately high temperature (~40°C -80°C) in which salts are insoluble, followed by recovery of the dissolved pure water from DS at a lower temperature by phase separation. Efficiency of this process requires that, the solubility of DS in the saline solution and in the resulting recovered pure water be negligible. The amphiphilic molecular structure of decanoic acid provides the properties of a suitable DS, which includes a polar and hydrophilic carboxylic acid group, which facilitates association with water, as well as a non-polar alkyl chain, which allows for the rejection of ions as well as the insolubility of the solvent in saline and pure water.

The objective in this project is to identify a DS with a 3X increase in effectiveness when compared to decanoic acid, which can make the DSE process more energetically efficient than a reverse osmosis process. Effectiveness is defined as the mass of pure water extracted per mass of directional solvent used in a DSE cycle and is calculated as a percentage. When using the temperature pair of 70Â°C and 34Â°C in a DSE cycle, decanoic acid has an effectiveness of approximately 0.9% to 1.7%. Our hypothesis is that effectiveness, as well as the solubility of water and salts in DS, and the solubility of DS in saline solution and pure water is tunable through varying the molecular structure of DS. As the first step, we performed a literature survey and analysis of existing solubility data of organic compounds with water, to identify key molecular factors that lead to DSs with high water extraction efficiency. Specifically, we compiled data for 28 ketones, 21 alcohols and 14 amines with varying â€“CH2â€“ backbones. The data includes the mutual solubilities of the organic solvents with pure water, over the temperature range of approximately 0°C-90°C. We converted the solubility data of pure water in bulk organic solvent to mass ratios, to calculate DS effectiveness for the temperature pair of 70°C and 30°C. In addition, we converted the solubility data of organic solvent in bulk pure water to parts per million at 70°C and 30°C. Allowing for the convenient analysis of the performance of the selected solvents in a DSE cycle.

The analysis performed supports a general trend, for organic compounds with an alcohol or ketone functional group and a non-polar component (which may be a normal alkyl chain, branched alkyl chain, or cycloalkyl group), that as temperature increases the solubility of water in an organic solvent increases. Conversely, with amines, we observe that as temperature increases the solubility of water in the organic bulk decreases. Thus, desalination with amines may be possible by reversing the DSE cycle. This is an encouraging result as this suggests that the type of behavior desired of a suitable DS is not unique to decanoic acid, but rather, very common to amphiphilic solvents. Furthermore, this study supports our central hypothesis that effectiveness, as well as the solubilities of water and salts in DS, and the solubility of DS in water is tuneable by varying the molecular structure of DS. This analysis concludes that, to minimize the solubility of a DS in bulk pure water, the optimal structure of the hydrophobic backbone is an n-alkyl chain. In addition, we conclude that the optimal hydrophilic functional group position is on the terminal carbon of a DS. Since alcohols show approximately a 3X greater effectiveness than ketones with similar structures, we hypothesize, that to increase the effectiveness of a DS, the hydrophilicity of the polar component of a DS must be increased. For example, we expect that replacing the carbonyl group on the base structure of decanoic acid with an alcohol, as with the structure of 1,1-decanediol, will result in a DS with greater effectiveness than decanoic acid while maintaining low solubility in pure water.

1. Bajpayee, A., Luo, T. F., Muto, A. & Chen, G. Very low temperature membrane-free desalination by directional solvent extraction. *Energ Environ Sci* **4**, 1672-1675 (2011).

Copyright © American Institute of Chemical Engineers. All rights reserved.

510185 Phase Transformation Kinetics of Suberic Acid Polymorphs and Amorphous Dispersions Formed from Monodisperse Droplet Evaporation

Abstract:

Recognizing the importance of the crystallization of organic molecules is important in many industries, ranging from pharmaceuticals to food products. These molecules have particle properties that are required for a specific product to have the desired characteristics. These properties include size, morphology which is the particle shape, and internal structure which is the arrangement of molecules within the particle. They can affect bioavailability, solubility, and ease of processing. The internal structure of a solid can be either crystalline (where there is a regular pattern to the molecules in the solid) or amorphous (where there is no long-range pattern). Polymorphs are different crystalline arrangements of a specific molecule. At any given temperature and pressure there is one polymorph that is thermodynamically stable, while other polymorphs are metastable. Finally, an amorphous form is typically the least stable in comparison to all available crystalline forms. Since amorphous particles have a less stable internal structure, they are able to break down easily and are therefore more soluble, which is often desired in the pharmaceutical industry.

In this work, I studied the crystallization and phase transformation kinetics of particles formed via monodisperse droplet evaporation of suberic acid solutions. A mixture of suberic acid and a polymer additive: polyvinylpyrrolidone (PVP) produces particles ranging from crystalline (formed from pure suberic acid) to partially and fully amorphous (both formed with the presence of PVP), depending on the suberic acid:PVP ratio. X-Ray Diffraction (XRD) data shows the transitioning of an initially fully amorphous suberic acid/PVP sample to one that is partly crystalline. Suberic acid has two published polymorphs: alpha and beta and evidence of a third polymorph, gamma, was recently discovered via production in a Vibrating Orifice Aerosol Generator (VOAG). A sample formed from pure suberic acid in isopropyl alcohol showed initial evidence of a mixture of alpha and gamma polymorphs, however over time the metastable gamma form showed evidence of transitioning to the stable alpha form.

Copyright © American Institute of Chemical Engineers. All rights reserved.

509814 Manganese Oxide Blended Polyethersulfone Ultrafiltration Membrane for Enriched Dye Removal Applications

Abstract:

Polyethersulfone (PES) is one of the most preferred polymers for ultrafiltration membrane synthesis, especially for industrial effluent treatment. However, pure PES membranes are susceptible to low permeate fluxes and membrane fouling due to their hydrophobic characteristics. Thus, modification of the pure PES membrane is an indispensable step to achieve blend systems with improved ultrafiltration performance and better antifouling abilities. In this current work, a series of manganese oxide (Mn_3O_4) blended polyethersulfone (base polymer) UF membranes were prepared using wet phase inversion by varying Mn_3O_4 from 0 to 4 wt%. Prepared membranes were analyzed for their morphology, thermal stability and membrane characterization. Membrane characterization results indicated that the PES/ Mn_3O_4 blend membranes possessed enhanced fluxes and better antifouling ability than the pristine PES membrane. Dye removal performances were more desirable for the composite membranes than the pristine PES membrane. A close analysis on the obtained result showed that the 2 wt% Mn_3O_4 blend PES membrane possessed desirable characteristics among the synthesized composite membrane series. Thus the manganese oxide blended polyethersulfone membranes seem to be more promising candidates for textile effluent treatment, especially for dye rejection.

Keywords: Membrane; Polyethersulfone; Manganese oxide; Ultrafiltration; Dye removal

References:

1. Crull A, Membranes for the Nineties: Highlighting Surface Modification Technology, Business Communications Co., Nonvalk, CT, 1990.

2. Zhao, Changsheng, et al., Modification of polyethersulfone membranesâ€"a review of methods, Progress in Materials Science,2013, 58,76-150.

3. Susanto, Heru; Mathias Ulbricht, Characteristics, performance and stability of polyethersulfone ultrafiltration membranes prepared by phase separation method using different macromolecular additives, Journal of Membrane Science, 2009, 327, 125-135.

4. Law Yong, et al., Polymeric membranes incorporated with metal/metal oxide nanoparticles: a comprehensive review, Desalination, 2013, 308, 15-33.

Copyright © American Institute of Chemical Engineers. All rights reserved.

510159 Separation of Aqueous Solutions By Sustainable Hybrid Process Design

Abstract:

Separation of Aqueous Solutions by Sustainable Hybrid Process Design

Nooshin Shahlari BS. Chemical Engineering University of Kansas ; Kyle V. Camarda PhD. University of Kansas, Lawrence, KS 66045; Rafiqul Gani PhD. Technical University of Denmark, S¿ltofts Plads Lyngby

Abstract

Distillation is an extremely energy intensive process, consuming about 10-15% of the worldÕs energy consumption. The rise of carbon dioxide in the atmosphere leads to the greenhouse effect, which traps the sunÕs energy on earth and results in an overall temperature increase. 90% of separation processes energy in the chemical and petroleum industries, which results in a large carbon footprint.[2] Replacement of energy intensive distillation processes with a hybrid membrane-distillation processes is proposed for aqueous separations in this paper to improve the energy efficiency the large energy consumption and greenhouse gas emissions. Three sets of aqueous separation systems simulated using Pro II, comparing traditional distillation with hybrid membrane distillation systems. Results show that an energy and CO_2 savings of 95% for acetic acid/water solutions are possible, 92% for acetone/water, while 42% for methanol/water with the hybrid systems. For the three separations investigated, the payback period for small scale systems was 0.13, 0.07, and 0.53 years respectively. A hybrid distillation process decreases energy usage by 63-99% to reach purities of 90-99.5% in these aqueous mixtures. This study showed that hybrid membrane distillation systems are economically attractive for important industrial separations.

Sources

1. Sholl, D. S., & Lively, R. P. (2016). Seven chemical separations to change the world. *Nature, 532*(7600), 435-437. doi:10.1038/532435a

2. Oak Ridge National Laboratory. *Materials for Separation Technologies: Energy and Emission Reduction Opportunities* (2005).

Copyright © American Institute of Chemical Engineers. All rights reserved.

510559 Dielectrophoretic Response of Condensed DNA Clusters in AC Fields

Abstract:

Insulator-based dielectrophoresis (iDEP) has been applied in analytical research as a method to separate and fractionate DNA in lab-on-a-chip devices. The development of similar continuous-flow, size-based separation techniques for DNA Next-Generation Sequencing (NGS) could advance the speed, efficiency, and accuracy of in situ analysis of genomic/biological data. When low-frequency and high-amplitude AC potentials are applied to homogeneously disperse Î»-DNA, reversible segregated clusters are observed in high frame rate imaging. However, the fundamental physics of such cluster phenomena are poorly understood and limited studies have been published referencing the aggregation formation realized under these conditions. This recently observed clustering phenomenon has gained the attention of colloidal physicists who wish to understand the physical parameters that influence clustering of Î»-DNA. We believe applied low-frequency potentials generate this clustering behavior, which influences the dielectrophoretic response of DNA molecules. As a result, this phenomenon could be used to improve DNA sample preparation and pre-concentration using microdevices for future NGS applications. Research investigating cluster formation through modeling and computer simulations has shown that this phenomenon may be due to electrohydrodynamic instability including by not limited to dipole-dipole interactions between molecules. However, these models often address simpler DNA systems and fail to predict the unexpected switch we observe from positive to negative DEP in DNA clustering. Clustering behavior has the potential to enhance the efficiency of DNA fractionation by exploiting strong AC electric fields to induce condensation and separation of DNA molecules, thus it is important that we understand the physics behind this phenomenon. Our recent work examines DNA cluster formation and migration characteristics using various electric field parameters in order to quantify the movement of DNA particles in solution. Preliminary evidence suggests that the electrokinetic mobility of DNA clusters changes negligibly with changes in frequency. Various amplitudes and frequencies of applied potential were examined to determine the extent of correlation in DNA clustering. DNA clustering was largely observed with frequencies ranging from 10 to 100 Hz and electric field strengths above 800 V/cm. Similar electric field parameters have produced some of the highest-observed sorting efficiencies in an iDEP constriction sorter for DNA molecules ranging in size from 10 to 50 kbp. Future work will aim to enhance techniques capable of automatically tracking DNA clusters throughout high-frame rate imaging to further understand clustering and transport mechanisms. Such techniques will significantly reduce biosensing and bioanalytical processing times.

Copyright © American Institute of Chemical Engineers. All rights reserved.

510799 Study of Enantiomeric Separation of Propylene Oxide Using L- or D-Penicillamine Modified Au Nanoparticles

Abstract:

Several pharmaceutical drugs synthesized in the laboratory are chiral. However, while one enantiomer of the drug is beneficial, the other enantiomer could be harmful. As a result, the enantiomers of chiral pharmaceutical drugs have to be separated before using the drug. The current method used for separation of enantiomers of chiral pharmaceuticals is High Pressure Liquid Chromatography (HPLC). This method is expensive and time-consuming. In response to the issue, we are exploring the idea of using chirally modified gold nanoparticles for enantiomeric separations of chiral pharmaceutical drugs. This work investigates the use of chirally modified urchin-like gold nanoparticles for enantioselective separation of racemic propylene oxide (PO) from the solution phase. Urchin-like gold nanoparticles were successfully synthesized according to Li, Jing, and *et.al.* [1]. Surface of the gold nanoparticles was modified with L- or D-Penicillamine in order to make them chiral. Gold nanoparticles were used to study separation of racemic mixture of PO. Propylene oxide is used as a probe molecule this work. Polarimetry was used to determine the optical rotation of the solutions and the extent of enantiomeric separations.

[1] Li, Jing, et al. â€œControllable Synthesis of Stable Urchin-like Gold Nanoparticles Using Hydroquinone to Tune the Reactivity of Gold Chloride.â€ The Journal of Physical Chemistry C, vol. 115, no. 9, Oct. 2011, pp. 3630â€“3637., doi:10.1021/jp1119074.

Copyright © American Institute of Chemical Engineers. All rights reserved.

510836 Predicting the Dynamic Binding Capacities for Adsorbents Using BSA As a Model Protein

Abstract:

Predicting the dynamic binding capacities for adsorbents using BSA as a model protein

Authorâ€™s name: Gaoshan Li

Presenterâ€™s name: Gaoshan Li

Faculty mentor: Dr. Heather Chenette

Abstract

Membrane adsorbers are one alternative to traditional bead-based chromatography that receive attention in the bioseparation process because of their ability to process material with lower residence times, and thus, higher product throughput. Another important metric is the dynamic binding capacity (DBC), which is defined as the mass of protein bound to the column per volume of the column adsorbent when the outlet stream reaches a certain concentration [2]. The equilibrium binding capacity (EBC) is the binding capacity when mass transfer is at equilibrium status such that the rate of protein adsorbing is equal and opposite to the rate of protein desorbing. While the DBC includes the effects of mass transfer under dynamic conditions, there are instances when it is more convenient to measure EBC, as it can be performed easily by measuring the equilibrium concentration. Along with determining the processing time and product resolution, determining the dynamic binding capacities (DBC) of innovative protein adsorbers as a performance characterization is also of importance. This is relevant for selecting appropriate materials for the purification process of biological products, such as monoclonal antibodies. Researchers have shown how EBC measurements can predict DBC values for specific cases when assuming that the Langmuir isotherm is rectangular and the intra-particle mass transfer is controlled by pore diffusion [1,2]. However, some properties of membranes, for example, the substrate morphology, make the assumptions that apply to the traditional bead-based chromatography invalid. Therefore, the model that predicts the DBC from the EBC cannot be applied to the membrane adsorbers. The goal of this research is to extend our analysis to determine DBC values using the existing theory and experimental EBC measurements using bovine serum albumin (BSA) as a model protein for commercial anion exchange resin (Q-Sepharose Fast Flow) adsorbers. In addition, this work explores existing literature on predicting DBC values for non-traditional adsorbent types, like those used in membrane chromatography, identifies the challenges with this method, and presents preliminary experimental EBC results for commercial anion exchange membrane adsorbers (Sartobind Q).

Reference

[1]Carta, Giorgio, and A. Jungbauer. *Protein Chromatography: Process Development and Scale-Up*. 2010.

[2]Carta, Giorgio. "Predicting protein dynamic binding capacity from batch adsorption

tests." *Biotechnology Journal* 7.10(2012):1216â€“1220.

Copyright © American Institute of Chemical Engineers. All rights reserved.

510979 Chemical-Free Electrostatic Separation As a Dry Approach for Enrichment of Plant Proteins

Abstract:

Processing of plant-based resources and fractionation of their major components (protein and starch) still relies on wet conventional technologies that involve the use of solvents, concentrated acids and alkali through a series of operations that include aqueous extraction, heating, centrifugation, isoelectric precipitation, and energy-intensive dehydration step. These harsh processing conditions result in protein denaturation by reducing the quality, solubility and functionality of the protein components. We have developed a dry tribo-electrostatic separation approach for producing protein-rich fractions from different agricultural flours including oat, yellow pea and navy bean flours. Our process consisted of a fluidized bed, polytetrafluoroethylene (PTFE) tribo-charging tube, and a rectangular separation chamber supplied with two electrode plates made of copper. For this approach, the milled flour particles were initially suspended in the fluidized bed before being passed through the PTFE tribo-charging tube. The milled flour particles were then charged in contact with the PTFE media, according to their charging behavior. Protein particles are characterized by high ionizability, and they can be charged upon physical contact with the tribo-charging media. However, starch particles are characterized by low ionizability, thus, they cannot be charged effectively using triboelectricity. The charged protein particles were finally separated from uncharged starch granules in the rectangular chamber under the influence of an applied electric field. The influence of process parameters including air/nitrogen flow rate, tribo-charger tube length and plate voltage on the protein enrichment level was evaluated. Using this approach, the protein content of the original flours were increased by at least 50% in the final protein-rich fractions collected from the electrode plates. This proposed process provides an efficient, inexpensive, and environmentally friendly alternative to conventional protein separation processes in a chemical-free environment. The approach also preserves the native functionality of the proteins and eliminates the costly treatment of proteinaceous effluent released as a by-product stream during wet processes.

Copyright © American Institute of Chemical Engineers. All rights reserved.

511146 Efficient Enrichment of 1,3-Propanediol from Fermentation Broths Using Imidazolium Dibutylphosphate Ionic Liquid Based Methacrylate Polymer Membranes

Abstract:

Glycerol, a platform chemical for the production of intermediate bioproducts, can be obtained as a low-value waste feedstock from a wide variety of industrial sources. Its fermentative conversion to bio-renewable chemicals such as 1,3-propanediol (1,3-PD) is an environmentally and economically sustainable recourse to its utilization. 1,3-PD is an important chemical intermediate in manufacturing of polyethers, polyurethanes, and polyesters. It is also widely used in food, cosmetic, and pharmaceutical industries. The principal impediment in the fermentative conversion of glycerol to 1,3-PD lies in the downstream enrichment of 1,3-PD concentration from very low starting concentrations. Conventional processes, like evaporation, chromatography, and reactive extraction have been used but are associated with high-energy usage and process limitations which can be mitigated with pervaporation. The challenge of pervaporative enrichment of 1,3-propanediol (1,3-PD) from dilute aqueous mixtures was addressed by the synthesis of a novel hydrophobic imidazolium alkyl phosphate ionic liquid structure based monomer. Vinylimidazole was â €œcoupledâ€ with mercaptoethanol, converted into an alkylphosphate ionic liquid followed by esterification to the final monomer structure. Characterization and batch pervaporation was carried out on several different polymer chemistry mediated membrane structures based on this monomer, including semi-interpenetrating polymeric network structures with hydrophobic polyglycerol sebacate acrylate (PGSA). The batch pervaporation exhibited high separation factors and fair 1,3-PD fluxes compared to other hydrophobic membranes. The monomer provides recourse to high performing materials by balancing its 1,3-PD affinity with its hydrophobic nature. The excellent cost-performance trade-off makes this an attractive candidate for possible scale-up and industrial applications in multilayer modules.

Copyright © American Institute of Chemical Engineers. All rights reserved.

511147 Ptfema-r-PMAA Functionalized Membranes for Metal and Dye Separations

Abstract:

Functionalized polymer membranes have exciting potential applications in chemically selective membranes. Here we explore the usage of a poly(methacrylic acid)-polytetrafluoroethylene random copolymer (PMAA-r-PTFEMA) as a charge-selective layer and for metal ion capture. PMAA-r-PTFEMA self-assembles in methanol to present negatively charged carboxylic acid groups on the pore surfaces. This has been demonstrated to allow for charge-based separation of dyes of similar sizes, rejecting negatively charged dyes and allowing positively charged dyes to pass. The carboxylic acid groups have also been shown to adsorb metals in solution. The membraneâ€™s applications in removal of heavy metal from water and their ability to regenerate is currently being studied.

Copyright © American Institute of Chemical Engineers. All rights reserved.

509579 Efficient Coacervate Extraction of Cationic Industrial Dye from Wastewater

Abstract:

Effluent wastewater containing dyes from textile, paint, and various other industries have long posed environmental issues. Functional Nano materials offer new opportunities to treat these effluent wastes in unprecedentedly high efficient and facile fashion due to their large surface area-to-volume ratio. In this work, we explore a new organic-inorganic hybrid nanomaterial based on the complexation of zwitterionic polyampholyte with anionic polyoxometalate (POM) nanocluster to demonstrate coacervate extraction of cationic dye, methylene blue (MB) as a model waste from its salted aqueous solution. MB wastewater samples were prepared at a concentration of 0.2 g/L and were used to subject the hybrid complexes to multiple rounds of extraction to determine the upper limit of adsorption of the material. Efficiency of coacervated complexes at varied POM concentrations and molar ratio to polyampholyte were measured using a UV-vis spectrophotometer. Our polyionic complexes exhibited ~99% removal efficiency for all initial MB wastewater samples with measured adsorption capacity as high as 352 mg/g, indicating extremely high efficiency. Our study also suggested that coacervate extraction could be a facile and robust method for ionic selective wastewater treatment with parameters such as temperature and pH showing no influence on removal efficiency.

Copyright © American Institute of Chemical Engineers. All rights reserved.

510607 Floating Membraneless PV-Electrolyzer Based on Buoyancy-Driven Product Separation

Abstract:

Although the sun is a promising energy source to meet societyâ€™s demands, the storage of solar energy during intermittent hours remains a significant challenge. One possible solution is to electrolyze water into hydrogen fuel. The ocean is an untapped source of abundant water and space, but modern proton exchange membrane electrolysis devices are ill-equipped to electrolyze seawater due to their intolerance to the impurities present in the ocean. Motivated by this challenge, our group has recently investigated the application of membraneless electrolyzers based on mesh flow-through electrodes for seawater electrolysis. In this study, we have demonstrated a stand-alone, floating PV electrolyzer capable of safe and reliable separation of H_2 and O_2 product gases via bubble buoyancy in the absence of forced convection. By selectively depositing the electrocatalyst on only one side of mesh flow-through electrodes, we were able to separate the product gases with high purity in an un-pumped, free-standing electrolyte. This electrode architecture was applied to a 3D printed floating laboratory-scale PV-electrolyzer that achieved a solar-to-hydrogen efficiency of ~5%.[1]

[1] Davis, *et al.* 2017 Floating Membraneless PV-Electrolyzer Based on Buoyancy-Driven Product Separation *Manuscript in Submission.*

Copyright © American Institute of Chemical Engineers. All rights reserved.

509767 The Development and Application of High Performance Liquid Chromatography Capabilities for Lanthanide Separations

Abstract:

Nuclear fission produces multiple lanthanide (Ln) isotopes, many of which yield valuable insights into the type and origin of nuclear material. The ability to carefully separate and analyze selected lanthanide radioisotopes can therefore be a key tool for nuclear forensic applications. An effective method for separation of lanthanide cations is High Performance Liquid Chromatography (HPLC). HPLC techniques have long been used to separate lanthanides, and our goal has been to modernize current capabilities through the development of both a new â€˜non-radâ€™ test system and an expansion of the current rad capabilities. The chemistry is well developed, selective stripping of trivalent lanthanide cations from a cation exchange resin using Î±-hydroxyisobutyric acid (HIBA), but instrumentation advances yield scope for process improvements. Recently, a new HPLC system has been installed comprising of quaternary pumps, column and a Diode Array detector that allows for post column on-line analysis. Addition of PAR (4(-Pyridylazo)resorcinol)) for post column derivatization facilitates accurate determination of lanthanide cation elution using the diode array detector. This new system then served as a template for the installation of a new multi-column radiological HPLC. The application of both capabilities towards the development of enhanced Ln separations will be described.

LA-UR-17-26126

Copyright © American Institute of Chemical Engineers. All rights reserved.

511067 CO2 Capture Characteristics of MgO Produced from Calcination of Metal Organic Frameworks

Abstract:

As the correlation between CO_2 emissions and global climate change has become increasingly clear, research has been conducted into methods to sequester Carbon Dioxide and store it instead of releasing it into the environment.[1,2] Many methods for carbon capture have been formulated, but this poster focuses on CO2 capture using solid adsorbents, more specifically, the use of Magnesium Oxides to adsorb CO_2. Sequestration of Carbon Dioxide using Magnesium Oxides is advantageous because of the wide availability of raw materials, low energy cost of regeneration, high adsorption capacity, and low toxicity of the adsorbents.[3,4] However, conventionally prepared Magnesium Oxides lose their porous structure and therefore their adsorption capacity significantly after multiple adsorption/desorption cycles.[5,6,7,8]

To solve this problem, a novel method of Magnesium Oxide nanoparticle synthesis via decomposition of metal organic frameworks (MOFs) containing Magnesium was developed. To synthesize these Magnesium Oxide particles, a pre-synthesized MOF containing Magnesium was calcined in an air atmosphere. This process broke the organic ligands holding the MOF structure together, while simultaneously oxidizing the magnesium metal groups to Magnesium oxides. The composition and structure of the resulting particles were characterized using, among other methods, XRD, SEM, TEM, and BET analysis. Furthermore, the CO2 absorption capacity of the synthesized material was determined using thermal gravimetric analysis. Resulting data shows that the resulting nanoparticles have high surface area and CO2 adsorption capabilities. The most significant conclusion that can be drawn from this study is that Magnesium Oxide nanoparticles synthesized using this method are more robust than conventionally prepared Magnesium Oxides, and are able to retain higher adsorption capacities after multiple adsorption and desorption cycles.[7,8]

References:

[1] C. F. Cogswell, H. Jiang, J. Ramberger, D. Accetta, R. J. Willey, and S. Choi, Langmuir,

2015, 31, 4534–4541.

[2] D. Andirova, C. F. Cogswell, Y. Lei, S. Choi, Micropor. Mesopor. Mat., 2016, 219, 276–305.

[3] S. Choi, J. H. Drese and C. W. Jones, ChemSusChem, 2009, 2, 796–854.

[4] S. Lee and S. Park, J. Ind. Eng. Chem., 2015, 23, 1–11

[5] G. Song, X. Zhu, R. Chen, Q. Liao, Y.-D. Ding and L. Chen, RSC Adv., 2016, 6,

19069–19077

[6] G. Song, Y.-D. Ding, X. Zhu and Q. Liao, Colloids Surf., A, 2015, 470, 39–45

[7] A. Hanif, S. Dasgupta and A. Nanoti, Ind. Eng. Chem. Res., 2016, 55, 8070–8078

[8] A. M. Ruminski, K.-J. Jeon and J. J. Urban, J. Mater. Chem., 2011, 21, 11486–11491

Copyright © American Institute of Chemical Engineers. All rights reserved.

510539 Modifying Adsorption Kinetics of Zeolites Using Self-Assembled Monolayers

Abstract:

Zeolites, due to their defined pore size and structure, are used to separate molecules based on their kinetic diameter in a process known as molecular sieving. Compared to their traditional polymer membrane counterparts, zeolite membranes show improved thermal and chemical stability in separation processes. This stability allows for the effective usage of zeolites in natural gas separations, such as the separation of CH_4 and CO_2. However, longer alkanes commonly found in natural gas feed streams such as propane and butane tend to adsorb in zeolite pores, blocking the pores and reducing separation selectivity. Thus, a method to selectively adsorb CO_2 during natural gas separations while reducing alkane adsorption is desirable.

The potential for self-assembled monolayers (SAMs) to control zeolite selectivity was evaluated in this study. SAMs have been used as surface modifications for catalytic applications. Specifically, self-assembled monolayers can tune the surface properties of catalysts and alter their selectivity. In principle, this same technique can be applied to tune zeolite surfaces to selectively control alkane fouling and ultimately reduce adsorption of these contaminants.

Zeolite surface modifications were applied on two commercially available zeolites, 5A and ferrierite, using alkylphosphonic acid self-assembled monolayers. The SAM monomers consisted of phosphonic acid head groups covalently bonded to the zeolite surface and *n*-alkane tail groups organized through Van der Waals attractions. The presence of well-ordered SAMs on the zeolite surfaces was confirmed using FT-IR analysis.

Alkane adsorption tests using butane and propane were also performed. The initial rates of alkane adsorption decreased on zeolites treated with the SAMs. This decrease in adsorption rate can be attributed to nonpolar interactions between the carbons in the alkanes and the carbons in the self-assembled monolayer. Additionally, the total adsorption and rate of adsorption of CO_2 were relatively unchanged, suggesting that these modified zeolites selectively adsorb CO_2 and decrease the rate of adsorption of contaminants such as propane and butane.

Copyright © American Institute of Chemical Engineers. All rights reserved.

510793 Engineering of Physical Aging to Optimize Carbon Molecular Sieve (CMS) Hollow Fiber Membranes

Abstract:

Carbon Molecular Sieve (CMS) hollow fiber membranes can transcend the permeability/selectivity upper bound for many important gas pair separations such as CO_2 vs CH_4 and C_3H_6 vs C_3H_8 [1 â€“ 2]. Such membranes are made by pyrolyzing polymeric precursors under inert atmospheres to form amorphously distributed carbon sheets. The amorphous CMS material has micropores between sheets and includes ultramicropores within the sheets themselves. There is a driving force for the carbon sheets to arrange themselves into more thermodynamically stable, denser packing states with fewer and smaller micropores. This rearrangement causes a reduction in the penetrant diffusion jump lengths and sorption capacities, thereby reducing permeance. These complex changes are henceforth referred to as *physical aging*. Physical aging can potentially limit the utility of CMS membranes. Fortunately, physical aging can be minimized by introducing condensable sorbent molecules in the CMS micropores. These temporary molecular pillars still allow structural rearrangement of the micropores to reflect the size of the sorbed molecules. Ideally, removing the sorbent molecules via a purging technique before placing in service provides the CMS with sustained attractive performance. In this poster, hollow fiber properties and several engineered storage techniques of CMS will be discussed, along with the fundamental factors associated with each.

1. Mayumi Kiyono. Carbon Molecular Sieve Membranes for Natural Gas Separations. Ph.D., Georgia Institute of Technology, Atlanta, GA, December 2010.
2. Liren Xu. Carbon Molecular Sieve Hollow Fiber Membranes for Olefin/Paraffin Separations. Ph.D., Georgia Institute of Technology, Atlanta, GA, December 2012.

Copyright © American Institute of Chemical Engineers. All rights reserved.

510626 Simulation of Aromatic Extraction Feedstock Separation By Dividing-Wall Column (DWC)

Abstract:

Problem

(1)The design and optimization of DWC is a question in process system engineering. With the development of computer simulation technology, processes can be simulated strictly. But the lack of reasonable initial parameters makes some process difficult to solve.

(2)Some factory of aromatic extraction uses naphtha(C3~C11) as feedstock, wasting extra energy and extractant to separate arene. DWC can be used to separate naphtha into C3~C5, C6~C8, C9~C11; C6~C8 can be used as source of aromatic extraction.

Work

In this study, the DWC is transformed into a model of 3 simple columns. Fenske-Gilliland-Underwood method is used for each column to shortcut design. With the classification of Underwood roots, the optimized solution can be found, using Matlab Programming. Also this method is applied in aromatic extraction feedstock separation, and some initial parameters has been provided to strict simulation. On the basis of shortcut design, Aspen Plus is used to simulate DWC strictly.

Copyright © American Institute of Chemical Engineers. All rights reserved.

510923 Separation of CO2 and N2 By Rapid PSA Using 13X Zeolite and a Dual Reflux Cycle

Abstract:

Rapid pressure swing adsorption (PSA) or rapid cycle PSA is identical to traditional PSA except that the cycle times and thus cycle step times are typically much shorter. Traditional PSA systems utilize cycle times on the order of minutes to tens of minutes, with individual step times no shorter than 0.5 min (30 s) or so. However, since the mid 80â€™s there has been a great deal of interest in trying to reduce the cycle time to make the columns smaller and thus increase the feed throughput. In fact, any reduction in cycle time results in a proportional decrease in column size, a gas separation process feature that is absolutely unique to PSA. In this study, a stream of 15 vol% CO_2 in N_2 was separated with a single bed rapid PSA system that approximated a 3-bed PSA process utilizing both heavy reflux (HR) and light reflux (LR) streams (i.e., a dual reflux (DR) cycle). The HR and LR streams were approximated with flows from cylinders containing 70 vol% CO_2 in N_2 and pure N_2, respectively. The 16.35 cm long by 3.18 cm inside diameter bed was filled with 0.3 mm diameter beads of 13X zeolite (18.88 g). The effects of cycle time and temperature on the recovery of CO_2 in the heavy product were examined for cycle times of 1 and 2 min at room temperature and at 70°C, with a high and low absolute pressures of 108 kPa and 5 kPa. For the two different cycle times, the amount of gas fed into the bed per cycle was constant. The corresponding feed throughputs were 2,100 L(STP)/h kg) for the 2 min cycle and 4,200 L(STP)/(h kg) for the 1 min cycle. The overall recovery of CO_2 in the heavy product ranged from 81.1 to 90.3%, with the most CO_2 recovered at 70°C and the 2 min cycle time. The CO_2 in the heavy product ranged from 72.5 to 79.1 vol%, with the highest occurring at room temperature with a 1 min cycle time. Overall, this preliminary study showed, possibly for the first time, that a HR step in a rapid PSA cycle is effective at concentrating the heavy component (CO_2) just like in a traditional PSA cycle.

Copyright © American Institute of Chemical Engineers. All rights reserved.

510996 Nanostructured Faujasite Zeolites for Carbon Dioxide Adsorption

Abstract:

Carbon capture is an essential way to reduce greenhouse gas emissions. One way to decrease the emissions is through the use of adsorbents such as zeolites. The experimental nanostructured faujasite (NaX) zeolite received from Dr.Dong-Kyun Seoâ€™s group (School of Molecular Sciences, Arizona State University) has a higher CO_2 capacity, 4.0 mol/kg at 298 K and 0.1 bar, than the conventional NaX zeolite. The new NaX zeolite was characterized and the physisorption properties were determined with ESEM. ASAP 2020 tests of the nano-zeolite pellets at 77K in a liquid N_2 bath determined the BET surface area of 617.0477 m^3/mol, the T-plot micropore volume of 0.279146 cm^3/g, and the adsorption average pore width of 2.07177 nm. The adsorption isotherm (equilibrium) of CH_4, N_2, and CO_2 were measured at 25Â°C. Breakthrough experiments were performed in an alumina fixed, packed bed of a length of 22 inch, 0.5 inch column i.d. at 1 atm and 298 K with nano-zeolite pellets. The gas chromatographer tested and recorded the data every two minutes with a flow rate of 10 cm^3/min for N_2 and 10 cm^3/min CO_2. breakthrough simulations of the powder and pellet form of the zeolite in an alumina tube fixed bed adsorber column were conducted on MATLAB. A linear isotherm was used for N_2, and CH_4, while the Langmuir and Freundlich adsorption isotherms were used for CO_2. It was found that the Langmuir isotherm was more accurate than the Freundlich isotherm for CO_2. The simulations determined the ideal adsorbed solution theory (ISAT) calculations, and the breakthrough curve at various CO_2, N_2, and CH_4 feed ratios. At 20Â°C the zeolite capacity for CO_2 is 5.729 mmol/g, 0.82265 mmol/g for CH_4, and 0.39603 mmol/g for N_2. The new NaX zeoliteâ€™s capacity for CO_2 is greater than the conventional NaX zeolite. The nanostructured faujasite zeolite appears to be a very promising adsorbent for CO_2/N2 capture from flue gas, CO_2/CH_4 natural gas purification, and the separation of CO_2/N_2 and CO_2/CH_4 .

Copyright © American Institute of Chemical Engineers. All rights reserved.

Post-Combustion Carbon Capture through Biphasic Absorbents

Monoethanolamine (MEA) is an industry standard absorbent for post-combustion carbon capture due to its fast reaction kinetics and high carbon dioxide (CO_2) absorption capacity. However, it requires a large regenerator and consumes significant energy during regeneration. A biphasic absorbent, e.g. a mixture of MEA-Alcohol-Water, on the other hand, requires a smaller regenerator and consumes less regeneration energy. This is possible because upon CO_2 absorption, the homogeneous absorbent turns into two immiscible phases; one is CO_2 rich, the other is CO_2 lean, and only the CO_2 lean phase is sent to regenerator. Therefore, it becomes valuable to investigate the reaction kinetics and absorption capacity of biphasic absorbents, and compare then with those of MEA. This study specifically looks at how water and alcohol, and increasing alcohol carbon chain length influence the initial reaction kinetic and final absorption capacity in biphasic absorbents. Miscibility tests were performed to ensure that all biphasic absorbents used for the CO_2 capture experiments were initially homogeneous. The CO_2 capture process for three absorbents, MEA-Water, MEA-Alcohol, and MEA-Water-Alcohol were conducted inside a microfluidic chip in a plug-slug flow experiment. Images of the CO_2 bubbles were recorded by the IDT Motion Studio and analyzed with ImageJ to determine the rate of absorption. The result for initial rate shows that MEA-Water has the fastest rate, followed by MEA-Propanol, while the biphasic absorbent, MEA-Propanol-Water has the slowest rate. However, MEA-Propanol-Water has the highest overall capacity, followed by MEA-Propanol- and MEA-Water. Raman spectra of the initial absorbent, and two separated liquid phases after the reaction indicate that the carbon capture was complete. We observed the formation of a viscous liquid for biphasic absorbents containing butanol. This phase was determined to be CO_2 rich. We also observed a color change when heat was applied to those liquids to regenerate them. The study concludes that adding water and alcohol in biphasic absorbent decreases initial kinetics but increases final absorption capacity. In practice, factors such as the amount of CO_2 in flue gas, the size of scrubber and regenerator, and the mass flow rate of absorbent should all be considered to choose the best absorbent. Special treatments should be applied to absorbents that form highly viscous liquid during the reaction in order to recover CO_2 gases for further storage. However, oxygen should be avoided during the recovery process because it may incur unexpected reactions. This study addresses the reaction efficiency problems in post-combustion carbon capture, which is one of the most promising technologies to mitigate climate change under current fossil dominant energy structure.

510920 Simulation and Optimization of Metal Organic Frameworks for CO2 Adsorption

Abstract:

Every year, emissions of carbon dioxide by the U.S. electric power sector exceeds 1800 million metric tons [1]. These gases contribute to the global warming trend, such as the rise in temperature and in water levels [2]. In the recent decades, new technologies have been introduced to capture carbon dioxide to reduce emission levels. Recently, metal-organic frameworks (MOFs) have been a promising material for CO2 capture due to its pore sizes, high-surface areas, and easy pore functionalization [3]. In this work, we will present the use of traditional computational modeling coupled with new simulation tools as a promising strategy to accelerate the commercialization of carbon capture technologies. While MOFs have been able to capture CO2, most of this work is done at the bench scale level. In this company sponsored project, we explore the simulation and optimization of MOFs for CO2 capture for large scale-up implementation. To protect the sponsor companyâ€™s technology, this project will benchmark the simulations using MOFâ€™s data already available in the literature. We developed Aspen Adsorption models of pressure and thermal swing adsorption to model CO2 separation. We modeled the separation bed to include the MOFs specifications. Once the models were validated, we varied different input parameters to analyze output recovery and rate of separation. Using the Department of Energyâ€™s Carbon Capture Simulation Initiative (CCSI), we were able to optimize these simulations to create an economic analysis for large scale implementation. From the Aspen simulations, we obtained a 90% yield of CO2 after separation. This work will present preliminary results on the CCSI tool to optimize our simulations for the economic analysis.

References

[1]"How much of U.S. carbon dioxide emissions are associated with electricity generation? - FAQ - U.S. Energy Information Administration (EIA)", Eia.gov, 2017. [Online]. Available: https://www.eia.gov/tools/faqs/faq.php?id=77&t=11. [Accessed: 12- Oct- 2017].

[2] N. MIMURA, "Sea-level rise caused by climate change and its implications for society", Proceedings of the Japan Academy, Series B, vol. 89, no. 7, pp. 281-301, 2013.

[3] A. Lu and S. Dai, Porous Materials for Carbon Dioxide Capture. Berlin, Heidelberg: Springer Berlin Heidelberg, 2014, pp. 79-113.

Copyright © American Institute of Chemical Engineers. All rights reserved.

510845 Utilization of Highly Reactive Calcium Oxide Nanofibers in the Sequestration of CO_2

Abstract:

Developing carbon sequestration methods is one of fourteen Grand Challenges to be faced by engineers this century. Evidence continues to mount implicating carbon dioxide (CO_2) emissions as a primary contributor to climate change, and if CO_2 levels continue to rise, further changes to the climate could have dire consequences, resulting in rising sea levels, agriculture disruptions, and more extreme weather striking more often. Therefore, it has become necessary for CO_2 to be captured and stored safely away from the atmosphere. Over the years, several studies have been conducted to identify and develop efficacious candidates for CO_2 sequestration and utilization. Calcium oxide (CaO) is an attractive material for capturing CO_2 from flue gas due to its energy efficiency, high reactivity, and low material cost; however, its capturing capacity and thus efficiency is strongly affected by the synthesis method.

The primary objective of this research was to synthesize CaO nanofibers via electrospinning and test their performance as sorbents in a multi-cycle process of carbonation and decarbonation. CaO nanoparticles were also synthesized via a hydrothermal approach to compare their performance to the nanofibers. Electrospinning was used because it is a contactless electrostatic synthesis technique that produces one-dimensional nanofibers that have high surface area-to-volume ratios, interconnected porous networks, and greater mechanical properties. It is theorized that compared to other synthesis methods, electrospinning results in the fabrication of nanomaterials that contain more accessible reactive sites.

In this work, the electrospinning solution was prepared by dissolving 1:1 weight ratio of polyvinylpyrrolidone (PVP) MW=1,300,000 g/mol and calcium nitrate tetrahydrate in a co-solvent system containing 70:30 volumetric ratio of ethanol to water. Electrospinning was conducted at 30 kV, 1 mL/h extrusion rate, and a 16-inch separation between the collector and the tip of the extrusion needle. PVP-Ca nanofibers were collected at room temperature on aluminum foil and heat-treated afterwards at 650°C (2°C/min ramp rate) for 8 hours to decompose the PVP chains and produce CaO nanofibers.

The results indicate a promising opportunity for the application of electrospun materials in CO_2 sequestration. CaO nanofibers and nanoparticles were characterized using X-ray diffraction (XRD), Brunauer-Emmett-Teller (BET) analysis, and scanning electron microscopy (SEM) to elucidate information about crystallite size, specific surface area, and nanofiber morphology, respectively. The reactivity of the electrospun nanofibers and hydrothermally synthesized nanoparticles were tested using CO_2 thermogravimetry at 550°C, 600°C, and 650°C. Sorbent degradation was also tested over the course of ten cycles of carbonation at 600°C and calcination. CaO nanofibers exhibited significantly higher rates and conversion (100% vs 78.4%) compared to traditionally synthesized CaO nanoparticles. Furthermore, the CaO nanofibers continued to produce higher or comparable conversions over repeated cycling. The increased rate and conversion of the CaO nanofibers result from their smaller, more reactive domains. Smaller, more reactive metal oxide domains allow for these materials to achieve stoichiometric CO_2 capacities. These improvements in CO_2 sequestration materials have the potential to reduce costs of CO_2 capture by offering greater cost advantage than current conventional technologies due to the high availability and low cost of CaO sorbents.

Copyright © American Institute of Chemical Engineers. All rights reserved.

AUTHOR INDEX

AUTHOR INDEX

AUTHOR INDEX

AUTHOR INDEX

AUTHOR INDEX

AUTHOR INDEX

AUTHOR INDEX

AUTHOR INDEX

AUTHOR INDEX